The Quest for Extraterrestrial Life
A BOOK OF READINGS

Donald Goldsmith

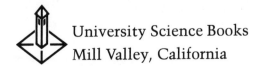
University Science Books
Mill Valley, California

To Rachel, and to all children of delight.

University Science Books
20 Edgehill Road
Mill Valley, CA 94941

Library of Congress Catalog Card Number: 79-57423

ISBN 0-935702-02-4 papercover
ISBN 0-935702-08-3 hardcover

Printed in the United States of America

10   9   8   7   6   5   4   3   2   1

The Quest for Extraterrestrial Life

Foreword

It has always seemed to me that life is unlikely to be confined to just this one small planet Earth. In 1950, when formulating a number of radio programs for the British Broadcasting Corporation, I wrote as follows:

> Will living creatures arise on every planet where favorable physical conditions occur? No certain answer can be given, but those best qualified to judge the matter, the biologists, seem to think that life would in fact arise wherever conditions were able to support it. Accepting this, we can proceed with greater assurance. The extremely powerful process of natural selection would come into operation and would shape the evolution of life on each of these distant planets. Would creatures arise having some sort of similarity to those on the Earth? The distinguished biologist, C. D. Darlington, suggests that this is by no means unlikely. To quote Darlington's own words: "There are such great advantages in walking on two legs, in carrying one's brain in one's head, in having two eyes on the same eminence at a height of five or six feet, that we might as well take quite seriously the possiblity of a pseudo man and a pseudo woman with some physical resemblance to ourselves. . . .
>
> A further question: Will travel between different planetary systems ever be possible? I am sorry to give an unpopular answer, but I believe this to be an uncompromising, no. Communication is a different matter. If living creatures at a high technological level exist on planets belonging to any of the nearest thousand stars it would be feasible to establish communication. A two-way interchange of information would take many centuries to develop. Even so, perhaps we should be starting now? (Hoyle, 1950, p. 103).

This was not popular stuff when it was written two decades ago, and it is still not popular with orthodox astronomical opinion today, probably because there seems at first sight to be nothing that one can do about it. But this seemingly respectable conservative position is in error. Although there is no direct hammer-and-tongs astronomical test of the validity or otherwise of such ideas, there are plenty of more subtle tests if one looks carefully for them.

It is also a mistake to think that life on many separate planets would need to have had a separate origin on each of them. The correct concept, I believe, is that life has had just one origin, and that it has invaded every planet in the form of protozoa, bacteria, and viruses, wherever it could gain a toehold. A straightforward calculation of probabilities shows this statement must almost surely be true.

Many enzymes and other biosubstances run across the whole spectrum of life, from bacteria to man. Many of these molecules permit variations, for instance, in the amino acids that make up the enzymes. But by no means all the amino-acid sites can be so varied. Typically, about one-third of them appear fixed. Indeed in some special cases essentially no variations at all have been permitted. The implication is that the fixed sites are unique to the functioning of the biosubstances in question. So how did the situation get the way it is in the first place?

The answer to this question turns out to be discouraging to the orthodox idea that life began on Earth. Biomolecules are so complex, and there is so much to be gotten right in their structures, that large-scale cosmic resources, not a minuscule planetary environment, were necessary for their origin. Indeed, it is surprising the early microbiologists did not realize that the amazing new world into which they had miraculously penetrated just had to be cosmic in its proportions.

But of course it was harder in the early days to see things clearly than it is today. Nowadays, the writing for the old life-originated-here-on-the-Earth prejudice is on the wall. A month or two ago, organic inclusions in rocks from the Isua region of Western Greenland were reported. These are the Earth's oldest rocks, with an age of about 3.8 billion years. The remarkable feature of the inclusions was that they imitated closely the highly distinctive shapes which occur in the budding of modern yeast cells.

Even in the face of this observation, indicating the presence of a complex cell almost at the outset of Earth's history, the old prejudice will doubtless grind on for a while with a certain lumbering inertia. But the situation is clearly there for the innocently unprejudiced to read. How far the tide will eventually run in the opposite direction has still to unfold. My bet is that the tide will run farther than even the most enthusiastic life-in-the-universe addict has yet dared to contemplate. For me, this is one of the most exciting prospects in astronomy. It will be especially interesting to see whether it is astronomy that absorbs biology, or the other way around.

Fred Hoyle

Preface

Nothing is so firmly believed as that which we least know.
—MONTAIGNE

Life on Earth has begun to yield its mysteries to scientific investigation. As detail upon detail of the workings of living beings comes to light, even the central mystery—the origin of life itself—may soon stand revealed. But until then, we cannot say with confidence just when life arose on this planet, or how. Great as is the importance of these questions to terrestrial biology, greater still will be the importance of their answers for exobiology, the study of life beyond the Earth.

This book contains selections from two millennia of thought on the possibility of extraterrestrial life, a time span that falls short of one millionth of the time that our sun and Earth have existed. With such a long stretch of time, and with so many possible sites for life in the universe, it might seem reasonable that life of all possible varieties must have appeared *somewhere*. This view has indeed held the majority—with important exceptions—so long as the Earth has not been considered unique. But the forms that life may take, and the ways in which we might find life elsewhere, or talk with other civilizations, have provoked a continuing debate, with no immediate resolution in sight.

Since science cannot proceed without debate, I have enjoyed making my own contribution by assembling some of the outstanding writings on the quest for extraterrestrial life. Huygens said almost three centuries ago that "we must not think that so great a diversity of minds were placed in different men to no end or purpose." Huygens had in mind a divine creator; I prefer to see the diversity of minds as the natural result of our evolution, and the key to our future progress. Today we stand at the dawn of the era in which humans can communicate with beings—if such exist!—in relatively nearby planetary systems. How long this era will last, no one knows. Perhaps the success or failure of our attempts to communicate will directly affect the answer to this question. In any case, the average lifetime for a communicating civilization in the Milky Way galaxy is a significant number which we do not yet know, nor do we know whether we shall exceed the average or fall short of it.

Our ignorance of the ways in which life may have developed elsewhere and in which communication may proceed remains profound but not total. The writings in this volume testify to the power (and the limitations) of scientific imagination and understanding in dealing with a subject that, at first glance, may seem fit only for pure speculation. Indeed the distinction between reasonable extrapolation and fanciful yearning may at times appear diffuse. Was Huygens being "scientific" when he demonstrated that beings on other planets must have hands, feet, eyes, and ears? Was Lowell "scientific" when he deduced the characteristics of Martian civilization from the canals he saw? In a way, yes. These scientists, and the others represented in this volume, have taken the best observations at their disposal and have drawn conclusions from them. By publishing their thoughts, they have allowed others to criticize or to support their analyses. Only in this way can a scientific debate occur, and only then can others appreciate or join this debate.

If human beings decide to invest heavily in a search for extraterrestrial intelligence, they will do so only after engaging in a political and scientific discussion of the merits of such a search. We might, of course, avoid this effort, as well as the expense itself, because some other civilization has already made it, and has taken the trouble to find us. Although many people consider this a probable turn of events, the writings collected here emphasize the reverse: Space is vast, and we have little extraordinary to offer. We must therefore almost certainly prepare for a long search if we are to find our neighbors in the Milky Way. In our own solar system, the search has already begun in earnest, but without finding any proof of life beyond the Earth. The quest for life beyond the sun's family of planets will take much more effort—if we choose to expend it. I favor such an effort, and to help the process, I present to you these readings. May you enjoy them, think them over, and wonder whether we should search more vigorously for extraterrestrial life.

In preparing this volume, I owe a debt of gratitude to Professor A. G. W. Cameron for his pioneering collection *Interstellar Communication* (Benjamin, 1963), and to all the other authors who kindly allowed the reprinting of their work, in particular to Professors Frank Drake, Sidney Fox, William Markowitz, and Carl Sagan for their assistance with my editing. It has been a pleasure to work with Aidan Kelly, Dick Palmer, and Bruce Armbruster on the production of this book. I would also like to thank Jerry Heymann, Paul Goldsmith, and Donald Kripke for their kind assistance with this project.

Donald Goldsmith

Contents

I
Historical Perspective

As long as human beings regarded Earth as the center of the cosmos, endowed by creation with a preeminent position among celestial objects that differed greatly from Earth, there was no obvious reason to imagine that other places should have conditions suitable to life, let alone life itself. The slow realization that our planet does not occupy a central position, nor represent a unique object, had to overcome our resistance to seeing ourselves adrift in an uncaring, possibly hostile universe.

Almost alone among ancient philosophers, Lucretius pursued the logical results of coupling his theory that life on Earth arose through natural processes with his belief that the universe must be infinite. The thread of Lucretius's reasoning provoked nothing further until the Copernican revolution finally dethroned Earth; then Giordano Bruno could present his well-reasoned, though apparently heretical, notion that God created innumerable worlds, all capable of bearing beings with souls. In the century after Bruno's execution, the idea of other inhabited worlds lost most of its power to shock, though Voltaire used it well to place the world-shaking conflicts of his time in a different setting.

1

Lucretius
On the Nature of the Universe

Titus Lucretius Carus, usually known simply as Lucretius, was a contemporary of Julius Caesar, born about 98 B.C. to an aristocratic family of ancient Rome. We know little about Lucretius's life, or about his writings other than his famous poem, *De Rerum Natura* (*On the Nature of the Universe*). In this epic work, Lucretius laid out his understanding of the heavens and Earth with a clarity and beauty that has probably not been surpassed since. The fact that much of what Lucretius concluded about the universe turned out to be wrong has far less importance than his ability to outline the paths along which scientific speculation could proceed. Lucretius showed his readers through the centuries where the power of reason could lead, and what startling conclusions could follow from a few bold hypotheses.

The chief motive of Lucretius's work was, apparently, to lay to rest a fear of the gods and their punishments by showing the natural causes of events. The great theme of the poem is that all of nature forms a single, interdependent whole, extremely subtle but open to human understanding. Lucretius is now best known for his exposition of the "atomistic" theory of nature, that all matter consists of eternal, unchanging atoms that can arrange and rearrange themselves into different, changing forms. But it is not surprising to find that this highly original Roman also speculated about the infinity of the universe, about other worlds than ours, and about the possibility of life distributed through the cosmos.

In this modern translation of a small part of Lucretius's poem, made by Prof. Mary-Kay Orlandi of the University of California, Santa Cruz, the reader can see the confidence and logic with which Lucretius analyzed the world around him. Even when Lucretius arrives at the wrong conclusions, as he often does, he arrives there marvelously, and when he is right, he is gloriously right. His statement that "such combinations of other atoms happen elsewhere in the universe to make worlds such as this one" provides the touchstone of this collection.

And so we are all sprung from heavenly seed. The pure upper air is father to all beings. When fecund mother earth receives drops of flowing liquid from it, she brings forth young—shining fruits, lush groves, the race of man, as well as all the kinds of animals. Then she proffers sustenance on which all these nourish their bodies, draw out sweet life, and generate offspring. Therefore the name Mother is rightly given to the earth. The moisture which came from the earth goes back into the earth, and that which was sent out from the edges of the upper air is carried back and the aether receives it again. In this way, death removes nothing by destroying the elements of its being, but it only disturbs their arrangement. So death joins one thing to another and brings it about that all creatures change their shapes and alter their colors, acquire sensations, and, at the right moment, give them up again.

You must understand that it matters with which other atoms, and in what position, the same atoms are conjoined, and what motions they make back and forth among each other. You must not think that something which we perceive as unstable like the surface of water rippling—at

From *De Rerum Natura*, by Lucretius (ca. 70 B.C.), translated by Prof. Mary-Kay Gamel Orlandi. Copyright © 1980 by Mary-Kay Gamel Orlandi.

one moment appearing, the next disappearing—is a permanent aspect of the everlasting atoms. In just the same way it is very important what the elements (letters) of my verses are, and in what position and order they occur. For the same letters indicate heaven, earth, sea, rivers, sun, fruits, plants, and living creatures. (Even if not all the letters are exactly the same, nonetheless the greater part is similar; they differ in the way they are placed.) And in just the same way in material objects, when the motions, configuration, placing, shape and juxtapositions of their component matter change, the composed objects also must change.

Now please pay close attention to a true fact. For something new is earnestly striving to reach your ears; a new aspect of the universe will reveal itself. No reality, of course, is so easy to understand that it does not seem at first difficult to believe. And yet nothing is so incredible and impossible that people don't stop being amazed by it in time. Take the pure bright hue of the sky, for example, and all the bodies it contains—the wandering stars, the moon, the splendor of the sun's dazzling light. Suppose all these appeared for the first time now to mortal sight, suddenly thrust forth without warning. What could possibly be called more wonderful than these? What would people have been less able to believe possible, before they actually saw them? Nothing, surely, the vision would have been so wonderful. And yet we are now so bored with this glorious sight that no one takes the time to gaze at heaven's shining sweep. Therefore, don't be frightened by the novelty of an idea, and spit it out of your mind, but thoroughly weight it with keener judgment, and then, if it seems true to you, side with it; if false, take up arms against it. Since the totality of space outside our own world's boundaries is infinite, the spirit seeks to understand what is out there where the mind longs to gaze, where the questing intellect can freely roam.

In the first place, then, in all parts round about, on both sides, above, below, throughout all things in the universe, there is no end. I have mentioned this before; and the truth proclaims itself and the nature of vast space makes it clear. For since infinite space stretches out on all sides, and atoms of numberless number and incalculable quantity fly about in all directions quickened by eternal movement, it can in no way be considered likely that this is the only heaven and earth created, and all those other atoms there beyond are doing nothing. For this world was created by Nature after atoms had collided spontaneously and at random in a thousand ways, driven together blindly, uselessly, without any results, when at last suddenly the particular ones combined which could become the perpetual starting points of things we know—earth, sea, sky, and the various kinds of living things. Therefore, we must acknowledge that such combinations of other atoms happen elsewhere in the universe to make worlds such as this one, held in the close embrace of the aether.

Moreover, when there is an abundance of matter available, when there is space vacant, and no object or reason delays the process, then certainly shapes of reality must be combined and created. For there is such a huge supply of atoms that all eternity would not be enough time to count them; there is the force which drives the atoms into various places just as they have been driven together in this world. So we must realize that there are other worlds in other parts of the universe, with races of different men and different animals.

It follows from this that in the totality of creation no thing is unique, the only one to be born and the only one to grow, without being part of a class of which there are many others. Consider first the living creatures: you will find that there are many examples of the races of beasts which range the mountains, the children of men are many, many also the herds of silent scaly fish, and the ranks of flying things. Therefore it must be acknowledged that in the same way the heaven, the earth, the sea, the sun and the moon and all other things which exist are not unique, but rather are part of an incredible number. For the perimeters on their life are firmly fixed, and they too are made of mortal clay, just like any race of creatures which here on earth abounds in specimens differing according to their kind.

If you keep these principles well in mind, you will see that Nature is totally autonomous, free from haughty overlords, able to accomplish all things by her own will—with no god's interference. In the name of the spirits of the gods, untouched in peaceful silence, who lead a life of repose and serenity! *Who* could have the power to rule the totality of the unmeasured universe, to hold firmly in his hands the staunch reins? *Who* could turn all the heavens at once, and warm the fruitful worlds with ethereal fire? Who could be in all places at once, make shadows by bringing on clouds, then—in a clear sky—crash with thunder, sending lightning to rock his own dwelling, then shifting to attack empty places, brandishing a bolt which often passes over the guilty, but kills those who have done nothing?

After the birth-date of the world, and the first appearance of the sea, the earth, and the sun, many bodies have been added from outer space, atoms added around, which the great totality has brought together by the process of hurling. From these the sea and the earth could increase, the air could rise in a mass, and the mansion of heaven could add to its extent, and raise its roofpeaks far from the earth. For because of the buffets they receive atoms are pushed out of anywhere they are and arranged with their own kind—water goes toward water, earth grows by means of earth, fire is forged out by fire, and air by air.

2

Giordano Bruno
On the Infinite Universe and Worlds

Giordano Bruno was born near Nola, close to Naples, in 1548, became a Dominican monk at the age of fifteen, was accused of impiety, and traveled widely (through Italy, Switzerland, France, and England) as he formed his own philosophical thoughts. The two years that Bruno spent in England (1583–1584) at the height of the Elizabethan era led to the writing of his best-known works, the *Cena de la Ceneri,* or *Ash Wednesday Conversation,* and *De l'Infinito, Universo, e Mondi,* or *On the Infinite Universe and Worlds.* The latter work in particular brought Bruno fatal difficulties.

Bruno heavily imbibed the attitudes of the Renaissance, and considered established religion a bar to joy and understanding. For this and for many other heresies Bruno was imprisoned by the Inquisition upon his return to Italy in 1593, and was kept in close confinement for seven years, the first two without a cloak or pillow. When Bruno refused to renounce his heretical views after several examinations by the Inquisition, he was burned at the stake (February 17, 1600) after having been excommunicated the week before. Three centuries later, a statue in Bruno's honor was erected at the Campo dei Fiori in Rome, where he died.

In accepting the Copernican view of the solar system, Bruno had, of course, to reject the natural philosophy of Plato and Aristotle, who considered Earth to be the center of the cosmos. This was a difficult mental process, since the Greek philosophers had dominated scholastic thought for centuries, but, as Bruno wrote, every well-regulated mind and alert judgment must decide what the truth is. Lucretius had concluded that the universe must be infinite because otherwise everything would fall toward the bottom; Bruno came to the same conclusion because "infinite perfection is far better presented in innumerable individuals than in those which are numbered and finite." In seeking absolute perfection, Bruno clearly remained heir to the tradition of Christian theology, mixed with its Greek forerunners.

At any rate, Bruno wrote of innumerable worlds, "no less inhabited and no less nobly" than our own. Such a concept struck hard at the established view of the relationship between God and human beings, and made a significant contribution to the "crimes" for which Bruno was punished.

The excerpt presented here, from the Third Dialogue of Bruno's *On the Infinite Universe and Worlds,* features a discussion between two philosophers named Burchio and Fracastorio.

FRACASTORIO: I would conclude as follows. The famous and received order of the elements and of the heavenly bodies is a dream and vainest fantasy, since it can neither be verified by observation of nature nor proved by reason or argued, nor is it either convenient or possible to conceive that it exists in such fashion. But we know that there is an infinite field, a containing space which does embrace and interpenetrate the whole. In it is an infinity of bodies similar to our own. No one of these more than another is in the centre of the universe, for the universe is infinite and therefore without centre or limit, though these appertain to each of the worlds within the universe in the way I have explained on other occasions, especially when we demonstrated that there are certain determined definite centres, namely, the suns, fiery bodies around which revolve all planets, earths and waters, even as we see the seven wandering planets take their course around our sun. Similarly we showed that each of these stars or worlds, spinning around his own centre, has the appearance of a solid and continuous world which takes by force all visible things which can become stars and whirls them around himself as the centre of their universe. Thus there is not merely one world, one earth, one sun, but as many worlds as we see bright lights around us, which are neither more nor less in one heaven, one space, one containing sphere than is this our world in one containing universe, one space or one heaven. So that the heaven, the infinitely extending air, though part of the infinite universe, is not therefore a world or part of worlds; but is the womb, the receptacle and field within which they all move and live, grow and render effective the several acts of their vicissitudes; produce, nourish and maintain their inhabitants and animals; and by certain dispositions and orders they minister to higher nature, changing the face of single being through countless subjects. Thus each of these worlds is a centre toward which converges every one of his own parts; toward it every kindred thing does tend just as the parts of this our star, even though at a certain distance, are yet brought back to their own field from all sides of the surrounding region. Therefore, since no part which flows thus outward from the great Body fails ultimately to return thereto; it happens that every such world is eternal though dissoluble; albeit if I mistake not, the inevitability of such eternity depends on an external maintaining and provident Being and not on intrinsic power and self-sufficiency. But I will explain you this matter with special arguments on other occasions.

BURCHIO: Then the other worlds are inhabited like our own?

FRACASTORIO: If not exactly as our own, and if not more nobly, at least no less inhabited and no less nobly. For it is impossible that a rational being fairly vigilant, can imagine that these innumerable worlds, manifest as like to our own or yet more magnificent, should be destitute of similar and even superior inhabitants; for all are either themselves suns or the sun doth diffuse to them no less than to us those most divine and fertilizing rays, which convince us of the joy that reigns at their source and origin and bring fortune to those stationed around who thus participate in the diffused quality. The innumerable prime members of the universe are then infinite [in number], and all have similar aspect, countenance, prerogative, quality and power.

BURCHIO: You will not admit any difference between them?

FRACASTORIO: [On the contrary]. You have heard more than once that some, in whose composition fire does predominate, are by their own quality bright and hot. Others shine by reflection, being themselves cold and dark, for water does predominate in their composition. On this diversity and opposition depend order, symmetry, complexion,[1] peace, concord, composition and life. So that the worlds are composed of contraries of which some, such as earth and water, live and grow by help of their contraries,[2] such as the fiery suns. This I think was the meaning of the sage who declared that God creates harmony out of sublime contraries;[3] and of that other who believed this whole universe to owe existence to the strife of the concordant and the love of the opposed.[4]

BURCHIO: In this way, you would put the world upside down.

FRACASTORIO: Would you consider him to do ill who would upset a world which was upside down?

BURCHIO: Would you then render vain all efforts, study and labours on such work as *De physico auditu* and *De coelo et mondo* wherein so many great commentators, paraphrasers, glossers, compilers, epitomizers, scholiasts, translators, questioners and logicians have puzzled their brains? Whereon profound doctors, subtle, golden, exalted, inexpugnable, irrefragable, angelic, seraphic, cherubic and divine, have established their foundation?

FRACASTORIO: Add the stonebreakers, the rocksplitters, horn-footed highkickers.[5] Add also the deep seers, know-alls,[6] the Olympians, the firmamenticians, celestial empirics, loud thunderers.

BURCHIO: Should we cast them all at your suggestion

From *On the Infinite Universe and Worlds*, by Giordano Bruno (1584), translated by Dorothea Waley Singer, in *Giordano Bruno: His Life and Thought* (New York: Henry Schuman Publishers, 1950).

---

[1] Of course in the Aristotelian sense.

[2] Cf. especially the closing passages of Fracastorio's unfinished work, *Fracastorius sive de anima*.

[3] This may refer either to Nicolaus Cusanus or to Pseudo-Dionysius the Areopagite.

[4] This view is attributed to Heraclitus by Aristotle in the *Nicomachean Ethics*, VIII, 2, 1155b 5-6.

[5] i.e., donkeys.

[6] *Palladii*. Florio gives *palladio professore*, one that professeth to know of Minerva's cunning.

into a cesspool? The world will indeed be ruled well if the speculations of so many and such worthy philosophers are to be cast aside and despised.

FRACASTORIO: It were not well that we should deprive the asses of their fodder, and wish them to adopt our own taste. Talent and intellect vary no less than temperaments and stomachs.

BURCHIO: You maintain that Plato is an ignorant fellow, Aristotle an ass and their followers insensate, stupid and fanatical?

FRACASTORIO: My son, I do not say these are foals and those asses, these little monkeys and those great baboons, as you would have me do. As I told you from the first, I regard them as earth's heroes. But I do not wish to believe them without cause, nor to accept those propositions whose anthitheses (as you must have understood if you are not both blind and deaf) are so compellingly true.

BURCHIO: Who then shall be judge?

FRACASTORIO: Every well-regulated mind and alert judgement. Every discreet person who is not obstinate when he recognizes himself convinced and unable either to defend their arguments or to resist ours.

3

Fontenelle
Conversations on the Plurality of Worlds

Bernard le Bovier de Fontenelle, born in Rouen, France, in 1657, showed early talent as a writer and an immense staying power, since he died nearly a hundred years later, in 1757. A poet, playwright, essayist, and philosopher, Fontenelle also became a writer of history and a popularizer of science.

The *Conversations on the Plurality of Worlds,* written in 1686, was the first book to present scientific knowledge in a way that invited the reader to enjoy the process of learning. At the very time that Fontenelle was writing his popular work, Isaac Newton was composing his monumental *Mathematical Principles of Natural Philosophy,* which was written (in Latin) for those who had the ability to understand the most complex mathematics of the era. In contrast, Fontenelle, who was no mathematician, invented a series of dialogues between himself and a "marchioness," or wife of a marquis, in the language of the people, complete with all the intriguing digressions of ordinary conversation. The book was a great success, testimony to the public interest in learning about astronomy.

The price Fontenelle paid, as so many other popularizers have, lay in being out of date: The theories he was expounding were those of Descartes, soon to be totally disproven by Newton's work. The "vortexes" Fontenelle discusses are Descartes' answer to Newton's law of gravity, whirlpools of motion that carry the planets around the sun with no need of gravitation. But in the supposition that our sun is a star like other stars, Fontenelle, Descartes, and Newton were in unamimous—and correct—agreement.

This love, replied the marchioness, laughing, is a strange thing; let the world go how it will, it is never in danger; there is no system can do it any harm. But, tell me freely, is your system true? Pray, do not conceal anything from me; I will keep your secret very faithfully; it seems to have for its foundation, but a slight probability, which is, that if a fixed star be in itself a luminous body, like the sun, then by consequence, it must, as the sun is, be the centre and soul of a world, and have its planets turning round about it. But is there an absolute necessity that it must be so? Hear me, madam, says I; since we are in the humour of mingling the follies of gallantry with philosophy, I must tell you, that in love and the mathematics, people reason much alike: allow ever so little to a lover, yet presently after you must grant him more; nay, more and more; and he will at last go a great way: in like manner, grant a mathematician but one minute principle, he immediately draws a consequence from it, to which you must necessarily assent; and from this consequence another, till he leads you so far (whether you will or no) that you have much ado to believe all he has proved, and what you already assented to. These two sorts of people, lovers and mathematicians, will always take more than you give them. You grant, that when two things are like one another in all visible respects, it is possible they may be like one another in those respects which are not visible, if you have not some good reason to believe otherwise: now this way of arguing have I made use of. The moon, says I, is inhabited, because she is like the earth; and the other planets are inhabited, because they are like the moon; I find the fixed stars resemble our sun; therefore I attribute

From *Conversations on the Plurality of Worlds,* by Bernard le Bovier de Fontenelle (1686); from a contemporary English translation.

to them what is proper to him: you have gone too far to be able to retreat, therefore you must go forward with a good grace. But, says the lady, if you build upon this resemblance, or likeness, which is between our sun and the fixed stars, then, to the people of another great vortex, our sun must appear no larger than a small fixed star, and can be seen only when it is night with them. Without doubt, madam, says I, it must be so: our sun is much nearer to us, than the suns of other vortexes, and therefore its light makes a much greater impression on our eyes than theirs do: we see nothing but the light of our own sun; and when we see him, it darkens and hinders us from seeing any other; but in another great vortex, there is another sun, which rules and governs; and, in his turn, extinguishes the light of our sun, which is never seen there but in the night, with the rest of the other suns, that is, the fixed stars; with them our sun is suspended in the great arched roof of heaven, where it makes a part of some constellation: the planets which turn round about it, (our earth for example) as they are not seen at so vast a distance, so nobody will so much as dream of them. All the suns that are day-suns in their own vortexes, are but night-suns or fixed stars in other vortexes: in his own world or sphere, every sun is single, and there is but one to be seen; but everywhere else they serve only to make up a number of stars. May not these worlds, replied she, notwithstanding this great resemblance between them, differ in a thousand other things; for though they may be somewhat alike in this one particular, they may greatly differ in others.

It is certainly true, says I; but the difficulty is to know wherein they differ. One vortex may have many planets that turn round about its sun, another may have but a few: in one there may be inferior or lesser planets, which turn about those that are greater; in another, perhaps, there may be no inferior planets; here all the planets are got round about their sun, in form of a little squadron; beyond which may be a large void space, which reaches even to the neighbouring vortexes: in others, the planets may make their revolutions towards the extremity of their vortex, and leave the middle void. I doubt not, but that there may be vortexes also quite void, without any planets at all; others may have their sun not exactly in their centre; and that sun may so move, as to carry its planets along with it: some may have planets, which, in regard of their sun, rise and set according to the change of their equilibrium, which keeps them suspended. In short, what farther variety can you wish for? But, I think, I have said enough for a man that was never out of his own vortex.

You have not said too much, replied the marchioness, considering what a multitude of worlds there are; what you have said is scarce sufficient for five or six; and from hence I see thousands, I may say, of millions.

4

Christiaan Huygens
Cosmotheoros, or New Conjectures Concerning the Planetary Worlds

Christiaan Huygens, born at The Hague, Holland, in 1629, was the outstanding Dutch astronomer of the seventeenth century; his father Constantijn had been the chief literary figure of Holland. Huygens showed an early aptitude for mathematics and science, which excused him from the legal career his father had planned for him. By the time he was thirty, Huygens had discovered the true nature of the rings of Saturn, had made the first accurate observations of the Orion Nebula, and had invented a practical pendulum clock that greatly improved scientists' time-keeping ability. The most famous king of France, Louis XIV, who was nine years younger than Huygens, invited him to live in France, an offer which Huygens accepted from 1665 to 1681. Huygens then returned to Holland and proceeded to build better telescopes and to work on his theories of optics and of gravitation.

Huygens died in 1695, leaving several unpublished works, of which the *Cosmotheoros* was the most purely speculative. This work was soon translated into English (and other languages), and we have reproduced part of this contemporary translation to give a sense of the vigorous joy of late 17th century English. The last section of the excerpt, not strictly speculation on life elsewhere, describes Huygens's brilliant method for estimating the distances to stars other than the sun, the first reasonably accurate estimate ever made.

That the Planets are not without Water, is made not improbable by the late Observations: For about *Jupiter* are observed some spots of a darker Colour than the rest of his Body, which by their continual change show themselves to be Clouds: For the Spots of *Jupiter* which belong to him, and never remove from him, are quite different from these, being sometimes for a long time not to be seen for these Clouds; and again, when these disappear, showing themselves. And at the going off of these Clouds, some Spots have been taken notice of in him, much brighter than the rest of his Body, which remained but a little while, and then were hid from our Sight. These Monsieur *Cassini* thinks are only the Reflection from the Snow that covers the Tops of the Hills in *Jupiter*: But I should rather think that it is only the Colour of the Earth, which happens to be free from those Clouds that commonly darken it.

Mars too is found not to be without his dark Spots, by means of which he has been observed to turn round his own Axis in 24 Hours and 40 Minutes; the Length of his Day: but whether he has Clouds or no, we have not had the same opportunity of observing as in *Jupiter*, as well because even when he is nearest the Earth, he appears to us much less than *Jupiter*, as that his Light not coming so far, is so brisk as to be an Impediment to exact Observations: And this Reason is as much stronger in *Venus* as its Light is. But since 'tis certain that the Earth and *Jupiter* have their Water and Clouds, there is no Reason why the other Planets should be without them. I can't say that they are exactly of the same nature with our Water; but that they should be liquid their Use requires, as their Beauty does that they should be clear. For this Water of ours, in *Jupiter* or *Saturn*, would be frozen up instantly by reason of the vast distance of the Sun. Every Planet therefore must have its Waters of such a temper, as to be proportioned to its Heat: *Jupiter's* and *Saturn's* must be of such a Nature as not to be liable to Frost; and *Venus's* and *Mercury's* of such, as not to be easily evaporated by the Sun. But in all of them, for a continual supply of Moisture, whatever

From *New Conjectures Concerning the Planetary Worlds, Their Inhabitants and Productions* (1698); from a contemporary English translation.

Water is drawn up by the Heat of the Sun into Vapours, must necessarily return back again thither. And this it cannot do but in Drops, which are caused as well there as with us, by their ascending into a higher and colder Region of the Air, out of that which, by reason of the Reflection of the Rays of the Sun from the Earth, is warmer and more temperate.

Here then we have found in these new Worlds Fields warm'd by the kindly Heat of the Sun, and water'd with fruitful Dews and Showers: That there must be Plants in them as well for Ornament as Use, we have shewn just now. And what Nourishment, what manner of Growth shall we allow them? Probably, there can be no better, nay no other, than what we here experience; by having their Roots fastned into the Earth, and imbibing its nourishing Juices by their tender Fibres. And that they may not be only like so many bare Heaths, with nothing but creeping Shrubs and Bushes, we may allow them some nobler and loftier Plants, Trees, or somewhat like them: These being the greatest, and, except Waters, the only Ornament that Nature has bestowed upon the Earth. For not to speak of those many uses that are made of their Wood, there's no one that is ignorant either of their Beauty or Pleasantness. Now what way can any one imagine for a continual Production and Succession of these Plants, but their bearing Seed? A Method so excellent, that it's the only one that Nature has here made use of, and so wonderful, that it seems to be designed not for this Earth alone. In fine, there's the same reason to think that this Method is observed in those distant Countries, as there was of its being followed in the remote Quarters of this same Earth.

'Tis much the same in Animals as 'tis in Plants, as to their manner of Nourishment, and Propagation of their Kind. For since all the living Creatures of this Earth, whether Beasts, Birds, Fishes, Worms, or Insects, universally and inviolably follow the same constant and fix'd Institution of Nature; all feed on Herbs, or Fruits, or the Flesh of other Animals that fed on them: since all Generation is performed by the impregnating of the Eggs, and the Copulation of Male and Female: Why may not the same Rule be observed in the Planetary Worlds? For *'tis certain that the Herbs and Animals that are there would be lost, their whole Species destroyed without some daily new Productions*: except there be no such thing there as Misfortune or Accident: except the Plants are not like other humid Bodies, but can bear Heat, Frost, and Age, without being dry'd up, kill'd or decay'd: except the Animals have Bodies as hard and durable as Marble; which I think are gross Absurdities. If we should invent some new Way for their coming into the World, and make them drop like Soland Geese from Trees, how ridiculous would this be to any one that considers the vast Difference between Wood and Flesh? Or suppose we should have new ones made every Day out of some such fruitful Mud as that of *Nile*, who does not see how contrary this is to all that's reason-

able? And that 'tis much more agreeable to the Wisdom of God, once for all to create of all sorts of Animals, and distribute them all over the Earth in such a wonderful and inconceivable way as he has, than to be continually obliged to new Productions out of the Earth? And what miserable, what helpless Creatures must these be, when there's no one that by his Duty will be obliged, or by that strange natural fondness, which God has wisely made a necessary Argument for all Animals to take care of their own, will be moved to assist, nurse or educate them?

As for what I have said concerning their Propagation, I cannot be so positive; but the other Thing, namely, that they have Plants and Animals, I think I have fully proved, *viz.* from hence, that otherwise they would be inferiour to our Earth. And by the same Argument, they must have as great a Variety of both as we have. What this is, will be best known to him that considers the different Ways our Animals make use of in moving from one Place to another. Which may be reduc'd, I think, to these; either that they walk upon two Feet or Four; or like Insects, upon Six, nay sometimes Hundreds; or that they fly in the Air bearing up, and wonderfully steering themselves with their Wings; or creep upon the Ground without Feet; or by a violent Spring in their Bodies, or paddling with their Feet, cut themselves a Way in the Waters. I don't believe, nor can I conceive, that there should be any other Way than these mentioned. The Animals then in the Planets must make use of one or more of these, like our amphibious Birds, which can swim in Water as well as walk on Land, or fly in the Air; or like our Crocodiles and Sea-Horses, must be Mongrels, between Land and Water. There can no other Method be imagined but one of these. For where is it possible for Animals to live, except upon such a solid Body as our Earth, or a fluid one like the Water or still a more fluid one than that, such as our Air is? The Air I confess may be much thicker and heavier than ours, and so, without any Disadvantage to its Transparency, be fitter for the volatile Animals. There may also be many sorts of Fluids ranged over one another in Rows as it were. The Sea perhaps may have such a fluid lying on it, which tho' ten times lighter than Water, may be a hundred Times heavier than Air; whose utmost Extent may not be so large as to cover the higher Places of their Earth. But there's no Reason to suspect or allow them this, since we have no such Thing; and if we did, it would be of no Advantage to them, for that the former Ways of moving would not be hereby at all increas'd: But when we come to meddle with the Shape of these Creatures, and consider the incredible Variety that is even in those of the different parts of this Earth, and that *America* has some which are nowhere else to be found, I must then confess that I think it beyond the Force of Imagination to arrive at any knowledge in the Matter, or reach to Probability concerning the Figures of these Planetary Animals. Altho' considering these Ways of Motion we e'en now recounted, they may perhaps be no more different from ours than ours (those of ours I mean that are most unlike) are from one another.

But still the main and most agreeable Point of the Enquiry is behind, which is the placing some Spectators in these new Discoveries, to enjoy these Creatures we have planted them with, and to admire their Beauty and Variety. And among all, that have never so slightly meddled with these Matters, I don't find any that have scrupled to allow them their Inhabitants: not Men perhaps like ours, but some Creatures or other endued with Reason. For all this Furniture and Beauty the Planets are stock'd with seem to have been made in vain, without any Design or End, unless there were some in them that might at the same time enjoy the Fruits, and adore the wise Creator of them. But this alone would be no prevailing Argument with me to allow them such Creatures. For what if we should say, that God made them for no other Design, but that he himself might see (not as we do 'tis true; but that he that made the Eye sees, who can doubt?) and delight himself in the Contemplation of them? For was not Man himself, and all that the whole World contains, made upon this very account? That which makes me of this Opinion, that those Worlds are not without such a Creature endued with Reason, is that otherwise our Earth would have too much the Advantage of them, in being the only part of the Universe that could boast of such a Creature so far above, not only Plants and Trees, but all Animals whatsoever: a Creature that has something Divine in him, that knows, and understands, and remembers such an innumerable number of Things, that deliberates, weighs and judges of the Truth: A Creature upon whose Account, and for whose Use, whatsoever the Earth brings forth seems to be provided. For every Thing here he converts to his own Ends. With the Trees, Stones, and Metals, he builds himself Houses: the Birds and Fishes he sustains himself with: and the Water and Winds he makes subservient to his Navigation; as he doth the sweet Smell and glorious Colours of the Flowers to his Delight. What can there be in the Planets that can make up for its Defects in the want of so noble an Animal? If we should allow *Jupiter* a greater Variety of other Creatures, more Trees, Herbs and Metals, all these would not advantage or dignify that Planet so much as that one Animal doth ours by the admirable Productions of his penetrating Wit. If I am mistaken in this, I do not know when to trust my Reason, and must allow my self to be but a poor Judge in the true Estimate of Things. . . .

Since it has so pleased God to order the Earth, and every Thing in it as we see it is (for it's absurd to say it happen'd against his Will or Knowledge) we must not think that so great a Diversity of Minds were placed in different Men to no End or Purpose: but that this mixture of bad Men with Good, and the Consequents of such a Mixture, as Misfortunes, Wars, Afflictions, Poverty, and the like, were permitted for this very good End, *viz.* the exercising our

Wits, and sharpening our Inventions; by forcing us to provide for our own necessary Defence against our Enemies. 'Tis fo the Fear of Poverty and Misery that we are beholden for all our Arts, and for that natural Knowledge which was the Product of laborious Industry; and which makes us that we cannot but admire the Power and Wisdom of the Creator, which otherwise we might have passed by with the same indifference as Beasts. And if Men were to lead their whole Lives in an undisturbed continual Peace, in no fear of Poverty, no danger of War, I doubt they would live little better than Brutes, without all knowledge or enjoyment of those Advantages that make our Lives pass on with Pleasure and Profit. We should want the wonderful Art of Writing, if its great Use and necessity in Commerce and War had not forced out the Invention. 'Tis to these we owe our Art of Sailing, our Art of Sowing, and most of those Discoveries of which we are Masters; and almost all the Secrets in experimental Knowledge. So that those very Things on account of which the Faculty of Reason seems to have been accused, are no small helps to its Advancement and Perfection. For those Virtues themselves, Fortitude and Constancy, would be of no use if there were no Dangers, no Adversity, no Afflications for their Exercise and Trial.

If we should therefore imagine in the Planets some such reasonable Creature as Man is, adorn'd with the same Virtues, and liable to the same Vices, it would be so far from degrading or vilifying them, that while they want such a one, I must think them inferior to our Earth.

But if we allow these Planetary Inhabitants some sort of Reason, must it needs, may some say, be the same with ours? Certainly it must; whether we consider it as applied to Justice and Morality, or exercised in the Principles and Foundations of Science. For Reason with us is that which gives us a true Sense of Justice and Honesty, Praise, Kindness and Gratitude: 'tis That that teaches us to distinguish universally between Good and Bad; and renders us capable of Knowledge and Experience in it. And can there be anywhere any other Sort of Reason that this? or can what we call just and generous, in *Jupiter* or *Mars* be thought unjust Villany? This is not at all, I don't say probable, but possible. . . .

But I perceive I am got somewhat too far: Let us first enquire a little concerning the bodily Senses of these Planetary Persons; for without such, neither will Life be any Pleasure to them, nor Reason of any Use. And I think it very probable, that all their Animals, as well their Beasts as rational Creatures, are like ours in all that relates to the Senses: For without the Power of Seeing we should find it impossible for Animals to provide Food for themselves, or be fore-warn'd of any approaching Danger, so as to guard themselves from it. So that where-ever we plant any Animals, except we wou'd have them lead the Life of Worms or Moles, we must allow them Sight; than which nothing can conduce more either to the Preservation or Pleasure of their Lives. . . .

It's likely then, and credible, that in these Things the Planets have an exact correspondence with us, and that their Animals have the same Organs, and use the same way of Sight that we do. They must have Eyes therefore, and two at least we must grant them, otherwise they would not perceive those Things close to them, nor hardly be able to walk about with Safety. And if we must allow them to all Animals for the Preservation of their Life, how much more must they that make more, and more noble Uses of them, not be deprived of the Blessing of so advantageous Members? For by them we view the various Flowers, and the elegant Features of Beauty: with them we read, we write, we contemplate the Heavens and Stars, and measure their Distances, Magnitudes, and Journeys: which how far they are common to the Inhabitants of those Worlds with us, I shall presently examine. But first I shall enquire whether now we have given them one, we ought also to give them the other four Senses. And indeed as to Hearing many Arguments perswade me to give it a Share in the Animals of those new Worlds. For 'tis of great consequence in defending us from Sudden Accidents; and, especially when Seeing is of no use to us, it supplies its Place, and gives us seasonable warning of any imminent Danger. Besides, we see many Animals call their Fellow to them with their Voice, which Language may have more in it than we are aware of, tho' we don't understand it. But if we do but consider the vast Uses and necessary Occasions of Speaking on the one side, and Hearing on the other, among those Creatures that make use of their Reason, it will scarce seem credible that two such useful, such excellent Things were designed only for us. For how is it possible but that they that are without these, must be without many other Necessaries and Conveniencies of Life? Or what can they have to recompense this Want? Then, if we go still farther, and do but meditate upon the neat and frugal Contrivance of Nature in making the same Air, by the drawing in of which we live, by whose Motion we sail, and by whose Means Birds fly, for a Conveyance of Sound to our Ears; and this Sound for the Conveyance of another Man's Thoughts to our Minds: Can we ever imagine that she has left those other Worlds destitute of so vast Advantages? That they don't want the Means of them is certain, for their having Clouds in *Jupiter* puts it past doubt that they have Air too; that being mostly formed of the Particles of Water flying about, as the Clouds are of them gathered into small Drops. And another Proof of it is, the necessity of breathing for the preservation of Life, a Thing that seems to be as universal a Dictate of Nature, as feeding upon the Fruits of the Earth.

As for Feeling, it seems to be given upon necessity to all Creatures that are cover'd with a fine and sensible Skin, as a Caution against coming too near those Things that may injure or incommode them: and without it they would be liable to continual Wounds, Blows and Bruises. Nature seems to have been so sensible of this, that she has not left the least place free from such a Perception. Therefore it's probable that the Inhabitants of those Worlds are not without so necessary a Defence, and so fit a Preservative against Dangers and Mishaps.

And who is there that doth not see the inevitable necessity for all Creatures that live by feeding to have both Taste and Smell, that they may distinguish those Things that are good and nourishing, from those that are mischievous and harmful? If therefore we allow the Planetary Creatures to feed upon Herbs, Seeds, or Flesh, we must allow them Taste and Smell, that they may chuse or refuse any Thing according as they find it likely to be advantageous or noxious to them.

I know that it hath been a Question with many, whether there might not have been more Senses than these five. If we should allow this, it might nevertheless be reasonably doubted, whether the Senses of the Planetary Inhabitants are much different from ours. I must confess, I cannot deny but there might possibly have been more Senses; but when I consider the Uses of those we have, I cannot think but they would have been superfluous. The Eye was made to discern near and remote Objects, the Ear to give us notice of what our Eyes could not, either in the Dark or behind our Back. Then what neither the Eye nor the Ear could, the Nose was made (which in Dogs is wonderfully nice) to warn us of. And if anything escapes the notice of the other four Senses, we have Feeling to inform us of the too near Approaches of it before it can do us any mischief. Thus has Nature so plentifully, so perfectly provided for the necessary preservation of her Creatures here, that I think she can give nothing more to those there, but what will be needless and superfluous. Yet the Senses were not wholly designed for use: but Men from all, and all other Animals from some of them, reap Pleasure as well as Profit, as from the Taste in delicious Meats; from the Smell in Flowers and Perfumes; from the Sight in the Contemplation of beauteous Shapes and Colours; from the Hearing in the Sweetness and Harmony of Sounds; from the Feeling in Copulation, unless you please to count that for a particular Sense by itself. Since it is thus, I think 'tis but reasonable to allow the Inhabitants of the Planets these same Advantages that we have from them. For upon this Consideration only, how much happier and easier a Man's Life is rendered by the enjoyment of them, we must be obliged to grant them these Blessings, except we would engross everything that is good to our selves, as if we were worthier and more deserving than any else. But moreover, that Pleasure which we perceive in Eating or in Copulation, seems to be a necessary and provident Command of Nature, whereby it tacitly compels us to the preservation and continuance of our Life and Kind. It is the same in Beasts. So that both for their Happiness and Preservation it's very probable the rest of the Planets are not without it.

Certainly when I consider all these Things, how great, noble, and useful they are; when I consider what an admirable Providence it is that there's such a Thing as Pleasure in the World, I can't but think that our Earth, the smallest part almost of the Universe, was never design'd to monopolize so great a Blessing. And thus much for those Pleasures which affect our bodily Senses, but have little or no relation to our Reason and Mind. But there are other Pleasures which Men enjoy, which their Soul only and Reason can relish. Some airy and brisk, others grave and solid, and yet nevertheless Pleasures, as arising from the Satisfaction which we feel in Knowledge and Inventions, and Searches after Truth, of which whether the Planetary Inhabitants are not partakers, we shall have an opportunity of enquiring by and by. . . .

There may arise another Question, whether there be in the Planets but one sort of rational Creatures, or if there be not several sorts possessed of different degrees of Reason and Sense. There is something not unlike this to be observed among us. For to pass by those who have human Shape (altho' some of them would very well bear that Enquiry too) if we do but consider some sorts of Beasts, as the Dog, the Ape, the Beaver, the Elephant, nay some Birds and Bees, what Sense and Understanding they are masters of, we shall be forced to allow, that Man is not the only rational Animal. For we discover somewhat in them of Reason independent on, and prior to all Teaching and Practice.

But still no Body can doubt, but that the Understanding and Reason of Man is to be preferr'd to theirs, as being comprehensive of innumerable Things, indued with an infinite memory of what's past, and capable of providing against what's to come. That there is some such Species of rational Creatures in the other Planets, which is the Head and Sovereign of the rest, is very reasonable to believe: for otherwise, were many Species endued with the same Wisdom and Cunning, we should have them always doing Mischief, always quarrelling and fighting one with another for Empire and Sovereignty, a Thing that we feel too much of where we have but one such Species. But to let that pass, our next Enquiry shall be concerning those Animals in the Planets which are furnished with the greatest Reason, whether it's possible to know wherein they employ it, and whether they have made as great Advances in Arts and Knowledge as we in our Planet. Which deserves most to be considered and examined of anything belonging to their Nature; and for the better Performance of it we must take our Rise somewhat higher, and nicely view the Lives and Studies of Men. . . .

What is it then after all that sets human Reason above all other, and makes us preferable to the rest of the Animal World? Nothing in my Mind so much as the Contemplation of the Works of God, and the Study of Nature, and the improving those Sciences which may bring us to some knowledge in their Beauty and Variety. For without

Knowledge what would be Contemplation? And what difference is there between a Man, who with a careless supine Negligence views the Beauty and Use of the Sun, and the fine golden furniture of the Heaven, and one who with a learned Niceness searches into their Courses; who understands wherein the Fix'd Stars, as they are call'd, differ from the Planets, and what is the Reason for the regular Vicissitude of the Seasons; who by sound Reasoning can measure the Magnitude and Distance of the Sun and Planets? Or between such a one as admires perhaps the nimble Activity and strange Motions of some Animals, and one that knows their whole Structure, understands the whole Fabrick and Architecture of their Composition? If therefore the Principle we before laid down be true, that the other Planets are not inferiour in Dignity to ours, what follows but that they have Creatures not to stare and wonder at the Works of Nature only, but who employ their Reason in the Examination and Knowledge of them, and have made as great Advances therein as we have? They do not only view the Stars, but they improve the Science of Astronomy: nor is there anything can make us think this improbable, but that fond Conceitedness of every Thing that we call our own, and that Pride that is too natural to us to be easily laid down. But I know some will say, we are a little too bold in these Assertions of the Planets, and that we mounted hither by many Probabilities, one of which, if it chance to be false, and contrary to our Supposition, would, like a bad Foundation, ruin the whole Building, and make it fall to the Ground. But I would have them to know, that all I have said of their Knowledge in Astronomy, has Proofs enough, antecedent to those we know produced. For supposing the Earth, as we did, one of the Planets of equal Dignity and Honour with the rest, who would venture to say, that no where else were to be found any that enjoy'd the glorious Sight of Nature's Theatre? Or if there were any Fellow-Spectators, yet we were the only ones that had dived deep into the Secrets and Knowledge of it? So that here's a Proof not so far fetch'd for the Astronomy of the Planets, the same which we used for their having rational Creatures, and enjoying the other Advantages we before talk'd of, which serves at the same time for the Confirmation of our former Conjectures. But if Amazement and Fear at the Eclipses of the Moon and Sun gave the first occasion to the Study of Astronomy, as probably they did, then it's almost impossible that *Jupiter* and *Saturn* should be without it; the Argument being of much greater force in them, by reason of the daily Eclipses of their Moons, and the frequent ones of the Sun to their Inhabitants. So that if a Person disinterested in his Judgment, and equally ignorant of the Affairs of all the Planets, were to give his Opinion in this Matter, I don't doubt he would give the Cause for Astronomy to those two Planets rather than us.

This Supposition of their Knowledge and Use of Astronomy in the Planetary Worlds, will afford us many new

Conjectures about their manner of Life, and their State as to other things.

For, First: No Observations of the Stars that are necessary to the Knowledge of their Motions, can be made without Instruments; nor can these be made without Metal, Wood, or some such solid Body. Here's a necessity of allowing them the Carpenters Tools, the Saw, the Ax, the Plane, the Mallet, the File: and the making of these requires the Use of Iron, or some equally hard Metal. Again, these Instruments can't be without a Circle divided into equal Parts, or a strait Line into unequal. Here's a necessity for introducing Geometry and Arithmetick. Then the Necessity in Such Observations of marking down the Epochas or Accounts of Time, and of transmitting them to Posterity, will force us to grant them the Art of Writing; perhaps very different from ours which is commonly used, but I dare affirm not more ingenious or easy. For how much more ready and expeditious is our Way, than by that multitude of Characters used in *China;* and how vastly preferable to Knots tied in Cords, or the Pictures in use among the barbarous People of *Mexico* and *Peru*? There's no Nation in the World but has some way or other of writing or marking down their Thoughts: So that it's no wonder if the Planetary Inhabitants have been taught it by that great Schoolmistress Necessity, and apply it to the Study of Astronomy and other Sciences....

What could we invent or imagine that could be so exactly accommodated to all the design'd Uses as the Hands are? Elephants can lay hold of, or throw anything with their Proboscis, can take up even the smallest Things from the Ground, and can perform such surprising Things with it, that it has not very improperly been call'd their Hand, tho' indeed it is nothing but a Nose somewhat longer than ordinary. Nor do Birds show less Art and Design in the Use of their Bills in the picking up their Meat, and the wonderful Composure of their Nests. But all this is nothing to those Conveniences the Hand is so admirably suited to; nothing to that amazing Contrivance in its Capacity of being stretched, or contracted, or turned to any Part as Occasion shall require. And then, to pass by that nice Sense that the Ends of the Fingers are endued with, even to the feeling and distinguishing most sorts of Bodies in the Dark, what Wisdom and Art is show'd in the Disposition of the Thumb and Fingers, so as to take up or keep fast hold of any Thing we please? Either then the Planetary Inhabitants must have Hands, or somewhat equally convenient, which it is not easy to conceive; or else we must say that Nature has been kinder not only to us, but even to Squirrels and Monkeys than them.

That they have Feet also scarce any one can doubt, that does not consider what we said but just now of Animals' different Ways of going along, which it's hard to imagine can be perform'd any other ways than what we there recounted. And of all those, there's none can agree so well with the state of the Planetary Inhabitants, as that that we

here make use of. Except (what is not very probable, if they live in Society, as I shall show they do) they have found out the Art of flying in some of those Worlds....

If their Globe is divided like ours, into Sea and Land, as it's evident it is (else whence could all those Vapours in *Jupiter* proceed?) we have great Reason to allow them the Art of Navigation, and not vainly ingross so great, so useful a Thing to our selves. Especially considering the great Advantages *Jupiter* and *Saturn* have for Sailing, in having so many Moons to direct their Course, by whose Guidance they may attain easily to the Knowledge that we are not Masters of, of the Longitude of Places. And what a Multitude of other Things follow from this Allowance? If they have Ships, they must have Sails and Anchors, Ropes, Pullies, and Rudders, which are of particular Use in directing a Ship's Course against the Wind, and the Sailing different Ways with the same Gale. And perhaps they may not be without the Use of the Compass too, for the magnetical Matter, which continually passes thro' the Pores of our Earth, is of such a Nature, that it's very probable the Planets have something like it. But there's no doubt but that they must have the Mechanical Arts and Astronomy, without which Navigation can no more subsist, than they can without Geometry....

Seeing then that the Stars, as I said before, are so many Suns, if we do but suppose one of them equal to ours, it will follow that its distance from us is as much greater than that of the Sun, as its apparent Diameter is less than the Diameter of the Sun. But the Stars, even those of the first Magnitude, though view'd through a Telescope, are so very small, that they seem only like so many shining Points, without any perceivable Breadth. So that such Observations can here do us no good. When I saw this would not succeed, I studied by what way I could so lessen the Diameter of the Sun, as to make it not appear larger than the Dog, or any other of the chief Stars. To this purpose I clos'd one End of my twelve-foot Tube with a very thin Plate, in the Middle of which I made a Hole not exceeding the twelfth Part of a Line, that is the hundred and forty fourth Part of an Inch. That End I turn'd to the Sun, placing my Eye at the other, and I could see so much of the Sun as was in Diameter about the 182d part of the Whole. But still that little piece of him was brighter much than the Dog-Star is in the clearest Night. I saw that this would not do, but that I must lessen the Diameter of the Sun a great deal more. I made then such another Hole in a Plate, and against it I plac'd a little round Glass that I had made use of in my Microscopes, of much about the same Diameter with the former Hole. Then looking again towards the Sun (taking care that no Light might come near my Eye to hinder my Observation) I found it appeared of much the same Clearness with *Sirius*. But casting up my account, according to the Rules of *Dioptricks,* I found his Diameter now was but $^1/_{152}$ part of that hundred and eighty second part of his whole Diameter that I saw through the former Hole. Multiplying $^1/_{152}$ and $^1/_{182}$ into one another, the Product I found to be $^1/_{27,664}$. The Sun therefore being contracted into such a Compass, or being removed so far from us (for it's the same thing) as to make his Diameter but the 27664 part of that we every Day see, will send us just the same Light as the Dog-Star now doth. And his distance then from us will be to his present distance undoubtedly as 27664 is to 1, and his Diameter little above four Thirds, $4'''$. Seeing then *Sirius* is supposed equal to the Sun, it follows that his Diameter is likewise $4'''$, and that his Distance to the Distance of the Sun from us is as 27664 to 1. And what an incredible Distance that is, will appear by the same way of reasoning that we used in measuring that of the Sun. For if 25 Years are required for a Bullet out of a Cannon, with its utmost Swiftness, to travel from the Sun to us; then by multiplying the Number 27664 into 25, we shall find that such a Bullet would spend almost seven hundred thousand Years in its Journey between us and the nearest of the fix'd Stars. And yet when in a clear Night we look upon them, we cannot think them above some few Miles over our Heads. What I have here enquir'd into, is concerning the nearest of them. And what a prodigious Number must there be besides of those which are placed in the vast Spaces of Heaven, as to be as remote from these as these are from the Sun! For if with our bare Eyes we can observe above a Thousand, and with a Telescope can discover ten or twenty times as many; what bounds of Number can we set to those which are out of the Reach even of these Assistances! especially if we consider the infinite Power of God. Really, when I have been reflecting thus with my self, methoughts all our Arithmetick was nothing, and we are vers'd but in the very Rudiments of Numbers, in comparison of this great Sum. For this requires an immense Treasury, not of twenty or thirty Figures only, in our decuple Progression, but of as many as there are Grains of Sand upon the Shore. And yet who can say, that even this Number exceeds that of the Fix'd Stars? Some of the Ancients, and *Jordanus Brunus* carry'd it further, in declaring the Number infinite: he would perswade us that he has prov'd it by many Arguments, tho' in my opinion they are none of them conclusive. Not that I think the contrary can every be made out. Indeed it seems to me certain, that the Universe is infinitely extended; but what God has been pleas'd to place beyond the Region of the Stars, is as much above our Knowledge, as it is beyond our Habitation. . . .

5

Voltaire
Micromégas (Chapters One and Two)

Francois Marie Arouet, who later took the name of Voltaire, was born in Paris in 1694, and lived long enough to hear of the American Revolution of 1776. Voltaire received an excellent education from the Jesuits, but became a famous satirist whose sharpest attacks were on those who felt they had the right to impose their beliefs on others. More than any man of his time, Voltaire was the champion of human freedom, justly seen as the opponent of orthodoxy in all forms. One of Voltaire's best-remembered campaigns was in defense of a family he had never met that had been accused of witchcraft; it helps to put Voltaire in proper historical context if we note that he was born two years after the famous witch trials in Salem, Massachusetts.

Micromégas is one of Voltaire's lesser works, published when he was almost sixty, three years after his far more successful satire *Candide*. In this work (whose title means "Littlebig" in Greek), Voltaire adopted the handy device of imagining two giants visiting our solar system to comment on its inhabitants—an approach made familiar to us by countless science-fiction novels of the past century. Today it may seem commonplace to remind the reader of how small Earth must be in relation to the rest of the cosmos, but in Voltaire's time this was still a remarkable and revolutionary thought, perfectly suited to Voltaire's attitude toward humanity.

This modern translation is by Dr. Theodore Schuker, Interprète de Conférence, Paris, France.

CHAPTER ONE

The Voyage of a Native of the Star Sirius to the Planet Saturn

On one of the planets that revolve around the star called Sirius, there was a bright young man whom I had the honor to be acquainted with during the last voyage he made to our little ant-hill. His name was Micromégas, a fitting moniker for all great men. He was eight leagues tall—by eight leagues, I mean twenty-four thousand geometrical paces of five feet each.

Right away some mathematicians, who engage in an activity that is always so useful for the broad public, may start grabbing for their pens to discover that since Mr. Micromégas, as a native of Sirius, measured twenty-four thousand paces from head to foot, which means one hundred and twenty thousand king's feet, whereas we citizens of Earth are scarcely five feet tall, and since our globe is nine thousand leagues in circumference—they will discover, I was saying, that the globe that produced him must necessarily have a girth twenty-one million six-hundred thousand times greater than our little Earth. Nothing in nature is more simple or ordinary. The territory of some German or Italian sovereign, which may be toured in half an hour, when compared to the empire of Turkey, Moscovia or China, gives one only an inkling of the prodigious differences that nature has placed in all things.

Since His Excellency's stature was as I have indicated, all our sculptors and painters will readily accept that his waistline could be fifty thousand king's feet, which means that he was very handsomely proportioned.

In addition, he had one of the most cultivated minds around. He knew a great deal, and he had invented a few things. When not yet two hundred and fifty years of age, and still a student, as was customary, at his planet's Jesuit school, he discerned by mere dint of mind more than fifty propositions of Euclid, which is eighteen more than Blaise Pascal, who, after coming up with thirty-two to amuse himself, according to his sister, went on to become a rather mediocre mathematician and a very poor metaphysician.

Around the age of four hundred and fifty, towards the end of childhood, he dissected many of those little insects less than one hundred feet in diameter which escape notice under ordinary microscopes, and he wrote a very curious book about them which got him into trouble. His country's mufti, a rather picayunish and ignorant man, found that his book contained suspect, objectionable, bold, heretical, virtually heretical propositions, and set about prosecuting him. The issue was whether or not fleas on Sirius had the same substantial form as snails. Micromégas defended himself intelligently, and won the women over to his side. The trial lasted two hundred and twenty years. In the end, the mufti got the book condemned by jurisconsults who had never read it, and the author was given the order not to appear in court for eight hundred years.

He was not deeply affected by being banished from a court full of petty annoyances. He composed an amusing ditty about the mufti, which scarcely bothered the latter, and then set out to journey from planet to planet so as to complete the education of his heart and mind, as the saying goes. People whose travel experience is limited to post-chaises and berlins will probably be surprised by the means of locomotion used up there, inasmuch as on the little mud-heap we live on, we are at a loss to conceive of anything that surpasses our methods. Our traveler was amazingly familiar with the laws of gravitation and all the forces of attraction and repulsion. He made such deft use of them, availing himself of a sun ray here and a comet there, that he and his retinue moved about from globe to globe like a bird flitting from branch to branch. He crossed the Milky Way in no time, and I must confess that though he peered through that array of stars, he never saw through them to that fair empyreal firmament that the illustrious vicar Derham boasts of having observed through a telescope. I'm not claiming that Mr. Derham's vision was bad. God forbid! But Micromégas did go there, he was a good observer, and I don't wish to contradict anybody.

After many a turn, Micromégas arrived at the planet Saturn. However accustomed he was to seeing new things, he could not keep from smiling with superiority, as happens occasionally to even the wisest of men, when he noticed how small this globe and its inhabitants are.

Saturn, it turns out, is only nine hundred times bigger than Earth, and its denizens are dwarfs only one thousand fathoms or so in height. In the beginning, he joked about it with his servants, in much the way an Italian musician laughs at the music of Lully when he first comes to France. But the Sirian had a fine mind, and he quickly understood that a thinking being is not necessarily ridiculous just because he is only six thousand feet tall. He got to know the Saturnians once their surprise wore off. He struck up a close friendship with the Secretary of the Saturn Academy, a man of fine parts who had not actually invented anything on his own but who did have a grasp of others' inventions, and who was tolerably good at making minor verse and major calculations. For the reader's edification, I shall record here a singular conversation that occurred one day between Micromégas and the Secretary.

CHAPTER TWO

The Conversation Between the Native of Sirius and the Inhabitant of Saturn

Once His Excellency had laid down and the Secretary had drawn close to his face, Micromégas said:

"You must admit that nature is very varied."

"Yes," the Saturnian replied, "nature is like a flower-bed whose blossoms . . ."

"Bah!" the other exclaimed. "Forget about your flower-bed."

"It is," the Secretary resumed, "like a gathering of blondes and brunettes whose jewels . . ."

"Hmmm! What would I do with your brunettes?" the other retorted.

"Then it is like a gallery of paintings the brush strokes of which . . ."

"No it's not!" the traveler said. "Once again, nature is like nature. Why bother with similes?"

"To please you," the Secretary answered.

"I don't want to be pleased," the traveler responded, "I want to be instructed. So start by telling me how many senses men on your globe have."

"We have seventy-two," the academician said, "and every day we complain how few they are. Our imagination surpasses our needs. We find that with our seventy-two senses, our ring and our five moons, we are too limited. In spite of all our curiosity and the rather large number of passions that stem from our seventy-two senses, we have plenty of time to be bored."

"I can believe that," Micromégas replied, "because in our globe, we have nearly one thousand senses yet we still have a kind of vague urge or uneasiness which keeps warning us of our own insignificance and pointing out that there are other beings much more perfect. I have traveled a bit, and I have seen mortals far beneath us as well as others far superior to us, yet I have not seen any whose desires were not greater than their genuine needs or whose needs did not exceed their satisfaction. Perhaps some day I'll come upon a land where nothing is lacking, but for the time being, I haven't heard of any such place."

The Saturnian and the Sirian then spent themselves in conjectures. After engaging in a great deal of highly ingenious and highly uncertain reasoning, they had to return to the facts.

"What is your life-span?" the Sirian inquired.

"Very short," the little man from Saturn replied.

"It's the same with us," the Sirian said. "We complain about how short it is. That must be a universal law of nature."

"Alas!" the Saturnian sighed. "We live only five hundred great revolutions of the sun (which in our way of reckoning amounts to approximately fifteen thousand years). So you see, death occurs almost right after birth: our existence is a mere point, our duration an instant, our globe an atom. Hardly have we started to get a little education that death overtakes us before we've had any experience. In my case, I don't dare to make any plans. I feel I'm like a drop of water in an immense ocean. I am ashamed, particularly in your presence, of the ridiculous figure I cut in this world."

Micromégas rejoined: "If you weren't a philosopher, I'd be afraid to distress you by informing you that our lifetime lasts seven hundred times longer than yours. You know full well, however, that when the time comes to relinquish one's body to the elements and reanimate nature in some other form—what is called death—when that moment of metamorphosis comes, there is no difference between having lived an eternity and just one day. I've been in countries where people live a thousand times longer than we do, and I found that they still grumble about it. But there are sensible people everywhere who know how to make the best of things and thank nature's author. He has scattered over this universe an abundance of varieties within a kind of admirable uniformity. For example, all thinking beings are different, yet all are basically similar by virtue of thier ability to have thoughts and desires. Matter extends everywhere, but it has different properties in each globe. How many of these diverse properties have you counted?"

"If you mean those properties," the Saturnian answered, "without which we believe this globe could not persist as it is, we reckon there are three hundred of them, such as extension, impenetrability, mobility, gravitation, divisibility, and so forth."

"Apparently," the traveler remarked, "that small number suffices for the Creator's designs for your humble abode. I admire his wisdom in all things. Everywhere I look I see differences, but there are also proportions everywhere. Your globe is tiny, but so too are your inhabitants. You have few sensations, but your matter has few properties. All that is the work of Providence. What color is your sun when examined closely?

"A yellowish white," the Saturnian replied. "When we break down one of its rays, we find it contains seven colors."

"Our sun has a reddish tint," the Sirian said, "and we have thirty-nine primary colors. Among all the suns I've come close to, there are not two alike, just as in your land there is no face that is not different from all the others."

After several questions of this nature, he inquired how many essentially different substances had been counted on Saturn. He learned that there were only about thirty, such as God, space, matter, extended sentient beings, extended sentient beings who think, thinking beings who do not have extension, those who are penetrable, those who are not penetrable, and the rest. The Sirian flabbergasted the philosopher from Saturn by informing him that his country had three hundred of them, and that he had discovered three thousand others during his travels. At last, having communicated to one another the little they knew and much of what they did not know, and having gone on reasoning for one complete revolution of the sun, they resolved to embark on a short philosophical voyage together.

II

The Origins of Life

To speculate about life beyond Earth, we must study the single example of life that we know, life on our own planet, and attempt to draw from this example some guiding principles that may determine what life elsewhere could or would be like. Thus, for example, when we realize that the four major elements found in Earth's living creatures—hydrogen, carbon, nitrogen, and oxygen—rank among the most abundant elements in the universe, we may reasonably conclude that life on Earth formed itself from the most common, chemically suitable elements, and that life on any planet, or in some interstellar cloud, would also be likely to consist of the cosmically most abundant elements. This does not mean that life based on such rare elements as hafnium and holmium could not exist, but that such forms of life should be immensely rarer than those based on the commonest elements.

The modern history of our understanding of how life arose on Earth begins with the thoughts of Charles Darwin and his colleagues, who saw how more complex organisms could evolve from less complex ones through the process of natural selection. Far harder to understand was how any living creatures could ever arise from nonliving matter. Twelve decades after Darwin's great work, we still are far from an answer to this puzzle, though most scientists think that within a far shorter time we shall be able to replicate some of the processes that brought forth life on Earth, to create from nonliving matter a system capable of self-replication.

The fundamental postulate, now in general adoption among those who try to lift the veil of life's origins, is that nothing particularly unusual had to occur for life to begin. Rather, the existence of a watery solution, laden with organic (carbon-based) molecules, provided the primordial broth in which countless chemical reactions eventually resulted in self-replicating molecules, precursors of the modern DNA (deoxyribonucleic acid) that accounts for all the reproduction on Earth. If this postulate is valid, then so long as water or an equivalent solvent exists, so long as carbon and other common elements are present, and so long as we have "world enough and time," we may expect life to appear.

The pioneering work by J. B. S. Haldane and (independently and still more obscurely) by Alexander Oparin in Russia showed as early as the 1920s that no logical barrier exists against the scenario for life's beginnings that we have outlined (now called the Haldane-Oparin theory). Interesting wrinkles on this theory come from the suggestion (first made by William Thomson, Lord Kelvin) that life arrived on Earth from other regions of space; but the questions and answers about the origin of life seem the same no matter where life began.

6

Charles Darwin on the Origin of Life

Charles Darwin's monumental work on his theory of evolution, *The Origin of Species,* appeared in 1859; it brought Darwin enough criticism to last a lifetime (he died in 1882). Perhaps understandably, Darwin never wrote his thoughts about the actual origin of life for publication, but his correspondence shows that he certainly thought about the puzzle of how life began.

In 1863, Darwin wrote to J. D. Hooker:

It will be some time before we see "slime, protoplasm, &c." generating a new animal. But I have long regretted that I truckled to public opinion, and used the Pentateuchal term of creation, by which I really meant "appeared" by some wholly unknown process. It is mere rubbish, thinking at present of the origin of life; one might as well think of the origin of matter.

Eight years later, Darwin wrote (also in a letter):

It is often said that all the conditions for the first production of a living organism are now present, which could ever have been present. But if (and oh! what a big if!) we could conceive in some warm little pond, with all sorts of ammonia and phosphoric salts, lights, heat, electricity, &c. present, that a protein compound was chemically formed ready to undergo still more complex changes, at the present day such matter would be instantly devoured or absorbed, which would not have been the case before living creatures had formed.

Darwin's "warm little pond," perhaps a set of tide pools filled and drained twice a day, remains the favored spot for the origin of life on this planet.

From the letters of Charles Darwin, as quoted by J. D. Bernal in *The Origin of Life* (Cleveland: World Publishing Company, 1967).

7

Simon Newcomb
Life in the Universe

Simon Newcomb (1835–1909) was the best-known American astronomer of his era, director for many years of the United States Naval Observatory in Washington, D.C. Newcomb was also a skilled popularizer of astronomy who wrote many books and articles on that subject as well as on questions of finance and economics. The article "Life in the Universe," published in *Harper's Magazine* in 1905, gives a cogent presentation of the idea that if Earth is a representative planet orbiting a representative star, then life must be abundant throughout the universe.

So far as we can judge from what we see on our globe, the production of life is one of the greatest and most incessant purposes of nature. Life is absent only in regions of perpetual frost, where it never has an opportunity to begin; in places where the temperature is near the boiling-point, which is found to be destructive to it; and beneath the earth's surface, where none of the changes essential to it can come about. Within the limits imposed by these pro-hibitory conditions—that is to say, within the range of temperature at which water retains its liquid state, and in regions where the sun's rays can penetrate and where wind can blow and water exist in a liquid form—life is the universal rule. How prodigal nature seems to be in its production is too trite a fact to be dwelt upon. We have all read of the millions of germs which are destroyed for every one that comes to maturity. Even the higher forms of life are found almost everywhere. Only small islands have ever been discovered which were uninhabited, and animals of a higher grade are as widely diffused as man.

If it would be going too far to claim that all conditions may have forms of life appropriate to them, it would be going as much too far in the other direction to claim that life can exist only with the precise surroundings which nurture it on this planet. It is very remarkable in this connection that while in one direction we see life coming to an end, in the other direction we see it flourishing more and more up to the limit. These two directions are those of heat and cold. We cannot suppose that life would develop in any important degree in a region of perpetual frost, such as the polar regions of our globe. But we do not find any end to it as the climate becomes warmer. On the contrary, every one knows that the tropics are the most fertile regions of the globe in its production. The luxuriance of the vegetation and the number of the animals continually increase the more tropical the climate becomes. Where the limit may be set no one can say. But it would doubtless be far above the present temperature of the equatorial regions.

It has often been said that this does not apply to the human race, that men lack vigor in the tropics. But human vigor depends on so many conditions, hereditary and otherwise, that we cannot regard the inferior development of humanity in the tropics as due solely to temperature. Physically considered, no men attain a better development than many tribes who inhabit the warmer regions of the globe. The inferiority of these regions in intellectual power is more likely the result of race heredity than that of temperature.

We all know that this earth on which we dwell is only one of countless millions of globes scattered through the wilds of infinite space. So far as we know, most of these globes are wholly unlike the earth, being at a temperature so high that, like our sun, they shine by their own light. In such worlds we may regard it as quite certain that no organized life could exist. But evidence is continually increasing that dark and opaque worlds like ours exist and revolve around their suns as the earth on which we dwell revolves around its central luminary. Although the

From *Harper's Magazine*, 111 (1905), 404.

number of such globes yet discovered is not great, the circumstances under which they are found lead us to believe that the actual number may be as great as that of the visible stars which stud the sky. If so, the probabilities are that millions of them are essentially similar to our own globe. Have we any reason to believe that life exists on these other worlds?

The reader will not expect me to answer this question positively. It must be admitted that, scientifically, we have no light upon the question, and therefore no positive grounds for reaching a conclusion. We can only reason by analogy and by what we know of the origin and conditions of life around us, and assume that the same agencies which are at play here would be found at play under similar conditions in other parts of the universe.

If we ask what the opinion of man has been, we know historically that our race has, in all periods of its history, peopled other regions with beings even higher in the scale of development than we are ourselves. The gods and demons of an earlier age all wielded powers greater than those granted to man—powers which they could use to determine human destiny. But, up to the time that Copernicus showed that the planets were other worlds, the location of these imaginary beings was rather indefinite. It was therefore quite natural that when the moon and planets were found to be dark globes of a size comparable with that of the earth itself, they were made the habitations of beings like unto ourselves.

The trend of modern discovery has been against carrying this view to its extreme, as will be presently shown. Before considering the difficulties in the way of accepting it to the widest extent, let us enter upon some preliminary considerations as to the origin and prevalence of life, so far as we have any sound basis to go upon.

A generation ago the origin of life upon our planet was one of the great mysteries of science. All the facts brought out by investigation into the past history of our earth seemed to show, with hardly the possibility of a doubt, that there was a time when it was a fiery mass, no more capable of serving as the abode of a living being than the interior of a Bessemer steel furnace. There must therefore have been, within a certain period, a beginning of life upon its surface. But, so far as investigations had gone—indeed, so far as it has gone to the present time,—no life has been found to originate of itself. The living germ seems to be necessary to the beginning of any living form. Whence, then, came the first germ? Many of our readers may remember a suggestion by Sir William Thompson, now Lord Kelvin, made twenty or thirty years ago, that life may have been brought to our planet by the falling of a meteor from space. This does not, however, solve the difficulty—indeed, it would only make it greater. It still leaves open the question how life began on the meteor; and granting this, why it was not destroyed by the heat generated as the meteor passed through the air. The popu-

lar view that life began through a special act of creative power seemed to be almost forced upon man by the failure of science to discover any other beginning for it. It cannot be said that even to-day anything definite has been actually discovered to refute this view. All we can say about it is that it does not run in with the general views of modern science as to the beginning of things, and that those who refuse to accept it must hold that under certain conditions which prevail, life begins by a very gradual process, similar to that by which forms suggesting growth seem to originate even under conditions so unfavorable as those existing in a bottle of acid.

But it is not at all necessary for our purpose to decide this question. If life existed through a creative act, it is absurd to suppose that that act was confined to one of the countless millions of worlds scattered through space. If it began at a certain stage of evolution by a natural process, the question will arise, what conditions are favorable to the commencement of this process? Here we are quite justified in reasoning from what, granting this process, has taken place upon our globe during its past history. One of the most elementary principles accepted by the human mind is that like causes produce like effects. The special conditions under which we find life to develop around us may be comprehensively summed up as the existence of water in the liquid state and the presence of nitrogen, free perhaps in the first place, but accompanied by every substance with which it may form combinations. Oxygen, hydrogen, and nitrogen are, then, the fundamental requirements. The addition of calcium or other forms of matter necessary to the existence of a solid world goes without saying. The question is whether these necessary conditions exist in other parts of the universe.

The spectroscope shows that, so far as the chemical elements go, other worlds are composed of the same substance as ours. Hydrogen especially exists everywhere, and we have reason to believe that the same is true of oxygen and nitrogen. Calcium, the base of lime, is almost universal. So far as chemical elements go, we may therefore take it for granted that the conditions under which life begins are very widely diffused in the universe. It is contrary to all the analogies of nature to suppose that life began only on a single world.

It is a scientific inference, based on facts so numerous as not to admit of serious question, that during the history of our globe there has been a continually improving development of life. As ages upon ages pass, new forms are generated, higher in the scale than those which preceded them, until at length reason appears and asserts its sway. In his last and well-known work Alfred Russell Wallace has argued that this development of life required the presence of such a rare combination of conditions that there is no reason to suppose that it prevailed anywhere except on our earth. It is quite impossible in an article like the present to follow his reasoning in detail; but it seems to me altogether inconclusive. Not only does life, but intelligence, flourish on this globe under a great variety of conditions as regards temperature and surroundings, and no sound reason can be shown why under certain conditions, which are frequent in the universe, intelligent beings should not acquire the highest development.

Now let us look at the subject from the view of the mathematical theory of probabilities. A fundamental tenet of this theory is that no matter how improbable a result may be on a single trial, supposing it at all possible, it is sure to occur after a sufficient number of trials,—and over and over again if the trials are repeated often enough. For example, if a million grains of corn, of which a single one was red, were all placed in a pile, and a blindfolded person were required to grope in the pile, select a grain, and then put it back again, the chances would be a million to one against his drawing out the red grain. If drawing it meant he should die, a sensible person would give himself no concern at having to draw the grain. The probability of his death would not be so great as the actual probability that he will really die within the next twenty-four hours. And yet if the whole human race were required to run this chance, it is certain that about fifteen hundred, or one out of a million, of the whole human family would draw the red grain and meet his death.

Now apply this principle to the universe. Let us suppose, to fix the ideas, that there are a hundred million worlds, but that the chances are 1000 to 1 against any of these taken at random being fitted for the highest development of life or for the evolution of reason. The chances would still be that 100,000 of them would be inhabited by rational beings whom we may call human. But where are we to look for these worlds? This no man can tell. We only infer from the statistics of the stars, and this inference is fairly well grounded, that the number of worlds which, so far as we know, may be inhabited, are to be counted by thousands, and perhaps by millions.

In a number of bodies so vast we should expect every variety of conditions as regards temperature and surroundings. If we suppose that the special conditions which prevail on our planet are necessary to the highest forms of life, we still have reason to believe that these same conditions prevail on thousands of other worlds. The fact that we might find the conditions in millions of other worlds unfavorable to life would not disprove the existence of the latter on countless worlds differently situated.

Coming down now from the general question to the specific one, we all know that the only worlds the conditions of which can be made the subject of observation are the planets which revolve around the sun, and their satellites. The question whether these bodies are inhabited is one which, of course, completely transcends not only our powers of observation at present, but every appliance of research that we can conceive of men devising. If Mars is inhabited, and if the people of that planet have equal powers with ourselves, the problem of merely producing an illumination which could be seen in our most powerful telescope would be beyond all the ordinary efforts of an entire nation. An unbroken square mile of flame would be invisible in our telescopes, but a hundred square miles might be seen. We cannot, therefore, expect to see any signs of the works of inhabitants even on Mars. All that we can do is to ascertain with greater or less probability whether the conditions necessary to life exist on the other planets of the system.

The moon being much the nearest to us of all the heavenly bodies, we can pronounce more definitely in its case than in any other. We know that neither air nor water exists on the moon in quantities sufficient to be perceived by the most delicate tests at our command. It is certain that the moon's atmosphere, if any exists, is less then the thousandth part of the density of that around us. The vacuum is greater than any ordinary air-pump is capable of producing. We can hardly suppose that so small a quantity of air could be of any benefit whatever in sustaining life; an animal that could get along on so little could get along on none at all.

But the proof of the absence of life is yet stronger when we consider the results of actual telescopic observation. An object such as an ordinary city block could be detected on the moon. If anything like vegetation were present on its surface, we should see the changes which it would undergo in the course of a month, during one portion of which it would be exposed to the rays of the unclouded sun, and during another to the intense cold of space. If men built cities, or even separate buildings the size of the larger ones on our earth, we might see some signs of them.

In recent times we not only observe the moon with the telescope, but get still more definite information by photography. The whole visible surface has been repeatedly photographed under the best conditions. But no change has been established beyond question, nor does the photograph show the slightest difference of structure or shade which could be attributed to cities or other works of man. To all appearances the whole surface of our satellite is as completely devoid of life as the lava newly thrown from Vesuvius.

We next pass to the planets. Mercury, the nearest to the sun, is in a position very unfavorable for observation from the earth, because when nearest to us it is between us and the sun, so that its dark hemisphere is presented to us. Nothing satisfactory has yet been made out as to its condition. We cannot say with certainty whether it has an atmosphere or not. What seems very probable is that the temperature on its surface is higher than any of our earthly animals could sustain. But this proves nothing.

We know that Venus has an atmosphere. This was very conclusively shown during the transits of Venus in 1874 and 1882. But this atmosphere is so filled with clouds or vapor that it does not seem likely that we ever get a view of the solid body of the planet through it. Some observers have thought they could see spots on Venus day after day, while others have disputed this view. On the whole, if intelligent inhabitants live there, it is not likely that they ever see sun or stars. Instead of the sun they see only an effulgence in the vapory sky which disappears and reappears at regular intervals.

When we come to Mars, we have more definite knowledge, and there seems to be greater possibilities for life there than in the case of any other planet besides the earth. The main reason for denying that life such as ours could exist there is that the atmosphere of Mars is so rare that, in the light of the most recent researches, we cannot be fully assured that it exists at all. The very careful comparisons of the spectra of Mars and the moon made by Campbell at the Lick Observatory failed to show the slightest difference in the two. If Mars had an atmosphere as dense as ours, the result could be seen in the darkening of the lines of the spectrum produced by the double passage of the light through it. There were no lines in the spectrum of Mars that were not seen with equal distinctness in that of the moon. But this does not prove the entire absence of an atmosphere. It only shows a limit to its density. It may be one-fifth or one-fourth the density of that on the earth, but no more.

That there must be something in the nature of vapor at least seems to be shown by the formation and disappearance of the white polar caps of this planet. Every reader of astronomy at the present time knows that, during the Martian winter, white caps form around the pole of the planet which is turned away from the sun, and grow larger and larger until the sun begins to shine upon them, when they gradually grow smaller, and perhaps nearly disappear. It seems, therefore, fairly well proved that, under the influence of cold, some white substance forms around the polar regions of Mars which evaporates under the influence of the sun's rays. It has been supposed that this substance is snow produced in the same way that snow is produced on the earth, by the evaporation of water.

But there are difficulties in the way of this explanation. The sun sends less than half as much heat to Mars as to the earth, and it does not seem likely that the polar regions can ever receive enough of heat to melt any considerable quantity of snow. Nor does it seem likely that any clouds from which snow could fall every obscure the surface of Mars.

But a very slight change in the explanation will make it tenable. Quite possibly the white deposits may be due to something like hoar frost condensed from slightly moist air, without the actual production of snow. This would produce the effect that we see. Even this explanation implies that Mars has air and water, rare though the former may be. It is quite possible that a density less than this would sustain life in some form. Life not totally unlike that on the earth may therefore exist upon Mars for any-

thing that we know to the contrary. More than this we cannot say.

In the case of the outer planets the answer to our question must be in the negative. It now seems likely that Jupiter is a body very much like our sun, only that the dark portion is too cool to emit much, if any, light. It is doubtful whether Jupiter has anything in the nature of a solid surface. Its interior is in all likelihood a mass of molten matter far above a red heat, which is surrounded by a comparatively cool, yet, to our measure, extremely hot, vapor. The beltlike clouds which surround the planet are due to this vapor combined with the rapid rotation. If there is any solid surface below the atmosphere that we can see, it is swept by winds such that nothing we have on earth could withstand them. But, as we have said, the probabilities are very much against there being anything like a surface. At some great depth in the fiery vapor there is a sold nucleus; that is all we can say.

The planet Saturn seems to be very much like that of Jupiter in its composition. It receives so little heat from the sun that, unless it is a mass of fiery vapor like Jupiter, the surface must be far below the freezing-point.

We cannot speak with such certainty of Uranus and Neptune; yet the probability seems to be that they are in much the same condition as Saturn. They are known to have very dense atmospheres, which shine only by the light of the sun. But nothing is known of the composition of these atmospheres.

To sum up our argument: the fact that, so far as we have yet been able to learn, only a very small proportion of the visible worlds scattered through space are fitted to be the abode of life does not preclude the probability that among hundreds of millions of such worlds a vast number are so fitted. Such being the case, all the analogies of nature lead us to believe that, whatever the process which led to life upon this earth—whether a special act of creative power or a gradual course of development—through that same process does life begin in every part of the universe fitted to sustain it. The course of development involves a gradual improvement in living forms, which by irregular steps rise higher and higher in the scale of being. We have every reason to believe that this is the case wherever life exists. It is, therefore, perfectly reasonable to suppose that beings, not only animated, but endowed with reason, inhabit countless worlds in space. It would, indeed, be very inspiring could we learn by actual observation what forms of society exist throughout space, and see the members of such societies enjoying themselves by their warm firesides. But this is, so far as we can now see, entirely beyond the possible reach of our race, so long as it is confined to a single world.

8

J. B. S. Haldane
The Origin of Life

The thread of Darwin's thought about the origin of life was too elusive to be picked up successfully in his lifetime, or indeed for several generations thereafter. Only when a deeper understanding of biochemistry had been achieved through years of patient effort could scientists begin to speculate in detail about the processes by which living organisms arose through chemical combination.

The 1920s brought two remarkable works on the origin of life on Earth (and by implication, elsewhere in the universe)—works so original that their thoughts remained basically unnoticed until after the Second World War. Working independently, the brilliant English biologist John Burton Sanderson Haldane and the equally remarkable Russian biochemist Alexander Oparin had reached the conclusion that, as Oparin put it, "Life is not characterized by any special properties but by a definite, specific combination of these [chemical] properties." It remained for scientists to work out, by detective work on the structure of living organisms and by experiments with simulated primitive conditions on Earth, just how life evolved from the "primordial soup."

Oparin's work, which appeared in 1924, went almost unnoticed outside of Russia—or for that matter inside Russia, which was recovering from the revolutions of 1917. Haldane's brief essay, published in 1929, points out a program of research that remains the basic plan, partly completed, of biochemical investigation into the origin of life. His essay now has mostly historical importance, but was so far in advance of contemporary thought that it deserves attention even today.

Until about 150 years ago it was generally believed that living beings were constantly arising out of dead matter. Maggots were supposed to be generated spontaneously in decaying meat. In 1668 Redi showed that this did not happen provided insects were carefully excluded. And in 1860 Pasteur extended the proof to the bacteria which he had shown were the cause of putrefaction. It seemed fairly clear that all the living beings known to us originate from other living beings. At the same time Darwin gave a new emotional interest to the problem. It had appeared unimportant that a few worms should originate from mud. But if man was descended from worms such spontaneous generation acquired a new significance. The origin of life on the Earth would have been as casual an affair as the evolution of monkeys into man. Even if the latter stages of man's history were due to natural causes, pride clung to a supernatural, or at least surprising, mode of origin for his ultimate ancestors. So it was with a sigh of relief that a good many men, whom Darwin's arguments had convinced, accepted the conclusion of Pasteur that life can originate only from life. It was possible either to suppose that life had been supernaturally created on Earth some millions of years ago, or that it had been brought to Earth by a meteorite or by microorganisms floating through interstellar space. But a large number, perhaps the majority, of biologists, believed, in spite of Pasteur, that at some time in the remote past life had originated on Earth from dead matter as the result of natural processes.

The more ardent materialists tried to fill in the details of this process, but without complete success. Oddly

From the *Rationalist Annual,* 1929. Reprinted by permission of the Rationalist Press Association, London.

enough, the few scientific men who professed idealism agreed with them. For if one can find evidences of mind (in religious terminology the finger of God) in the most ordinary events, even those which go on in the chemical laboratory, one can without much difficulty believe in the origin of life from such processes. Pasteur's work therefore appealed most strongly to those who desired to stress the contrast between mind and matter. For a variety of obscure historical reasons, the Christian Churches have taken this latter point of view. But it should never be forgotten that the early Christians held many views which are now regarded as materialistic. They believed in the resurrection of the body, not the immortality of the soul. St. Paul seems to have attributed consciousness and will to the body. He used a phrase translated in the revised version as 'the mind of the flesh', and credited the flesh with a capacity for hatred, wrath, and other mental functions. Many modern physiologists hold similar beliefs. But, perhaps unfortunately for Christianity, the Church was captured by a group of very inferior Greek philosophers in the third and fourth centuries A.D. Since that date views as to the relation between mind and body which St. Paul, at least, did not hold, have been regarded as part of Christianity, and have retarded the progress of science.

It is hard to believe that any lapse of time will dim the glory of Pasteur's positive achievements. He published singularly few experimental results. It has even been suggested by a cynic that his entire work would not gain a Doctorate of Philosophy today! But every experiment was final. I have never heard of any one who has repeated any experiment of Pasteur's with a result different from that of the master. Yet his deductions from these experiments were sometimes too sweeping. It is perhaps not quite irrelevant that he worked in his latter years with half a brain. His right cerebral hemisphere had been extensively wrecked by the bursting of an artery when he was only forty-five years old; and the united brain-power of the microbiologists who succeeded him has barely compensated for that accident. Even during his lifetime some of the conclusions which he had drawn from his experimental work were disproved. He had said that alcoholic fermentation was impossible without life. Buchner obtained it with a cell-free and dead extract of yeast. And since his death the gap between life and matter has been greatly narrowed.

When Darwin deduced the animal origin of man, a search began for a 'missing link' between ourselves and the apes. When Dubois found the bones of Pithecanthropus some comparative anatomists at once proclaimed that they were of animal origin, while others were equally convinced that they were parts of a human skeleton. It is now generally recognized that either party was right, according to the definition of humanity adopted. Pithecanthropus was a creature which might legitimately be described either as a man or an ape, and its existence showed that the distinction between the two was not absolute.

Now the recent study of ultramicroscopic beings has brought up at least one parallel case, that of the bacteriophage, discovered by d'Herelle, who had been to some extent anticipated by Twort. This is the case of a disease, or, at any rate, abnormality of bacteria. Before the size of the atom was known there was no reason to doubt that

> Big fleas have little fleas
> Upon their backs to bite 'em;
> The little ones have lesser ones,
> And so ad infinitum.

But we now know that this is impossible. Roughly speaking, from the point of view of size, the bacillus is the flea's flea, the bacteriophage the bacillus' flea; but the bacteriophage's flea would be of the dimensions of an atom, and atoms do not behave like fleas. In other words, there are only about as many atoms in a cell as cells in a man. The link between living and dead matter is therefore somewhere between a cell and an atom.

D'Herelle found that certain cultures of bacteria began to swell up and burst until all had disappeared. If such cultures were passed through a filter fine enough to keep out all bacteria, the filtrate could infect fresh bacteria, and so on indefinitely. Though the infective agents cannot be seen with a microscope, they can be counted as follows. If an active filtrate containing bacteriophage be poured over a colony of bacteria on a jelly, the bacteria will all, or almost all, disappear. If it be diluted many thousand times, a few islands of living bacteria survive for some time. If it be diluted about ten million fold, the bacteria are destroyed round only a few isolated spots, each representing a single particle of bacteriophage.

Since the bacteriophage multiplies, d'Herelle believes it to be a living organism. Bordet and others have taken an opposite view. It will survive heating and other insults which kill the large majority of organisms, and will multiply only in presence of living bacteria, though it can break up dead ones. Except perhaps in presence of bacteria, it does not use oxygen or display any other signs of life. Bordet and his school therefore regard it as a ferment which breaks up bacteria as our own digestive ferments break up our food, at the same time inducing the disintegrating bacteria to produce more of the same ferment. This is not as fantastic as it sounds, for most cells while dying liberate or activate ferments which digest themselves. But these ferments are centainly feeble when compared with the bacteriophage.

Clearly we are in doubt as to the proper criterion of life. D'Herelle says that the bacteriophage is alive, because, like the flea or the tiger, it can multiply indefinitely at the cost of living beings. His opponents say that it can multiply only as long as its food is alive, whereas the tiger certainly, and the flea probably, can live on dead products of life. They suggest that the bacteriophage is like a book or a work of art, which is constantly being copied by living beings, and is therefore only metaphorically alive, its real life being in its copiers.

The American geneticist Muller has, however, suggested an intermediate view. He compares the bacteriophage to a gene—that is to say, one of the units concerned in heredity. A fully coloured and a spotted dog differ because the latter has in each of its cells one or two of a certain gene, which we know is too small for the microscope to see. Before a cell of a dog divides this gene divides also, so that each of the daughter-cells has one, two, or none according with the number in the parent cell. The ordinary spotted dog is healthy, but a gene common among German dogs causes a roan colour when one is present, while two make the dog nearly white, wall-eyed and generally deaf, blind or both. Most of such dogs die young, and the analogy to the bacteriophage is fairly close. The main difference between such a lethal gene, of which many are known, and the bacteriophage, is that the one is only known inside the cell, the other outside. In the present state of our ignorance we may regard the gene either as a tiny organism which can divide in the environment provided by the rest of the cell; or as a bit of machinery which the 'living' cell copies at each division. The truth is probably somewhere in between these two hypotheses.

Unless a living creature is a piece of dead matter plus a soul (a view which finds little support in modern biology) something of the following kind must be true. A simple organism must consist of parts A, B, C, D and so on, each of which can multiply only in presence of all, or almost all, of the others. Among these parts are genes, and the bacteriophage is such a part which has got loose. This hypothesis becomes more plausible if we believe in the work of Hauduroy, who finds that the ultramicroscopic particles into which the bacteria have been broken up, and which pass through filters that can stop the bacteria, occasionally grow up again into bacteria after a lapse of several months. He brings evidence to show that such fragments of bacteria may cause disease, and d'Herelle and Peyre claim to have found the ultramicroscopic form of a common staphylococcus, along with bacteriophage, in cancers, and suspects that this combination may be the cause of that disease.

On this view the bacteriophage is a cog, as it were, in the wheel of a life-cycle of many bacteria. The same bacteriophage can act on different species and is thus, so to say, a spare part which can be fitted into a number of different machines, just as a human diabetic can remain in health when provided with insulin manufactured by a pig. A great many kinds of molecule have been got from cells, and many of them are very efficient when removed from it. One can separate from yeast one of the many tools which it uses in alcoholic fermentation, an enzyme called invertase, and this will break up six times its weight of cane-sugar per second for an indefinite time without wearing out. As it does not form alcohol from the sugar, but only a sticky mixture of other sugars, its use is permitted

in the United States in the manufacture of confectionery and cake-icing. But such fragments do not reproduce themselves, though they take part in the assimilation of food by the living cell. No one supposes that they are alive. The bacteriophage is a step beyond the enzyme on the road to life, but it is perhaps an exaggeration to call it fully alive. At about the same stage on the road are the viruses which cause such diseases as smallpox, herpes, and hydrophobia. They can multiply only in living tissue, and pass through filters which stop bacteria.

With these facts in mind we may, I think, legitimately speculate on the origin of life on this planet. Within a few thousand years from its origin it probably cooled down so far as to develop a fairly permanent solid crust. For a long time, however, this crust must have been above the boiling-point of water, which condensed only gradually. The primitive atmosphere probably contained little or no oxygen, for our present supply of that gas is only about enough to burn all the coal and other organic remains found below and on the Earth's surface. On the other hand, almost all the carbon of these organic substances, and much of the carbon now combined in chalk, limestone, and dolomite, were in the atmosphere as carbon dioxide. Probably a good deal of the nitrogen now in the air was combined with metals as nitride in the Earth's crust, so that ammonia was constantly being formed by the action of water. The Sun was perhaps slightly brighter than it is now, and as there was no oxygen in the atmosphere the chemically active ultra-violet rays from the Sun were not, as they now are, mainly stopped by ozone (a modified form of oxygen) in the upper atmosphere, and oxygen itself lower down. They penetrated to the surface of the land and sea, or at least to the clouds.

Now, when ultra-violet light acts on a mixture of water, carbon dioxide, and ammonia, a vast variety of organic substances are made, including sugars and apparently some of the materials from which proteins are built up. This fact has been demonstrated in the laboratory of Baly of Liverpool and his colleagues. In this present world, such substances, if left about, decay—that is to say, they are destroyed by micro-organisms. But before the origin of life they must have accumulated till the primitive oceans reached the consistency of hot dilute soup. Today an organism must trust to luck, skill, or strength to obtain its food. The first precursors of life found food available in considerable quantities, and had no competitors in the struggle for existence. As the primitive atmosphere contained little or no oxygen, they must have obtained the energy which they needed for growth by some other process than oxidation—in fact, by fermentation. For, as Pasteur put it, fermentation is life without oxygen. If this was so, we should expect that high organisms like ourselves would start life as anaerobic beings, just as we start as single cells. This is the case. Embryo chicks for the first two or three days after fertilization use very little oxygen, but obtain the energy which they need for growth by fermenting sugar into lactic acid, like the bacteria which turns milk sour. So do various embryo mammals, and in all probability you and I lived mainly by fermentation during the first week of our pre-natal life. The cancer cell behaves in the same way. Warburg has shown that with its embryonic habit of unrestricted growth there goes an embryonic habit of fermentation.

The first living or half-living things were probably large molecules synthesized under the influence of the Sun's radiation, and only capable of reproduction in the particularly favourable medium in which they originated. Each presumably required a variety of highly specialized molecules before it could reproduce itself, and it depended on chance for a supply of them. This is the case today with most viruses, including the bacteriophage, which can grow only in presence of the complicated assortment of molecules found in a living cell.

The unicellular organisms, including bacteria, which were the simplest living things known a generation ago, are far more complicated. They are organisms—that is to say, systems whose parts co-operate. Each part is specialized to a particular chemical function, and prepares chemical molecules suitable for the growth of the other parts. In consequence, the cell as a whole can usually subsist on a few types of molecule, which are transformed within it into the more complex substances needed for the growth of the parts.

The cell consists of numerous half-living chemical molecules suspended in water and enclosed in an oily film. When the whole sea was a vast chemical laboratory the conditions for the formation of such films must have been relatively favourable; but for all that life may have remained in the virus stage for many millions of years before a suitable assemblage of elementary units was brought together in the first cell. There must have been many failures, but the first successful cell had plenty of food, and an immense advantage over its competitors.

It is probable that all organisms now alive are descended from one ancestor, for the following reason. Most of our structural molecules are asymmetrical, as shown by the fact that they rotate the plane of polarized light, and often form asymmetrical crystals. But of the two possible types of any such molecule, related to one another like a right and left boot, only one is found throughout living nature. The apparent exceptions to this rule are all small molecules which are not used in the building of the large structures which display the phenomena of life. There is nothing, so far as we can see, in the nature of things to prevent the existence of looking-glass organisms built from molecules which are, so to say, the mirror-images of those in our own bodies. Many of the requisite molecules have already been made in the laboratory. If life had originated independently on several occasions, such organisms would probably exist. As they do not, this event probably occurred only once, or, more probably, the descendants of the first living organism rapidly evolved far enough to overwhelm any later competitors when they arrived on the scene.

As the primitive organisms used up the foodstuffs available in the sea some of them began to perform in their own bodies the synthesis formerly performed haphazardly by the sunlight, thus ensuring a liberal supply of food. The first plants thus came into existence, living near the surface of the ocean, and making food with the aid of sunlight as do their descendants today. It is thought by many biologists that we animals are descended from them. Among the molecules in our own bodies are a number whose structure resembles that of chlorophyll, the green pigment with which the plants have harnessed the sunlight to their needs. We use them for other purposes than the plants—for example, for carrying oxygen—and we do not, of course, know whether they are, so to speak, descendants of chlorophyll or merely cousins. But since the oxygen liberated by the first plants must have killed off most of the other organisms, the former view is the more plausible.

The above conclusions are speculative. They will remain so until living creatures have been synthesized in the biochemical laboratory. We are a long way from that goal. It was only this year that Pictel for the first time made cane-sugar artificially. It is doubtful whether any enzyme has been obtained quite pure. Nevertheless I hope to live to see one made artificially. I do not think I shall behold the synthesis of anything so nearly alive as a bacteriophage or a virus, and I do not suppose that a self-contained organism will be made for centuries. Until that is done the origin of life will remain a subject for speculation. But such speculation is not idle, because it is susceptible of experimental proof or disproof.

Some people will consider it a sufficient refutation of the above theories to say that they are materialistic, and that materialism can be refuted on philosophical grounds. They are no doubt compatible with materialism, but also with other philosophical tenets. The facts are, after all, fairly plain. Just as we know of sight only in connection with a particular kind of material system called the eye, so we know only of life in connection with certain arrangements of matter, of which the biochemist can give a good, but far from complete, account. The question at issue is: 'How did the first such system on this planet originate?' This is a historical problem to which I have given a very tentative answer on the not unreasonable hypothesis that a thousand million years ago matter obeyed the same laws that it does today.

This answer is compatible, for example, with the view that pre-existent mind or spirit can associate itself with certain kinds of matter. If so, we are left with the mystery as to why mind has so marked a preference for a particular type of colloidal organic substances. Personally I regard all attempts to describe the relation of mind to matter as rather clumsy metaphors. The biochemist knows no more, and no less, about this question than anyone else. His ignorance disqualifies him no more than the historian or the geologist from attempting to solve a historical problem.

9

Svante Arrhenius
The Propagation of Life in Space

One of the more intriguing notions concerning the first appearance of life on Earth is that of "panspermia" ("all-seeding" in Greek), first proposed by the Swedish chemist Svante Arrhenius in 1903 in an article in the popular German magazine *Die Umschau*. Arrhenius suggested that life can travel through interstellar space, driven by radiation pressure from the sun and other stars, in the form of spores that would seed one planet from another. Of course, this theory does not purport to explain the ultimate origin of life, but it does make important predictions about the ways in which life on different planets should show a resemblance. The article from *Die Umschau* has been translated by Donald Goldsmith with the assistance of Raymond Goldsmith and Annemarie Kleinert.

In 1873, James Clerk Maxwell published his epochal work on the nature of magnetism and electricity. He demonstrated therein that light, or in general any radiation, must exert a pressure on irradiated objects; for our purposes, we may call this pressure "radiation pressure." In recent years this pressure has played a very important role in the explanation of celestial phenomena, and it appears that almost all the cosmic mysteries which could previously not be clarified may find an explanation through radiation pressure. According to Maxwell's calculation, the radiation pressure at the sun's surface is 2.75 milligrams per square centimeter, which means no more than 1/400,000 of the pressure of the atmosphere at the Earth's surface. In the Earth's vicinity the radition pressure from sunlight is 46,000 times weaker, thus only 1/18,400,000,000 of an atmosphere. This is an extraordinarily small value and one must not be surprised that the experimental confirmation of this pressure was successfully achieved only recently, by the Russian scientist Lebedev. . . .

The effect of radiation pressure can also be observed on comet tails, which are pushed away from the sun, and on the solar corona, which has a radial structure and consists mainly of small hard particles, as one can conclude from its spectrum.

This radiation pressure, which flings the tiniest particles into space with enormous velocity, must have played an important part in the transmission of living organisms from one planet to another. Already in the dawn of our culture the question was raised: How did living organisms reach the Earth? Did they originate here or did they reach us from elsewhere? Today the attitude toward so-called "spontaneous generation" is the more common one. It is generally believed, that with a suitable temperature and humidity, which no longer occur on the Earth, but which perhaps once did exist, organic living bodies can form. This contradicts all our experiences: Attempts in this direction have all failed until now; I simply recall Pasteur's work. Therefore I share Lord Kelvin's belief that this hypothesis is untenable. Lord Kelvin, one of the greatest physicists now living, assumes that fragments of foreign bodies occasionally reached the Earth and brought small organic germs with them, which continued to develop on Earth. And what holds true for the Earth also holds true for the other planets which are accessible to life.

This view, however, presents many difficulties. For a terrestrial object to escape from the solar system, it must possess a velocity of about 45 kilometers per second, or about 70 times more than our fastest projectiles. As is well known, projectiles become red-hot while rushing through the air. An object that had a velocity ten times as great would become much hotter still. Therefore, one can scarcely assume that a terrestrial object could escape the solar system without all the germs on its surface being burnt. We also know that meteorites which fall from the

From *Die Umschau,* 7 (1903), 481. Translated by Donald Goldsmith.

sky have a melted crust. To overcome these difficulties, Lord Kelvin supposes that the germs somehow came here in the cracks of fragments thrown out by volcanic force. How? Of this he has given no details. It also cannot be well understood. Moreover the enormous speeds would likewise be incomprehensible.

It is therefore a true relief that one can transform Kelvin's idea with the help of radiation pressure so that these difficulties are avoided. We know that there are living beings of such small dimensions that the radiation pressure from the sun exceeds their weight. Just at the limit of the magnifying power of our microscopes, a host of small bacteria with diameters of 0.2 to 0.3 microns exist, and it can hardly be doubted that still smaller beings than this limit exist, which cannot be discovered with the microscope. We also know that there are infectious outbreaks which are doubtless the result of organisms, although these cannot be discovered with a microscope. For example, the carriers of rabies, of foot-and-mouth disease, as well as a tobacco sickness in Sumatra, are caused by such small living beings.

Still smaller must be the spores that are those enduring forms of lower plant life, which can last a longer time without dying off. Spores are exceptionally resistant to outside influences, and it is especially interesting to know that one can subject them to the lowest temperatures yet attained without harming them. They have been preserved in liquid hydrogen, which means at temperatures of -252° C, for about a week, without losing their power of germination. . . .

[Spores] of 0.15 microns diameter will need 3,000 years to travel [from the solar system] to the nearest star, and about half a year to reach the farthest planet in our own solar system.

Now we can ask: *Can spores remain capable of germination for so long?* We know of many plants, grain for example, which retain their powers of germination for a long period. It has been even reported that grain found in the graves of Egyptian kings has germinated upon planting. A closer investigation, however, has shown that these reports are not very trustworthy. The same applies to spores, which can live unbelievably long under ordinary conditions; they are also subject to metabolism and this metabolism cannot be prolonged to infinity. Yet of metabolism doubtless the same is true as of chemical reactions. All reaction speeds decrease when the temperature is lowered, and we can reduce the speeds in this way almost indefinitely, for example to one part in 10^{50} by lowering the temperature from 17° C to -223° C (the temperature in the vicinity of the orbit of Neptune). Now if spores obey, with respect to metabolism, the law of usual reaction speeds, and if, furthermore, they are put at extremely low temperatures, then they will retain their powers of germination practically forever, and will be able to withstand at trip of 3,000 years very well, and thus will be able to reach the nearest stars, in condition able to germinate.

Now they reach another object in the universe that is still free of organisms. Those objects also pass through different epochs. At the beginning they are red-hot and gaseous; later they are covered with a hard crust. After this has occurred, the temperature falls rapidly. In one hundred years, according to Lord Kelvin, the temperature has fallen below 100° C, so that living beings can appear there. In a shorter or longer time every heavenly body will probably be covered with a crust and be accessible to life some time later. We know that algae live in geysers in New Zealand which must withstand a temperature up to more than 90° C. It is therefore not incorrect to assume that as soon as liquid water and a carbon-dioxide–containing atmosphere are at hand on a celestial object, organisms (plants) can exist. Of course these are only quite low forms of life, but according to the Darwinian theory they can evolve until the highest organisms arise.

According to this view it is quite conceivable that the living beings on all planets are related, and that a planet, as soon as it can shelter organic life, is soon occupied by such organic life.

In this way the question of the origin of life is not shunted aside, since the question of the origin of the first living beings is, in my opinion, at the same level as that of the origin of matter. We must gradually accustom ourselves to think that living beings have endured for eternity, and thus have no origin in time, that they originate through germs which come from other celestial bodies, that they die out when conditions have become unfavorable, but that they then live on elsewhere in the universe where organic life is possible. Such heavenly bodies accessible to life have beyond doubt always existed. In the solar system the outer planets are certainly gaseous and can support no organic living beings of the same nature as those on Earth. But they will in future times develop so far that they cool down, and then they can possibly obtain living beings from our Earth and will develop their lives further, as life on Earth dies out.

10

F. H. C. Crick and L. E. Orgel
Directed Panspermia

It might seem that Arrhenius's idea has been disproven by the discovery of just how hostile to life the interstellar environment appears to be. But as the 1972 article by the brilliant biologists Francis Crick and Leslie Orgel shows, the idea of panspermia is too good to abandon entirely. These authors propose that *directed panspermia*—life sent on its way deliberately from one planet to another—cannot be entirely rejected as an explanation for the origin of life on Earth.

It was not until the middle of the nineteenth century that Pasteur and Tyndall completed the demonstration that spontaneous generation is not occurring on the Earth nowadays. Darwin and a number of other biologists concluded that life must have evolved here long ago when conditions were more favourable. A number of scientists, however, drew a quite different conclusion. They supposed that if life does not evolve from terrestrial nonliving matter nowadays, it may never have done so. Hence, they argued, life reached the earth as an "infection" from another planet (Oparin, 1957).

Arrhenius (1908) proposed that spores had been driven here by the pressure of the light from the central star of another planetary system. His theory is known as Panspermia. Kelvin suggested that the first organisms reached the Earth in a meteorite. Neither of these theories is absurd, but both can be subjected to severe criticism. Sagan (Shklovski and Sagan, 1966; Sagan and Whitehall, 1973) has shown that any known type of radiation-resistant spore would receive so large a dose of radiation during its journey to the Earth from another Solar System that it would be extremely unlikely to remain viable. The probability that sufficiently massive objects escape from a Solar System and arrive on the planet of another one is considered to be so small that it is unlikely that a single meteorite of extrasolar origin has ever reached the surface of the Earth (Sagan, private communication). These arguments may not be conclusive, but they argue against the

"infective" theories of the origins of life that were proposed in the nineteenth century.

It has also been argued that "Infective" theories of the origins of terrestrial life should be rejected because they do no more than transfer the problem of origins to another planet. This view is mistaken; the historical facts are important in their own right. For all we know there may be other types of planet on which the origin of life *ab initio* is greatly more probable than on our own. For example, such a planet may possess a mineral, or compound, of crucial catalytic importance, which is rare on Earth. It is thus important to know whether primitive organisms evolved here or whether they arrived here from somewhere else. Here we reexamine this problem in the light of more recent biological and astronomical information.

Our Present Knowledge of the Galaxy

The local galactic system is estimated to be about 13×10^9 yr old (Metz, 1972). The first generation of stars, because they were formed from light elements, are unlikely to have been accompanied by planets. However, some second generation stars not unlike the Sun must have formed within 2×10^9 yr of the origin of the galaxy (Blaauw and Schmidt, 1965). Thus it is quite probable that planets not unlike the Earth existed as much as 6.5×10^9 yr before the formation of our own Solar System.

We know that not much more than 4×10^9 yr elapsed between the appearance of life on the Earth (wherever it came from) and the development of our own technological

Reprinted from *Icarus*, 19, (1973), 341. Copyright 1973 by Academic Press, Inc.

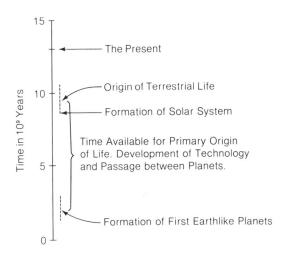

Figure 10.1. An approximate time-scale for the events discussed in the paper. To simplify illustration the age of the galaxy has been somewhat arbitrarily taken as 13×10^9 yr.

society. The time available makes it possible, therefore, that technological societies existed elsewhere in the galaxy *even before the formation of the Earth*. We should, therefore, consider a new "infective" theory, namely that a primitive form of life was deliberately planted on the Earth by a technologically advanced society on another planet.

Are there many planets which could be infected with some chance of success? It is believed, though the evidence is weak and indirect, that in the galaxy many stars, of a size not dissimilar to our Sun, have planets, on a fair fraction of which temperatures are suitable for a form of life based on carbon chemistry and liquid water, as ours is. Experimental studies of the production of organic chemicals under prebiotic conditions make it seem likely that a rich prebiotic soup accumulates on a high proportion of such Earthlike planets. Unfortunately, we know next to nothing about the probability that life evolves within a few billion years in such a soup, either on our own special Earth, or still less on other Earthlike planets.

If the probability that life evolves in a suitable environment is low we may be able to prove that we are likely to be alone in the galaxy (Universe). If it is high the galaxy may be pullulating with life of many different forms. At the moment we have no means at all of knowing which of these alternatives is correct. We are thus free to postulate that there have been (and still are) many places in the galaxy where life could exist but that, in at least a fraction of them, after several billion years the chemical systems had not evolved to the point of self-replication and natural selection. Such planets, if they do exist, would form an excellent breeding ground for external microorganisms. Note that because many if not all such planets would have

a reducing atmosphere they would not be very hospitable to the higher forms of life as we know them on Earth.

Our Proposal

The possibility that terrestrial life derives from the deliberate activity of an extraterrestrial society has often been considered in science fiction and more or less light-heartedly in a number of scientific papers. For example, Gold (1960) has suggested that we might have evolved from the microorganisms inadvertently left behind by some previous visitors from another planet (for example, in their garbage). Here we wish to examine a very specific form of Directed Panspermia. Could life have started on Earth as a result of infection by microorganisms sent here deliberately by a technological society on another planet, by means of a special long-range unmanned spaceship? To show that this is not totally implausible we shall use the theorem of detailed cosmic reversibility; if we are capable of infecting an as yet lifeless extrasolar planet, then, given that the time was available, another technological society might well have infected our planet when it was still lifeless.

The Proposed Spaceship

The spaceship would carry large samples of a number of microorganisms, each having different but simple nutritional requirements, for example blue-green algae, which could grow on CO_2 and water in "sunlight." A payload of 1000 kg might be made up of 10 samples each containing 10^{16} microorganisms, or 100 samples each of 10^{15} microorganisms.

It would not be necessary to accelerate the spaceship to extremely high velocities, since its time of arrival would not be important. The radius of our galaxy is about 10^5 light years, so we could infect most planets in the galaxy within 10^8 yr by means of a spaceship traveling at only one-thousandth of the velocity of light. Several thousand stars are within a hundred light years of the Earth and could be reached within as little as a million years by a spaceship travelling at only 60,000 mph, or within 10,000 yr if a speed of one-hundredth of that of light were possible.

The technology required to carry out such an act of interstellar pollution is not available at the present time. However, it seems likely that the improvements in astronomical techniques will permit the location of extrasolar planets within the next few decades. Similarly, the problem of sending spaceships to other stars, at velocities low compared with that of light, should not prove insoluble once workable nuclear engines are available. This again is likely to be within a few decades. The most difficult problem would be presented by the long flight times; it is not clear how long it will be before we can build components that would survive in space for periods of thousands or millions of years.

Although there are some techological problems associated with the distribution of the microorganisms in viable form after a long journey through space, none of them seems insuperable. Some radiation protection could be provided during the journey. Suitable packaging should guarantee that small samples, including some viable organisms, would be widely distributed. The question of how long microorganisms, and in particular bacterial spores, could survive in a spaceship has been considered in a preliminary way by Sneath (1962). He concludes "that life could probably be preserved for periods of more than a million years if suitably protected and maintained at temperatures close to absolute zero." Sagan (1960) has given a comparable estimate of the effects of radiation damage. We conclude that within the foreseeable future we could, if we wished, infect another planet, and hence that it is not out of the question that our planet was infected.

We can in fact go further than this. It may be possible in the future to send either mice or men or elaborate instruments to the planets of other Solar Systems (as so often described in science fiction) but a rocket carrying microorganisms will always have a much greater effective range and so be advantageous if the sole aim is to spread life. This is true for several reasons. The conditions on many planets are likely to favour microorganisms rather than higher organisms. Because of their extremely small size vast numbers of microorganisms can be carried, so much more wastage can be accepted. The ability of microorganisms to survive, without special equipment, both storage for very long periods at low temperatures and also an abrupt change back to room temperatures is also a great advantage. Whatever the potential range for infection by other organisms, microorganisms can almost certainly be sent further and probably much further.

It should be noted that most of the earliest "fossils" so far recognized are somewhat similar to our present bacteria or blue-green algae. They occur in cherts of various kinds and are estimated to be up to 3×10^9 yr old. This makes it improbable that the Earth was ever infected merely by higher organisms.

Motivation

Next we must ask what motive we might have for polluting other planets. Since we would not derive any direct advantage from such a programme, presumably it would be carried through either as a demonstration of technological capability or, more probably, through some form of missionary zeal.

It seems unlikely that we would deliberately send terrestrial organisms to planets that we believed might already be inhabited. However, in view of the precarious situation on Earth, we might well be tempted to infect other planets if we became convinced that we were alone in the galaxy (Universe).[1] As we have already explained we cannot at the moment estimate the probability of this. The hypothetical senders on another planet may have been able to prove that they were likely to be alone, and to remain so, or they may have reached this conclusion mistakenly. In either case, if they resembled us psychologically, their motivation for polluting the galaxy would be strong, if they believed that all or even the great majority of inhabitable planets could be given life by Directed Panspermia.

The psychology of extraterrestrial societies is no better understood than terrestrial psychology. It is entirely possible that extraterrestrial societies might infect other planets for quite different reasons than those we have suggested. Alternatively, they might be less tempted than we would be, even if they thought that they were alone. The arguments given above, together with the principle of cosmic reversibility, demonstrate the possibility that we have been infected, but do not enable us to estimate the probability.

Possible Biological Evidence

Infective theories of the origins of terrestrial life could be taken more seriously if they explained aspects of biochemistry or biology that are otherwise difficult to understand. We do not have any strong arguments of this kind, but there are two weak facts that could be relevant.

The chemical composition of living organisms must reflect to some extent the composition of the environment in which they evolved. Thus the presence in living organisms of elements that are extremely rare on the Earth might indicate that life is extraterrestrial in origin. Molybdenum is an essential trace element that plays an important role in many enzymatic reactions, while chromium and nickel are relatively unimportant in biochemistry. The abundance of chromium, nickel, and molybdenum on the Earth are 0.20, 3.16, and 0.02%, respectively. We cannot conclude anything from this single example, since molybdenum may be irreplaceable in some essential reaction—nitrogen fixation, for example. However, if it could be shown that the elements represented in terrestrial living organisms corelate closely with those that are abundant in some class of star—molybdenum stars, for example—we might look more sympathetically at "infec-

---

[1]In a somewhat different context the seeding of Venus and other solar planets has been suggested by C. Sagan (1961b), and T. Gold, private communication.

tive" theories.

Our second example is the genetic code. Several orthodox explanations of the universality of the genetic code can be suggested, but none is generally accepted to be completely convincing. It is a little surprising that organisms with somewhat different codes do not coexist. The universality of the code follows naturally from an "infective" theory of the origins of life. Life on Earth would represent a clone derived from a single extraterrestrial organism. Even if many codes were represented at the primary site where life began, only a single one might have operated in the organisms used to infect the Earth.

Conclusion

In summary, there is adequate time for technological society to have evolved twice in succession. The places in the galaxy where life could start, if seeded, are probably very numerous. We can foresee that we ourselves will be able to construct rockets with sufficient range, delivery ability, and surviving payload if microorganisms are used. Thus the idea of Directed Panspermia cannot at the moment be rejected by any simple argument. It is radically different from the idea that life started here *ab initio* without infection from elsewhere. We have thus two sharply different theories of the origin of life on Earth. Can we choose between them?

At the moment it seems that the experimental evidence is too feeble to make this discrimination. It is difficult to avoid a personal prejudice, one way or the other, but such prejudices find no scientific support of any weight. It is thus important that both theories should be followed up. Work on the supposed terrestrial origin of life is in progress in many laboratories. As far as Directed Panspermia is concerned we can suggest several rather diverse lines of research.

The arguments we have employed here are, of necessity, somewhat sketchy. Thus the detailed design of a long-range spaceship would be worth a careful feasibility study. The spaceship must clearly be able to home on a star, for an object with any appreciable velocity, if dispatched in a random direction, would in almost all cases pass right through the galaxy and out the other side. It must probably have to decelerate as it approached the star, in order to allow the safe delivery of the payload. The packets of microorganisms must be made and dispersed in such a way that they can survive the entry at high velocity into the atmosphere of the planet, and yet be able to dissolve in the oceans. Many useful feasibility studies could be carried out on the engineering points involved.

On the biological side we lack precise information concerning the life-time of microorganisms held at very low temperatures while traveling through space at relatively high velocities. The rocket would presumably be coasting most of the time so the convenient temperature might approximate that of space. How serious is radiation damage, given a certain degree of shielding? How many distinct types of organism should be sent and which should they be? Should they collectively be capable of nitrogen fixation, oxidative phosphorylation and photosynthesis? Although many "soups" have been produced artificially in the laboratory, following the pioneer experiments of Miller, as far as we know no careful study has been made to determine which present-day organisms would grow well in them under primitive Earth conditions.

At the same time present-day organisms should be carefully scrutinized to see if they still bear any vestigial traces of extraterrestrial origin. We have already mentioned the uniformity of the genetic code and the anomalous abundance of molybdenum. These facts amount to very little by themselves but as already stated there may be other as yet unsuspected features which, taken together, might point to a special type of planet as the home of our ancestors.

These enquiries are not trivial, for if successful they could lead to others which would touch us more closely. Are the senders or their descendants still alive? Or have the hazards of 4 billion years been too much for them? Has their star inexorably warmed up and frizzled them, or were they able to colonise a different Solar System with a short-range spaceship? Have they perhaps destroyed themselves, either by too much aggression or too little? The difficulties of placing any form of life on another planetary system are so great that we are unlikely to be their sole descendants. Presumably they would have made many attempts to infect the galaxy. If the range of their rockets were small this might suggest that we have cousins on planets which are not too distant. Perhaps the galaxy is lifeless except for a local village, of which we are one member.

One further point deserves emphasis. We feel strongly that under no circumstances should we risk infecting other planets at the present time. It would be wise to wait until we know far more about the probability of the development of life on extrasolar planets before causing terrestrial organisms to escape from the solar system.

Acknowledgments

We are indebted to the organisers of a meeting on Communication with Extraterrestrial Intelligence, held at Byurakan Observatory in Soviet Armenia in September 1971, which crystallized our ideas about Panspermia. We thank Drs. Freeman Dyson, Tommy Gold, and Carl Sagan for discussion and important comments on our argument.

The Origin of Life: The Search Begins

We still have no real knowledge of the actual processes that led to the first living organisms, and are unlikely to obtain such knowledge without many more years of investigation. Meanwhile, we may be proud that our ability to understand where we came from has advanced to the point that we can generalize with some confidence about likely conditions for life in other locales throughout the universe.

Twenty years ago, Joshua Lederberg introduced the space age to the question of "exobiology" at the First International Space Science Symposium. His comments have borne up well under the new information of two decades; only the limitations placed on the possibility of life elsewhere in the solar system by space vehicles have changed the assessments Lederberg made. In a more recent article, Richard Dickerson provides a masterly summary of what we now know about the origin of life on Earth, and the key time of "pollution" through the liberation of life-produced oxygen.

In a minority view on the origin of life, Fred Hoyle and Chandra Wickramasinghe have suggested that interstellar clouds are good sites in which to form organic molecules, perhaps amino acids. Hoyle and Wickramasinghe have gone on to suggest that life itself may have often developed in such clouds; that comets, the most primitive members of our solar system, may carry viruses and simple forms of life; that encounters with comets may have produced plagues on Earth in past eras; and that such encounters explain the widespread human superstition that comets are harbingers of evil. These views, although highly significant if true, have been judged unlikely in the extreme by other scientists who consider the origin and cosmic distribution of life. But the idea that comets may carry organic material, and that such material could give the origin of life a "head start" on planetary surfaces, seems worth considering seriously.

11

Joshua Lederberg
Exobiology: Approaches to Life beyond the Earth

It is a privilege to discuss some basic problems in biology with an audience whose special concern is for the recent striking advances in the physics of the earth in the solar system. However, many of us are looking forward to the close investigation of the planets, and few inquisitive minds can fail to be intrigued by what these studies will tell of the cosmic distribution of life. To conform to the best of our contemporary science, much thoughtful insight, meticulous planning, and laboratory testing must still be invested in the experimental approaches to this problem. This may require international cooperation and also—perhaps more difficult—mutual understanding among scientific disciplines as isolated as biochemical genetics and planetary astronomy.

Many discussions of space exploration have assumed that exobiological studies might await the full development of the technology for manned space flight and for the return of planetary samples to the terrestrial laboratory. To be sure, these might be preceded by some casual experiments on some instrumented landings. One advantage of such a program is that time would allow exobiological experiments to be planned with composure and deliberation. Undoubtedly, this planning would be more rigorous insofar as it was based on improved knowledge, from closer approaches, of the chemistry and physics of planetary habitats. Unfortunately, this orderly and otherwise desirable program takes insufficient account of the capacity of living organisms to grow and spread throughout a new environment. This unique capacity of life which engages our deepest interest also generates our gravest concerns in the scientific management of missions beyond the earth. On account of these, as well as of the immense costs of interplanetary communication, we are obliged to weigh the most productive experiments that we can do by remote instrumentation in early flights, whether or not manned space flight eventually plays a role in scientific exploration.

Reprinted from *Science*, 132 (1960), 393. Copyright 1960 by the American Association for the Advancement of Science.

Motivations for Exobiological Research

The demons which lurk beyond the Pillars of Hercules have colored the folklore and literature of ages past and present, not always to the benefit of fruitful exploration and dispassionate scientific analysis. Apart from such adventuresome amusements and the amateur delights of a cosmically enlarged natural history, how does exobiology relate to contemporary science and culture? The exploration of space may seem to have very little to do with fundamental questions in biology or medicine, with the role of genes in embryological development, protein synthesis, the biology of viruses, and the evolution of species. The physical sciences may sharpen our perspective. Twenty-five centuries of scientific astronomy have widened the horizons of the physical world, and the casual place of the planet Earth in the expanding universe is a central theme in our modern scientific culture. The dynamics of celestial bodies, as observed from the earth, is the richest inspiration for the generalization of our concepts of mass and energy throughout the universe. The spectra of the stars likewise testify to the universality of our concepts in chemistry. But biology has lacked tools for such extension, and "life" until now has meant only terrestrial life. This disparity in the domains of the physical versus the biological sciences attenuates most of our efforts to construct a theoretical biology as a cognate of theoretical physics or chemistry. For the most part, biological science has been the rationalization of particular facts, and we have had all too limited a basis for the construction and testing of meaningful axioms to support a theory of life. At present, perhaps the only potentially universal principle in biology is the Darwinian concept of evolution through the natural selection of random hereditary fluctuations.

Some chemical attributes of terrestrial life might support a claim to be basic principles: for example, polyphosphates (adenylpyrophosphate) occur in all organisms as coupling agents for the storage and transfer of metabolic energy. But, at least in principle, we can imagine that organisms may have found alternative solutions to the same problem. Only the perspective of comparative biology on a cosmic scale could tell whether this device is an indispensable element of all life or a particular attribute of its local occurrence on this planet.

An important aim of theoretical biology is an abstract definition of life. Our only consensus so far is that such a definition must be arbitrary. If life has gradually evolved from inanimate matter, the demarcation of chemical from biological evolution is one of useful judgment. For a working principle, we might again rely upon the evolutionary concept: a living system has those properties (of self-replication and metabolism) from which we may with more or less confidence deduce an evolutionary scheme that would encompass self-evidently living organisms. But I do not propose this as a rote formula for the assessment of life on other celestial bodies, and certainly not before we have some empirical knowledge of the diversities of chemical evolution.

From this standpoint, the overriding objective of exobiological research is to compare the over-all patterns of chemical evolution of the planets, stressing those features which are globally characteristic of each of them.

We are all thinking of the question: "Is there life on Mars?" To answer it may require a careful reassessment of our meaning of "life" and matching this with the accumulation of hard-won evidence on the chemical composition of that planet. On the other hand, we might be confronted with an object obviously analogous to an earthly plant, animal or microbe. But even this abrupt answer would be trivial in deference to a biochemical analysis of the organism and of its habitat for comparison with the fundamentals of terrestrial life.

In our first approaches to the nearby planets we will wish to design experiments which have some tangible foundation in the present accumulation of biochemical knowledge. The aqueous environment, and its corollary of moderate temperatures in which large carbonaceous molecules are reasonably stable, are implicit in terrestrial biochemistry. This is not to reject the abstract possibility of nonaqueous life, or noncarbonaceous molecules that might characterize temperatures of $< 200°$ or $> 500°$ K. However, we can defer our concern for such exotic biological systems until we have got full value from our searches for the more familiar, and have learned enough of the exotic chemistry to judge how to proceed.

Within the bounds of its aqueous environment, what are the most nearly universal features of terrestrial life? In fact, our plants, animals, and bacteria share a remarkable list of biochemical components, and a biochemist cannot easily distinguish extracts of yeast cells and beef muscle. Among these components, the nucleic acids warrant first attention. Although they constitute the hereditary material, so that all the variety of terrestrial life can be referred to subtle differences in the nucleic acids, the same basic

structure is found in the nuclei of all cells. This is a long, linear polymer fabricated from a sugar-phosphate repeating unit:

where R is a purine side group:

Adenine   Guanine

or a pyrimidine side group:

Thymine   Cytosine

The meaningful variety of nucleic acids depends on the specific order of the side group attached to each sugar on this monotonous backbone, a linear message written in a language of four letters, A, G, T, and C. The bacteria, which are the simplest free-living organisms, contain nucleotide sequences about 5 million units long; man contains sequences about 5000 million units long—this content being one of our best objective measures of biological complexity. On the other hand, the simplest viruses, which can multiply only inside living cells and come close to being single genes, have about 2500 units per particle. Playing a central role in the unification of terrestrial biology, nucleic acids underlie both heredity and (through their control of protein synthesis) development. Are they the only linear polymers which can subsume these functions, or have many other fundamental types evolved, to be found on other celestial bodies?

Equally general among the constituents of living cells are the proteins, which are also polymers, but with a more diverse set of constituents, some 20 amino acids. The fundamental backbone of a protein is a poly-amino acid chain:

$$H-NH-CH-CO-NH-CH-CO\ldots$$
$$\mid \qquad\qquad \mid$$
$$R \qquad\qquad R$$
$$\ldots NH-CH-CO-OH$$
$$\mid$$
$$R$$

where R may be any of 20 different groups, distinguishing a like number of amino acids found in natural proteins. Proteins assume a wide variety of three-dimensional shapes, through coiling and cross-linking of the polymer chains. They are in this way suited to perform such diverse functions as those of enzymes, structural elements, and antibodies. Not only do we find just the same 20 amino acids among the proteins of all terrestrial organisms but these are all the levo- isomers, although dextro- amino acids are found to have other metabolic functions. Next only to the incidence of nucleic acids, we would ask whether exobiota make analogous use of proteins, comprising the same amino acids, in hopes of understanding what seem to be random choices in the sculpture of our own living form.

Common to all forms of terrestrial life are also a number of smaller molecules which are involved in the working metabolism of the cells; for example, most of the B vitamins have a perfectly general distribution. They are vitamins for us only because we have learned, in our evolutionary history, to rely on their production by green plants, rather than to synthesize them within our own cells. But once formed, these vitamins, and similar categories of substances such as porphyrins, play entirely analogous roles in the metabolism of all cells.

A few substances, such as the steroid hormones, do play special roles in the metabolism of higher organisms, and testify to some progress in biochemical evolution. In fact, most objective evidence points to a loss of specific functions—microorganisms are certainly more versatile and less dependent than man is on a specific nutrient milieu. The main trend of biochemical evolution, from microbe to man, has been far less the innovation of new unit processes than the coordination of existing processes in time and space.

While we propose to give first priority to these most general questions, they by no means exhaust our interest in the peculiarities of extraterrestrial organisms, any more than they would for a newly discovered phylum of the earth's own repertoire. Nor should we preclude the possibility of finding new organisms that might be economically useful to man, just as new organisms were among the most fruitful yields of geographic exploration. However, the enlargement of our understanding, rather than of our zoos and botanical gardens, is surely our first objective.

Theories of the Origin of Life

At this point, a consideration of contemporary theory on the origin of life is justified for two reasons: (i) exobiological research gives us a unique, fresh approach to this problem, and (ii) we can find some basis to conclude that life need not be so improbable an evolutionary development as had once been supposed.

The interval between Pasteur's work on spontaneous generation and the recent past has been especially difficult for the mechanistic interpretation of the origin of life. Before Pasteur's time, many investigators could believe that simple microorganisms arose spontaneously in nutrient media. His demonstration that such media remained sterile if properly sterilized and protected seemed to rule out any possibility of "spontaneous generation." His conclusion was, of course, overdrawn, since life must have evolved at least once, and the event could still occur, though very much less frequently than had been supposed before. Meanwhile, the problem was compounded by the growth of biological knowledge. We now realize that bacteria, small as they are, are still extremely complex, well-ordered, and representative organisms. The first organisms must have been far simpler than present-day free-living bacteria.

With the growth of genetics since 1900, and the recognition of the self-replicating gene as the elementary basis of life, the question could focus on the origin of the first genetic molecule: given the power of self-replication, and incidents of stochastic variation, Darwin's principle could account for the eventual emergence of any degree of biological complexity.

An immense amount of fruitful genetic work was done in a period when "genetic molecule" was an abstraction and "self-replication" was an axiomatic principle whose chemical basis seemed beyond the grasp of human understanding. Now we recognize that the nucleic acids are the material basis of heredity, and we can begin to construct mechanistic models of their replication. The first principle, as already stated, is that the gene is a string of nucleotides, each position in the string being marked by one of the four nucleotide units A, T, C, and G. The polymerization of such strings by the union of the monomeric units presents no fundamental problems, but self-replication would necessitate the assembly of the units in a specific order, the one dictated by the order of the nucleotides in the parent molecule. The key to the solution of this problem was the realization by Watson and Crick that the complete nucleic acid molecule is a rigid, duplex structure in which two strings are united. In that rigid structure, as can be shown by suitable molecular models, adenine occupies a space which is just complementary to that of thymine, and cytosine is likewise complementary to guanine. A string can therefore replicate—that is, direct the assembly of another daughter string—in the following way. The nutrient mix of the cell contains all four nucleotide units. However, at any position of the parent nucleic acid molecule only one of these four can make a suitable fit and will therefore be accepted. After being accepted, the daughter units are firmly bound together by new chemical linkages giving a well-defined daughter string. Kornberg has reconstructed most of these events in some detail, by means of extracts from bacteria, to the very verge of proving duplication of genes in a chemically defined system in the test tube.

However, the media in which such syntheses can occur, in the cell or even in the test tube, are extremely complex. Knowing that the simplest organisms would be the most dependent on their environments for raw materials, where did these precursors come from before living organisms had evolved the enzymes to manufacture them?

Thanks to the insight of Haldane, Oparin, Horowitz and others, we now realize that this paradox is a false one, though it dates to the confusion between "carbon chemistry" and "organic chemistry" which still exists in English terminology. In fact, in 1828, Wöhler had already shown that an organic compound, urea, could be formed experimentally from an inorganic salt, ammonium cyanate. A hundred years later, a number of routes for synthesis of geochemically significant amounts of complex organic materials were pointed out, for example, the hydrolysis of metallic carbides, and subsequent reactions of olefins with water and ammonia. More recently, Miller and Urey demonstrated the actual production of amino acids by the action of electric discharges on gas mixtures containing the hydrides NH_3, OH_2, and CH_4. This demonstration converges with the other argument that the primitive atmosphere of the earth had just such a reduced composition, becoming oxidized secondarily (and in part through photosynthetic separation of carbon from oxygen).

An alternative theory of origin of carbonaceous molecules is even more pervasive. Perhaps we associate carbon with life, and rocks and metals with physical phenomena; beyond doubt we tend to connote the latter with the predominant substance of the universe. In fact, as a glance at tables of cosmic abundance will show, the lighter elements are by far the most prevalent, and after the dispersed hydrogen and helium these are carbon, oxygen, and nitrogen. The primitive condensation of free atoms to form the interstellar smoke, and eventually the stars themselves, must entail the molecular aggregation $H + C + O + N$; that is, a large fraction of the condensed mass of the universe must consist, or once have consisted, of organic macromolecules of great complexity. The chief problem for their synthesis is in fact not a source of chemical energy but how to dissipate the excess energy of reactions of free atoms and radicals.

This aspect of astrophysics may have place for a remote biological analogy: Once a few molecules have formed, the energy of subsequent impacts can be dissipated among the vibrational degrees of freedom. That is, such molecules can function as nuclei of condensation. As seeds for further condensation, those molecules will be favored which (i) most readily dissipate the energy of successive impacts and (ii) can undergo molecular fission to increase the number of nuclei. The actual molecular chemistry of the interstellar (or prestellar) smoke is thus subject to a kind of natural selection and cannot be a purely random sampling of available atoms.

Whether the earth has retained remnants of this chemistry is hard to say. There is at least some evidence of it in the spectra of comets, and fragments from these continue to form part of the meteoroidal infall. These particles, unless associated with larger meteorites, would be unrecognizable after traversing the earth's atmosphere; they are among the possible treasures to be found buried in protected crevices on the moon.

Light traversing the interstellar smoke has been found to be polarized. If primitive aggregation plays some role in furnishing precursors for biological evolution, this polarization furnishes at least one bias for a decision between levo- and dextro- isomers.

At any rate, possible sources of probiotic nutrition no longer pose a problem. Before the appearance of voracious organisms, organic compounds would accumulate until they reached equilibrium with thermal and radiative decomposition, from which the oceans would furnish ample protection. Locally, the concentration of the soup would be augmented by selective evaporation, and by adsorption onto other minerals. The main gap in the theory, not yet bridged by any experiment, is the actual formation of a *replicating* polymer in such a morass. We are beginning to visualize the essential conditions for chemical replication, and its ultimate realization is foreshadowed both by biochemical studies of nucleic acids and by industrial syntheses of stereospecific polymers.

There is some controversy over whether nucleic acids were the first genes, partly because they are so complex, partly because their perfection hints at an interval of chemical evolution rather than one master stroke. The advantage of the nucleic acid hypothesis is that no other self-replicating polymers have so far been found. But, as an alternative speculation, a simplified protein might replicate by the complementary attachment of acidic versus basic units, perhaps the crudest possible method of assembly. The nucleic acids would be perfections on this theme for replication. The existent proteins do not replicate; with their variety of amino acids, they would have evolved as better adaptations for assuming specific shapes. A comparative view of independent evolutionary systems may at least serve to check such speculations.

Although many steps in the generation of living molecules remain to be re-created, we can state this as a relevant problem for exobiological study, with considerable optimism for the prevalence of life elsewhere. But a sterile planet, too, would be of extraordinary interest to biology for the insight it should give on the actual progress of probiotic chemical evolution.

Natural and Artificial Panspermia

In the foregoing discussion it was tacitly assumed that the evolution of planetary life was a local phenomenon, independent of the incidence of life elsewhere. But, at a time when *de novo* generation seemed less plausible than it does now, Arrhenius defended another hypothesis: *panspermia,* the migration of spores through space from one planet to another. The credibility of the panspermia hypothesis has been eroded mainly for two reasons: (i) the lack of a plausible natural mechanism for impelling a spore-bearing particle out of the gravitational field of a planet as large as the earth, or any planet large enough to sustain a significant atmosphere, and (ii) the vulnerability of such a particle to destruction by solar radiation. In any case, the panspermia hypothesis could be disparaged for evading the fundamental problem by transposing it to an unknown, perhaps scientifically unknowable, site. These difficulties have impeached the standing of panspermia as an experimentally useful hypothesis, but not its immense significance for cosmic biology. In its defense, it might be indicated that, in view of the dormancy of microorganisms in high vacuum and at low temperatures and of their relatively low cross section for ionizing radiations, the hazards of exposure to space may have been exaggerated. The chief hazard to microorganisms might come from solar ultraviolet radiation and the proton wind, but a thin layer of overlying material would shield a spore from these. For the impulsion of particles we might possibly appeal to impacts with other heliocentric bodies, be they grazing meteorites or planetoids in cataclysmic encounters—suggestions not more remote than those invoked for other astronomical phenomena. Nor can we be sure that all the electrokinetic mechanisms which Arrhenius may have had in mind can be excluded from applying to any single particle. In testing for panspermia, we would be concerned first of all for evidence of interplanetary transport of any material. The moon suggests itself as a nearby trap for particles of terrestrial origin, among which living spores or biochemical fragments of them, might be the most characteristic markers. At one spore per kilogram of sample (a weight ratio of 10^{-15}), the sensitivity of easy biological detection would partly compensate for the vulnerability of spores to physical hazards.

The development of rocket-impelled spacecraft has, of course, furnished a mechanism for producing artificial panspermia. Several authors have recently revived Haldane's passing suggestion that life might even have been disseminated by intelligent beings from other stellar sys-

tems. That another century of productive science and technology could give the human species this capability would be hard to dispute. The hypothesis is connected with the age or agelessness of the universe, and until we have a basis for decision on this point, and can make independent tests for intelligent life elsewhere, it must join natural panspermia in the limbo of irrefutable, untestable scientific hypotheses. The technique for attempted radio communication with nearby stars has been detailed recently by Cocconi and Morrison [Reading 20].

These new tools for the exploration of the universe have caught many of us unawares, and few can pretend to have recaptured their equilibrium in dealing with these concepts. Irrefutable notions have little scientific value unless they lead to attempts at verification. A priori arguments for the presence or absence of intelligent life on the planets or in nearby stellar systems are equally unconvincing. The skepticism of most scientists is justified not by conviction but by the consistency of negative evidence in the limited scientific data that have so far been collected.

Planetary Targets

The suitability for life of the accessible bodies of the solar system has already received ample attention. Mars is, of course, the likeliest target, most nearly resembling the habitat of the earth. The indicated scarcity of free moisture and oxygen would severely limit the habitability of Mars by man or most terrestrial animals. However, there seems little doubt that many simpler, earthly organisms could thrive there. Indeed, many students have concluded that Mars does have a biota of its own. The most pertinent evidence is perhaps the infrared reflection spectrum recorded by Sinton which indicates an accumulation of hydrocarbonaceous materials in the dark areas. This is complemented by Dollfus' report (1960) on the seasonal changes of granularity of these areas. The main reservation that must be registered is that these might be meteorological phenomena involving masses of material which may be carbonaceous but not necessarily living. Most such material on the earth's surface is associated with life. However, this may be connected with the greedy utilization of such compounds by organisms rather than their production by vital synthesis. However, the most plausible explanation of the astronomical data is that Mars is a life-bearing planet. (The term *vegetation* is often used; this should be discouraged if it implies that the Martian biota will necessarily fall into the taxonomic divisions that we know on earth.)

The habitability of Venus is connected with its temperature, a highly controversial subject. Perhaps the most useful first contribution to the exobiology of Venus would be a definitive measurement of its temperature profile. Even should the surface be unbearably hot, this need not preclude a more temperate layer at another level.

The exposure of the moon's surface to solar radiation and the absence of a significant atmosphere have led scientists to discount the possibility of a lunar biology. However, the composition of the moon's deeper layers, below even a few meters beneath the surface, is very much an open question (see Urey, 1960) particularly in the light of Kozyrev's recent reports of gaseous emissions. Realistic plans for the biological study of the moon probably must await the results of chemical analyses. Apart from the remote possibility of indigenous life, the moon is a gravitational trap for meteoroidal material. We may eventually be able to screen large quantities of this virgin material for what Haldane called astroplankton, in an empirical test of the panspermia hypothesis. While exposed deposits would be subject to solar degradation, shaded refuges must also exist. Mercury may be analogous to the moon, except insofar as its dark side may furnish an even more reliable, though much more remote, refuge of this kind.

It may be academic to discuss the exploration of the major planets, in view of their distance and the difficulty of deceleration in the Jovian field. However, their wealth of light elements, subject to solar irradiation at temperatures and in gravitational fields very different from the earth's, offers the most exciting prospects for novel biochemical systems.

Experimental Approaches

A realistic view of our limitations requires that our treatment of this topic be one of utmost humility. Useful landings on planetary targets are fraught with difficulties and hazards, and experiments done at a distance must not be overlooked in the excitement of planning for more adventurous missions. Balloon- and satellite-mounted telescopes can tell much about planetary chemistry, and hence biology, and probes to the vicinity of a planet can furnish additional information prior to actual landing.

It is instructive to ask ourselves how we might diagnose the existence of life on the earth from distant observations. If we may judge from the photographs so far obtained from high-altitude flights, we could hope to detect only large-scale manifestations of organized culture—cities, roads, rockets. This reserve may not give due credit to the possibilities of high-resolution photography and sensitive infrared spectrometry, and reasonable implications from seasonal changes in the color and texture of terrain. However, we may conclude that distant approaches will be invaluable for deriving preliminary chemical information but probably will not be decisive for exobiological inferences. Even if we could more surely decide that the Martian cycle involved living organisms rather than inanimate chemical transformations, we would still have

little insight into the intimate biochemical details which are a major objective of exobiological research. On the other hand, a planet could harbor an extensive biota that would defy detection from a distance, like the biota of our own extensive deserts and deep waters.

Microorganisms, for many reasons, are the best prospects on which to concentrate marginal capabilities. They are more likely to flourish in a minimal environment than larger organisms. The microbes must also precede the macrobes in evolutionary sequence, though we must not suppose that present-day bacteria are necessarily very primitive. The earth is well endowed with both kinds of organisms; we can imagine another world with only microbes, but we cannot conceive of one lacking microbes if it bears any form of life at all. Likewise, taking the earth as a whole, we find that large organisms occupy only a small fraction of the surface. However, we can reasonably expect to find evidence of microscopic life in any drop of water, pinch of soil, or gust of wind. Given a limited sample for study, microbiological analysis will certainly give the most reliable evidence for the presence of life anywhere on the planet. By the same odds the greatest diversity of biochemical mechanisms will be represented among the microbiota of a small sample.

Microbiological probes also offer distinct advantages for the collection and analysis of living material. From a single particle, microbes can easily be cultivated within the confines of an experimental device. In this they remain accessible to physiological and chemical experiments that would be extremely cumbersome with larger organisms. (Compare, for example, the automatic instrumentation that would be needed to catch a mouse or an elephant and then to determine its nutritional requirements!) The techniques of cytochemistry already developed for the chemical analysis of microscopic cells and organisms appear to be the most readily adaptable to automation and telemetric recording, an important advantage under the existing pressure of time, talent, and cost. Important issues of policy cannot be decisively settled without factual information on the growth capacity of the microorganisms that might be exchanged among the planets. Accordingly, methodological precedents in terrestrial science for exobiology are most evident in microbial biochemistry. The conceptual aims are equally close to those of biochemical genetics. Needless to say, no other resource or objective of serious biological science can be neglected in the development of an experimental program.

Aside from experimental designs, the pace of exobiological research may be regulated by advances in vehicular and guidance capabilities and data communication. In the expectation that these will remain in reasonable balance—for static or real time television communication with the planetary probe—the microscope may be the most efficient sensory instrument. The redundancy of a pictorial image would not be altogether wasted: would we put our trust in a one-bit pulse from an efficient black box to answer our queries about the cosmos?

According to this experimental concept, the terminal microscope-Vidicon chain must be supported by three types of development: (i) for collection and transport of the specimen to the aperture of the microscope; (ii) for cytochemical processing of the samples; (iii) for protection of the device against environmental hazards, for appropriate location after landing, and for illumination, focusing, and perhaps preliminary image selection. Detailed studies of these problems are only just under way, and the following suggestions are only tentative.

The easiest specimens to obtain may be atmospheric dust and samples of surface soil, once the device has been landed. These would be collected on a traveling ribbon of transparent tape which would be thrown out and then rewound into the device. Larger samples, collected by a soil auger, could be subjected to a preliminary concentration of nonmineral components by flotation in a dense liquid. The use of such a tape would simplify the problem of treating the samples with a succession of reagents—for example, specific enzymes and fluorescent stains for the detection of nucleic acids and proteins. Microscopy with ultraviolet light, particularly at 2600 and 2800 angstroms, owing to its selectivity for nucleic acids and proteins, may be the most direct way to distinguish microorganisms from mineral particles. Generally speaking, the microscope can be adapted to many simple analytical procedures whose construction on a larger scale would present formidable problems for automatic technique.

The adaptation of the microscope system to a payload can be undertaken more realistically when laboratory prototypes have been built and tested. For example, we will have to decide between accurate prefocusing of a microscope whose lenses and entry slit are mounted in a rigid structure and continuous control of focus by an optically controlled servo system (an innovation that would be far from useless in the biological laboratory). Fluorescent staining may facilitate automatic discrimination for conservation of radio power; the traveling ribbon can be stopped and the Vidicon-transmitter activated just when a stained object is in view.

These preliminary experiments can indicate some of the general features of the planetary microbiota. The data they furnish will support more intensive studies of the growth characteristics, chemical composition, and enzymatic capabilities of organisms cultivated on a larger scale. The interaction of these organisms with tissue cultures of animal cells can also be considered. From the results of these initial probes we can better deduce how to anticipate the long-range consequences of the intercourse of planetary biota.

Conservation of Natural Resources

A corollary of interplanetary communication is the artificial dissemination of terrestrial life to new habitats. History shows how the exploitation of newly found resources has enriched human experience; equally often we have seen great waste and needless misery follow from the thoughtless spread of disease and other ecological disturbances. The human species has a vital stake in the orderly, careful, and well-reasoned extension of the cosmic frontier, how we react to the adventuresome and perplexing challenges of space flight will be a crucial measure of the maturity of our national consciences and our concern for posterity.

The introduction of microbial life to a previously barren planet, or to one occupied by a less well-adapted form of life, could result in the explosive growth of the implant, with consequences of geochemical scope. With a generation time of 30 minutes and easy dissemination by winds and currents, common bacteria could occupy a nutrient medium the size of the earth in a few days or weeks, being limited only by the exhaustion of available nutrients. It follows that we must *rigorously* exclude terrestrial contaminants from our spacecraft. This stricture must hold until we have acquired the factual information from which we can assess with assurance the detrimental effects of free traffic and determine whether these are small enough to warrant the relaxation of these controls.

At the present time, the values that would most obviously be threatened by contamination are scientific ones. The overgrowth of terrestrial bacteria on Mars would destroy an inestimably valuable opportunity for understanding our own living nature. Even if an intemperate mission has not contaminated a planet, the threat of its having done so will confuse later studies, if earth-like organisms are found. However, other values are also involved. Quite apart from strictly scientific concerns, would we not deplore a heedless intrusion on other life systems? It would be rash to predict too narrowly the ways in which undisturbed planetary surfaces, their indigenous organisms, or their molecular resources may ultimately serve human needs. If we have cause to prejudice these values, we surely would not wish to do so by inadvertence.

To guard effectively against contamination requires a nice appreciation of the ubiquity and durability of bacterial spores, which are well preserved in high vacua and at low temperatures and are rapidly destroyed only when kept at temperatures over 160°C. It is probable that spacecraft can be disinfected by the conscientious application of gaseous disinfectants, especially ethylene oxide, but this will succeed only if the procedure is carried out meticulously and with controlled tests of its effectiveness. Sealed components, if found to be potential sources of contamination, can be disinfected by chemicals prior to sealing, or subsequently by heat, or by irradiation at very high doses. The technology of disinfection is an expert one, and personnel already experienced in it should be delegated supervisory control.

The assessment of this problem involves a concept of risk that has not always been perceptively realized. The hazards of space flight itself, or of hard impact, or the planetary environment *might* suffice to neutralize any contaminants, but can we afford to rely on uncertain suppositions when the stakes are so high, and when we have practical means at hand for conservative protection? We must be especially sensitive to the extreme variations in the environments of spacecraft or of planetary surfaces which might furnish refuges for microbe survival no matter how hostile the *average* conditions.

The indication by agencies both in the United States and the U.S.S.R. that adequate precautions will be exercised on all relevant missions is an important step in the realization of constructive exobiology.

Scientists everywhere will call for the application of these measures with the same care and enthusiasm as the more positive, exciting, and patently rewarding aspects of space research. Scientific microbiology in the laboratory is absolutely dependent on the rigorous application of the special technique of pure culture with aseptic control. If we do not exercise the same rigor in space science, we might as well save ourselves the trouble of thinking about, and planning for, exobiological research.

While early traffic to the planets will be one-way, we must anticipate round-trip, and even manned, space flight. Undoubtedly, planetary samples can be analyzed for any scientific purpose more conveniently and more exactly in the terrestrial laboratory than by remote devices. For each step of analysis, special devices can be used (or if need be, newly designed and constructed), and a constant give-and-take between human judgment and instrumental datum is possible. However, the return of such samples to the earth exposes *us* to a hazard of contamination by foreign organisms. Since we are not yet quite certain of the existence of planetary (that is, Martian) organisms, and know nothing of their properties, it is extremely difficult to assess the risk of the event. The most dramatic hazard would be the introduction of a new disease, imperiling human health. What we know of the biology of infection makes this an extremely unlikely possibility; most disease-producing organisms must evolve very elaborate adaptations to enable them to resist the active defenses of the human body, to attack our cells, and to pass from one person to another. That a microorganism should have evolved such a capacity in the absence of experience with human hosts or similar organisms seems quite unlikely. However, a converse argument can also be made, that we have evolved our specific defenses against terrestrial bacteria and that we might be less capable of coping with organisms that lack

the proteins and carbohydrates by which they could be recognized as foreign. Furthermore, a few diseases are already known (for example, psittacosis, botulism, aspergillosis) whose involvement of man seems to be a biological accident. These arguments can only be resolved by more explicit data. Nonetheless, if they are harmful at all, exobiota are more likely to be weeds than parasites, to act on our agriculture and the general comfort of our environment, and to be pervasive nuisances than acute aggressors. However, even the remotest risk of pandemic disease, and the greater likelihood of serious economic nuisance, must dictate a stringent embargo on the premature return of planetary samples, or of craft that might inadvertently carry them. Again, our preliminary experiments must give us the foundation of knowledge to cope with exoorganisms, even to select those which may be of economic benefit. A parallel development of techniques for disinfection may mitigate some of these problems; at present the prospects for treating a returning vehicle to neutralize any possible hazard are at best marginal by comparison with the immensity of the risks.

Of the possible payloads for interplanetary travel, living man, of course, excites the widest popular interest. In due course, he may be supported by a sufficient payload to accomplish useful tasks in exploration beyond the capacities of instrumentation. However, he is a teeming reservoir of microbial contamination, the most difficult of all payloads to neutralize, and he is an especially suitable vehicle for infectious organisms. In view of these difficulties, and insofar as manned space flight is predicated on the return of the crew, a sound basis of scientific knowledge from instrumented experiments is a *sine qua non* for the planning of such missions.

Timely effort now to devise and build instrumented experiments is essential to keep pace with the technical capacities of space vehicles.

Conclusion

Many of the ideas presented in this article are not new. In the scientific literature they have been treated only occasionally, for example in a remarkable article by J.B.S. Haldane [Reading 8]. They are also anticipated in the classic works of science fiction—for example, H. G. Wells' *War of the Worlds*—and in a flood of derivative fantasies of less certain quality either as science or as fiction. This kind of attention has not necessarily contributed to realistic evaluation of the biological aspects of space travel, discussion of which may still be dismissed as overimaginative by some of our colleagues. However, exobiology is no more fantastic than the realization of space travel itself, and we have a grave responsibility to explore its implications for science and for human welfare with our best scientific insights and knowledge.

Notes

The principles embodied in this article reflect the judgment of one among several of the scientific groups advisory to the Space Science Board of the U.S. National Academy of Sciences. However, they do not necessarily represent any official policy of the committed views of each consultant. The continued interest and advice of M. Calvin, R. Davies, N. Horowitz, S. E. Luria, A. G. Marr, D. Mazia, A. Novick, C. Sagan, G. Stent, H. C. Urey, C. B. van Niel, and H. Weaver, among many others, have been indispensable.

12

Richard E. Dickerson
Chemical Evolution and the Origin of Life

Perhaps the most striking aspect of the evolution of life on the earth is that is happened so fast. Various radioactive-isotope methods of dating stony meteorites all give approximately the same age: 4.6 billion years. If one assumes that the sun, the planets, the meteorites and other debris of the solar system all formed from the same primordial dust cloud at about the same time, then 4.6 billion years is the age of our planet as well. Much of the earth's early geological history has been erased by later events. Some of the most ancient sedimentary rocks known are in the Fig Tree and Onverwacht deposits of South Africa, respectively 3.2 and 3.4 billion years old. Both deposits contain microfossils that resemble bacteria. Evidently some kind of primitive life had appeared on our planet a little more than a billion years after its formation.

Twice as much time was then required before the emergence of eukaryotic cells (cells with nuclei) and of multicelled organisms. The step from nonbiological organic matter to life seems to have been easier than one night have expected, and the step from one-celled bacteria to multicelled organisms seems to have been harder. The later processes are better understood because the results in the fossil record are apparent. The first billion pages in the book of the earth's history are almost completely missing. One must try to reconstruct them from information in the later pages and from what one knows about the other planets and about chemistry in general. This article describes the attempts of investigators to reconstruct the evolution of life during the missing first billion years.

Plausible mechanisms have been demonstrated for synthesizing under primitive terrestrial conditions most of the monomers, or simple molecules, needed by the living cell. Some of these monomer units are assembled into two broad classes of polymers: nucleic acids, which embody and transmit the hereditary material, and proteins, of which some serve as structural materials and others as enzymes for catalyzing the scores of complex chemical

Reprinted from *Scientific American,* 239 (September 1978), 70–86. Reprinted with permission.

reactions that underlie both metabolism and reproduction. The problem of showing how the monomers might have linked up into biologically effective polymers has proved to be more difficult, but a number of plausible pathways have been demonstrated. Moreover, in some experiments droplets with a membranelike boundary surface or skin have exhibited the capacity to catalyze rudimentary reactions resembling those observed in living cells, and they have demonstrated the survival advantage of being isolated from the surrounding medium. All of this is a long way from creating "life in a test tube," but that is hardly the objective. The broad goal is to arrive at an intellectually satisfying account of how living forms could have emerged step by step from inanimate matter on the primitive earth. That goal appears to be in sight.

It is possible, of course, that life did not arise on the earth at all. According to the theory of panspermia, which was popular in the 19th century, life could have been propagated from one solar system to another by the spores of microorganisms. Francis H. C. Crick and Leslie E. Orgel recently made the more venturesome suggestion that the earth, and presumably other sterile planets, might have been deliberately seeded by intelligent beings living in solar systems whose stage of evolution was some billions of years ahead of our own. The process, which Crick and Orgel called directed panspermia, might explain, for example, why molybdenum, which is quite scarce on the earth, is essential for the functioning of many key enzymes.

One can neither prove nor disprove theories of panspermia, but they are not really relevant to the inquiry of interest here. The earth is hospitable to the kind of life found on it. If that kind of life did not evolve on the earth, it must surely have evolved on a planet not drastically different from the earth in its temperature and composition. The question really is: How might life have evolved on an earthlike planet?

Assuming that terrestrial life did evolve on the earth, what was the planet like when the process began? One thing is certain: the atmosphere contained little or no free oxygen and hence was not strongly oxidizing as it is today. The organic matter that must accumulate as the raw materials from which life could evolve is not stable in an oxidizing atmosphere. One tends to forget that oxygen is a dangerously corrosive and poisonous gas, from which human beings and other organisms are protected by elaborate chemical and physical mechanisms. Many bacteria and all higher forms of life "burn" their food by combining it with oxygen, because this process yields far more energy per gram of fuel than simple anaerobic (nonoxygen) fermentation. Enzymes such as catalase, peroxidase and superoxide dismutase have evolved to protect oxygen-using organisms from toxic side effects. Anaerobic bacteria lack these protective systems; for them oxygen is both useless and lethal.

J. B. S. Haldane, the British biochemist, seems to have been the first to appreciate that a reducing atmosphere, one with no free oxygen, was a requirement for the evolution of life from nonliving organic matter. Without oxygen in the atmosphere there would have been no high-altitude ozone to block most of the ultraviolet radiation from the sun as there is today. The unblocked ultraviolet radiation reaching the surface of the planet could have then provided the energy for the synthesis of a great many organic compounds from molecules such as water, carbon dioxide and ammonia. Without free oxygen in the atmosphere to destroy them again such compounds would have accumulated in the oceans until, in Haldane's words, "the primitive oceans reached the consistency of hot dilute soup."

Haldane's ideas appeared in *Rationalist Annual* in 1929, but they elicited almost no reaction. Five years earlier the Russian biochemist A. I. Oparin had published a small monograph proposing rather similar ideas about the origin of life, to equally little effect. Orthodox biochemists were too convinced that Louis Pasteur had disproved spontaneous generation once and for all to consider the origin of life a legitimate scientific question. They failed to appreciate that Haldane and Oparin were proposing something very special; not that life evolves from nonliving matter today (the classical theory of spontaneous generation, which was untenable after Pasteur) but rather that life once evolved from nonliving matter under the conditions prevailing on the primitive earth and in the absence of competition from other living organisms.

Charles Darwin was on the right track, as he so frequently was, when he wrote to a friend in 1871: "It has often been said that all the conditions for the first production of a living organism are now present which could ever have been present. But it (and oh! what a big if!) we could conceive in some warm little pond, with all sorts of ammonia and phosphoric salts, light, heat, electricity, etc., present, that a protein compound was chemically formed ready to undergo still more complex changes, at

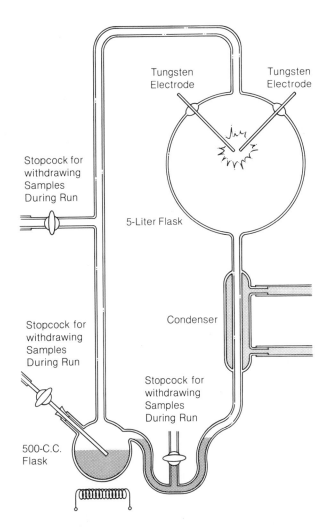

Figure 12.1. Organic compounds were synthesized in an apparatus designed by Stanley L. Miller and Harold C. Urey at the University of Chicago to simulate conditions in the atmosphere of the primitive earth. Various mixtures of gases presumed to have been present in that atmosphere were admitted to the apparatus through the stopcock in the middle of the vertical tube at the left. Water in the 500-cubic-centimeter flask at the bottom of the tube was boiled to drive gases in a closed circuit through the apparatus. In the five-liter flask at the upper right the gases were subjected to a spark discharge simulating energy inputs also presumed to have been present in the primitive atmosphere. The various compounds that were formed in the discharge (see Figure 12.6) accumulated in solution at bottom of apparatus.

the present day such matter would be instantly devoured or absorbed, which would not have been the case before living creatures were formed."

Harold C. Urey restated the Oparin-Haldane thesis in 1952 in his book *The Planets,* and he and Stanley L. Miller began conducting actual laboratory experiments at the University of Chicago to see whether or not energy sources available on the primitive earth could have induced the synthesis of organic compounds from gases that would have been present in the primitive atmosphere. They demonstrated that spark discharges in mixtures of hydrogen, methane, ammonia and water gave rise to aldehydes, carboxylic acids and amino acids. Other mixtures of gases including carbon monoxide, carbon dioxide and nitrogen, were equally productive, provided that no free oxygen was present. Those experiments, and the beginnings of space exploration in the 1960's, reawakened interest in the origin of life and in the possible existence of life elsewhere in the universe.

From the outset Oparin and Haldane held divergent views about the initial conditions most important for the evolution of life, a disagreement that continues among origin-of-life theorists today. A living cell has two central talents: a capacity for metabolism and a capacity for reproduction. The cell survives in the short run by rearranging the atoms of the compounds it ingests into molecules needed for its own maintenance. It survives vicariously over the long run by being able to reproduce itself and give rise to offspring with similar biochemical talents. Which came first, a functioning metabolism, protected by some kind of membrane against dilution and destruction by its surroundings, or a large molecule that survived by making copies of itself from materials in its surroundings? In other words, which is older, the "protobiont" or the "naked gene"? Haldane favored the latter idea. Oparin has always been more interested in the chemical reactions that can proceed inside droplets segregated from the bulk medium, and in the question of competition for survival among such droplets. (At the age of 84 Oparin perseveres in Moscow as the grand old man of origin-of-life research.) To Oparin the reproductive machinery and DNA are only the ultimate biochemical subtleties that turned metabolically competing protobionts into living cells.

The metabolism v. reproduction (or protein v. nucleic acid) argument should ultimately turn out to be as sterile as the chicken v. egg or heredity v. environment arguments of earlier generations. Today nucleic acids cannot replicate without enzymes, and enzymes cannot be made without nucleic acids. To the question, "Which came first, enzymes or nucleic acids?" the answer must be, "They developed in parallel." The catalysts needed to encourage reactions that favored the survival of particular droplets within the primitive Haldane soup, and the copying machinery to ensure that the catalysts were not lost as the droplets were broken up and dispersed by wave action and other mechanical forces, must have evolved together. The older systems did not survive because they could not compete with later improvements for their raw materials. Enzymic catalysis and DNA replication today are so thoroughly interwoven in living cells that it is hard to see what a simpler system might have been like. But as the

British physicist J. D. Bernal wrote: "The picture of the solitary molecule of DNA on a primitive seashore generating the rest of life was put forward with slightly less plausibility than that of Adam and Eve in the Garden."

The step from aldehydes and amino acids nonbiologically into a living cell is a giant one. It is one thing to propose scenarios for the origin of life that might have been; it is another thing entirely to demonstrate that such scenarios are either possible or probable. As evidence there is a meager record of fossil microorganisms, a geological history of the planet, laboratory experiments that can demonstrate what primitive reactions might have been possible, extraterrestrial evidence for organic matter in meteorites and in spectra of interstellar dust, and the hope of detecting life that evolved independently on other planets.

We can divide the problem of the evolution of living cells from nonliving matter into five steps: (1) the formation of the planet, with gases in the atmosphere that could serve as raw materials for life; (2) the synthesis of biological monomers such as amino acids, sugars and organic bases; (3) the polymerization of such monomers into primitive protein and nucleic acid chains in an aqueous environment where depolymerization is thermodynamically favored; (4) the segregation of droplets of Haldane soup into protobionts with a chemistry and an identity of their own, and (5) the development of some type of reproductive machinery to ensure that the daughter cells have all the chemical and metabolic capabilities of the parent cells. Stated concisely, these are the problems of raw materials, monomers, polymers, isolation and reproduction.

The universe as a whole consists almost entirely of hydrogen (92.8 percent) and helium (7.1 percent), with minor impurities such as nitrogen, oxygen, neon and all the other elements. Two important features should be noted in Figure 12.2: the abundance of an element decreases in general with an increase in its atomic number (equal to the number of protons in its nucleus), and atoms with even atomic numbers are more abundant than their neighbors with odd atomic numbers. The reason is that the heavier elements are synthesized from the lighter ones in the interior of stars and that this synthesis, at least of the elements with an atomic number up to that of iron, involves the capture of alpha particles, or helium nuclei, which have two protons. The even-numbered elements are more abundant because they lie in the mainstream of synthesis, the odd-numbered elements are less abundant because they are synthesized by side reactions.

It was once thought that the sun and planets of the solar system were formed by the aggregation and cooling of a cloud of hot gas. It now seems more likely that the starting point was a cloud of cold gas, dust particles and debris that became flattened by rotation and developed a protosun or

Element	Symbol	Atomic Number	Entire Universe	Entire Earth	Crust of Earth	Ocean Water	Human Body
HYDROGEN	H	1	92,714	120	2,882	66,200	60,563
HELIUM	He	2	7,185	—	—	—	—
LITHIUM	Li	3	—	—	9	—	—
BERYLLIUM	Be	4	—	—	—	—	—
BORON	B	5	—	—	—	—	—
CARBON	C	6	8	99	56	1.4	10,680
NITROGEN	N	7	15	0.3	7	—	2,440
OXYGEN	O	8	50	48,880	60,425	33,100	25,670
FLUORINE	F	9	—	3.8	77	—	—
NEON	Ne	10	20	—	—	—	—
SODIUM	Na	11	0.1	640	2,554	290	75
MAGNESIUM	Mg	12	2.1	12,500	1,784	34	11
ALUMINUM	Al	13	0.2	1,300	6,251	—	—
SILICON	Si	14	2.3	14,000	20,475	—	—
PHOSPHORUS	P	15	—	140	79	—	130
SULFUR	S	16	0.9	1,400	33	17	130
CHLORINE	Cl	17	—	45	11	340	33
ARGON	Ar	18	0.3	—	—	—	—
POTASSIUM	K	19	—	56	1,374	6	37
CALCIUM	Ca	20	0.1	460	1,878	6	230
SCANDIUM	Sc	21	—	—	—	—	—
TITANIUM	Ti	22	—	28	191	—	—
VANADIUM	V	23	—	—	4	—	—
CHROMIUM	Cr	24	—	—	8	—	—
MANGANESE	Mn	25	—	56	37	—	—
IRON	Fe	26	1.4	18,870	1,858	—	—
COBALT	Co	27	—	—	1	—	—
NICKEL	Ni	28	0.1	1,400	3	—	—
COPPER	Cu	29	—	—	1	—	—
ZINC	Zn	30	—	—	2	—	—
			99,999.5	99,998.1	99,999	99,994.4	99,999

Figure 12.2. Distribution of the major elements varies widely according to the nature of the sample. This table shows the abundance of the first 30 elements in the periodic table in atoms per 100,000 for the entire universe, for the entire earth, for the crust of the earth, for ocean water and for the human body. The blanks indicate that the abundance is less than .1 atom per 100,000. It is apparent from the table that the earth is a highly unrepresentative sample of the elements present in the universe as a whole. The composition of the human body is fairly typical of the composition of all living organisms. Twenty-four of the elements are now known to be essential for the processes of life; the 20 that are shaded together with selenium (Atomic No. 34), molybdenum (No. 42), tin (No. 50) and iodine (No. 53).

concentrated core at the center. The cloud was subsequently heated by the release of gravitational energy and to a lesser extent by the natural radioactivity of some of its atoms. As the sun coalesced at the center of the revolving flat cloud other local inhomogeneities at varying distances from the center became aggregation points for the formation of the planets. The large outer planets—Jupiter, Saturn, Uranus and Neptune—probably represent a fair sample of the composition of the original cloud, since their elemental makeup is close to that of the universe at large. They are composed mainly of hydrogen, helium, methane, ammonia and water. The small inner planets—Mercury, Venus, the earth and Mars—are richer in the heavier elements and poorer in such gases as helium and neon, which could escape from the weak gravitational pull of these planets.

A combination of low gravity and high temperature led to a loss of most of the earth's volatile constituents to interplanetary space soon after the planet coalesced. Oxygen was greatly increased in relative abundance because it was locked into the nonvolatile silicate minerals; much nitrogen was lost because the nitrides are less stable and

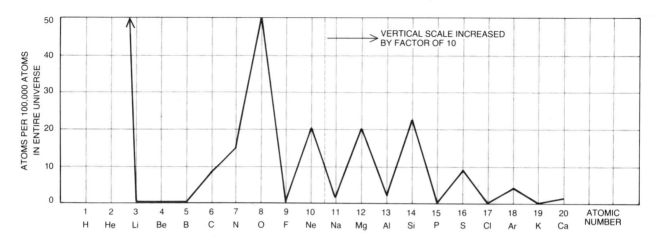

Figure 12.3. Graphical representation of the abundance of elements in the universe reveals the predominance of elements of even atomic number, that is, elements whose nuclei contain an even number of protons. The explanation is that the stellar nuclear reactions synthesizing the heavier elements, at least those up to iron, favor the helium nucleus as a building block. Beryllium is scarce because the fusion of three helium nuclei to form carbon is favored over the reaction that stops with the fusion of two helium nuclei to form beryllium.

more easily converted into volatile gases. In general terms the earth consists of an iron-nickel core and a mantle that corresponds roughly in composition to the silicate mineral olivine ($FeMgSiO_4$). Only about .034 percent of the earth is carbon.

The earth became stratified into a core, a mantle and a crust as a result of the heat released as the planet was built up by accretion. The original surface may have been too hot for water to remain liquid, but as soon as the temperature had fallen below the boiling point water released from the interior by outgassing processes such as volcanism would have condensed to form the original oceans. Outgassing would have given rise to a secondary atmosphere composed of water vapor from the water of hydration of minerals, methane (CH_4), carbon dioxide (CO_2), carbon monoxide (CO) from the decomposition of metal carbides, ammonia (NH_3) and nitrogen from nitrides, and hydrogen sulfide (H_2S) from sulfides. It is from this secondary atmosphere, the character of which was reducing rather than oxidizing, that life presumably arose. As Haldane pointed out, the oxygen in the atmosphere today was put there mainly by the earliest living organisms, which succeeded in harnessing the energy of sunlight to split water molecules and fix carbon dioxide to make glucose ($C_6H_{12}O_6$), releasing oxygen as a by-product. Having once appeared on the earth, life changed the planet and destroyed the conditions that made the original appearance of life possible.

What molecules would have to be synthesized in the primitive atmosphere and oceans as the precursors of life? The list would have to include amino acids for proteins; sugars, phosphates and organic bases for nucleic acids;

lipids for membranes and a number of other special-purpose organic molecules such as flavins. If polymeric chains of proteins and nucleic acids are to be forged out of their precursor monomers, a molecule of water must be removed at each link in the chain. It is therefore hard to see how polymerization could have proceeded in the aqueous environment of the primitive ocean, since the presence of water favors depolymerization rather than polymerization. We shall have to face up to this difficulty, but first let us see how the monomers could have arisen.

The formation of monomers from the gases of the primitive atmosphere is the step about which the most is known, since the reactions can be simulated and studied in the laboratory. In Miller and Urey's original experiments they worked with an artificial atmosphere consisting of hydrogen and the fully reduced forms of carbon, nitrogen and oxygen: methane, ammonia and water. Miller considered working with ultraviolet radiation in the first experiments, but difficulties with providing windows in the

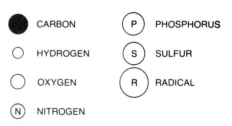

Figure 12.4. Six kinds of atoms in the molecules in the following figures are identified in this key. "Radical" refers to side chains that go through a series of reactions without change.

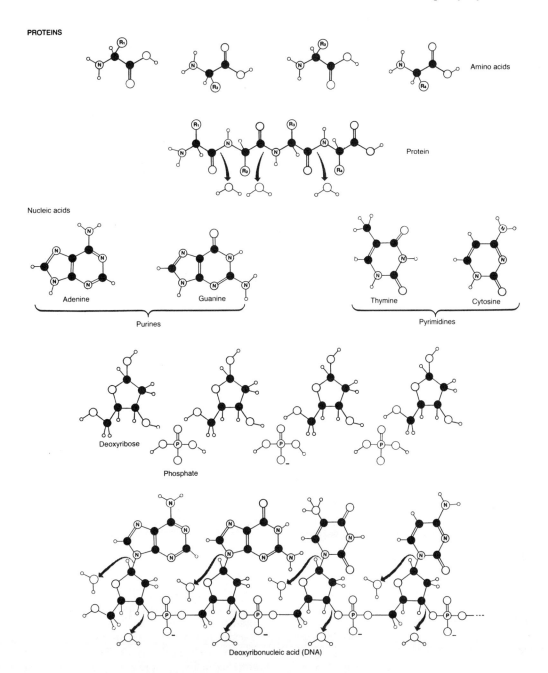

Figure 12.5. Central polymers of life are proteins and nucleic acids. Proteins are polymers that serve both as structural materials and as catalysts for directing the course of biochemical reactions. Nucleic acids are polymers that embody the genetic code in which the specification for each of an organism's proteins is recorded. Deoxyribonucleic acid (DNA) has a backbone assembled from alternating units of deoxyribose (a sugar) and phosphate. Attached to each deoxyribose unit is one of four different organic bases: adenine (*A*), guanine (*G*), thymine (*T*) or cytosine (*C*). The genetic code is written in sequences of three bases (for example *ATC, GCA* or *GTA*), which represent one of 20 different amino acid monomers of a protein polymer. A sequence of triplet bases in DNA therefore specifies the sequence of amino acids in a protein. The side chains that distinguish one amino acid from another are indicated here by R_1, R_2, R_3 and so on. When amino acids are linked into polymers, a molecule of water must be removed at each linkage point, creating a peptide bond. The polymer is thus known as a polypeptide. A protein is a polypeptide of biological origin. The construction of nucleic acid chains also proceeds by removal of water molecules at the critical linkage points, a step that would require special conditions on the primitive earth.

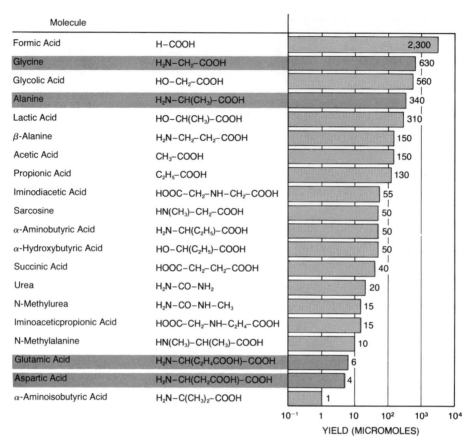

Figure 12.6. Molecules associated with life were produced in the experiments conducted by Miller and Urey. The compounds listed here were created in one such experiment by repeatedly passing the spark discharge through a gaseous mixture of hydrogen, methane, ammonia and water. The mixture originally held 710 milligrams, or 59,000 micromoles, of carbon in the form of methane gas, of which about 15 percent was converted into the compounds listed. A substantially larger percentage of the carbon was deposited as a tarry residue that could not be analyzed. This particular experiment yielded four of the 20 amino acids that are commonly present in proteins, identified here by shading. Structures of these compounds appear in Figure 12.17.

reaction vessel and keeping them clean of polymerized organic matter led him to conduct the first trials with an electric spark discharge, a laboratory simulation of a lightning flash.

In a typical experiment the gases were circulated past the spark for a week. The progress of synthesis was monitored by taking samples out of the boiling flask for analysis. It was surprising at first to find that the synthesized substances included several common amino acids and other molecules that are constituents of living matter. Many variations of the experiment have since been tried by Miller and by others, substituting carbon monoxide or carbon dioxide for methane, nitrogen for ammonia and ultraviolet radiation for the spark discharge. Many naturally occurring amino acids have been found, including leucine, isoleucine, serine, threonine, asparagine, lysine, phenylalanine and tyrosine. It is clear that amino acids would have been readily synthesized in the primitive atmosphere.

Two cautionary comments are necessary. Although the simulations yield many of the amino acids found in the proteins of living organisms, they also yield at least as many related molecules that are not present. For example, experiments of the Miller type synthesize three isomeric forms of an amino acid with the formula $C_3H_7NO_2$: alanine, beta-alanine and sarcosine. Yet only alanine has been incorporated into the proteins of living organisms. Of the three isomers valine, isovaline and norvaline only valine appears in proteins today. Seven amino acid isomers with the formula $C_4H_9NO_2$ are created in spark-discharge experiments, none of which is designated as a protein constituent by the universal genetic code of ter-

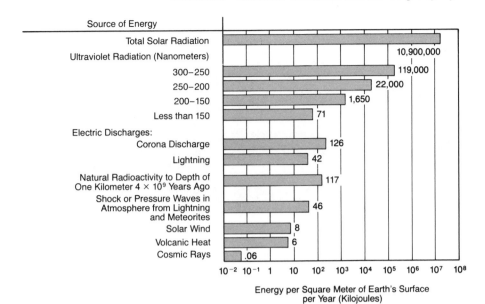

Figure 12.7. Sources of energy for chemical synthesis in the primitive atmosphere of the earth embrace a wide variety of phenomena. More than 98 percent of the energy in solar radiation arrives in the form of photons too weak to make and break chemical bonds. Only ultraviolet photons with wavelengths shorter than about 200 nanometers (less than 1.2 percent of the total ultraviolet) would have been highly effective in triggering chemical reactions. Value for energy in electric discharges, a total of 168 kilojoules per square meter of surface per year, is based on present-day weather conditions and could have been larger on the primitive earth.

restrial life. It is obvious that the choice of the 20 amino acids in the genetic code was not foreordained by the availability of a particular set of molecules on the primitive earth. One of the fascinating side issues of origin-of-life biochemistry is why the present set of 20 amino acids was chosen. Were there false starts, with genetic codes that specified different sets of amino acids, in lines of development that died out without a trace because they could not compete with the lines that survived? There probably were.

The other cautionary observation is that these laboratory simulations of prebiological reactions give rise to equal numbers of both forms of optically active molecules: molecules that rotate polarized light in opposite directions because the molecules exist in two configurations that are mirror images of each other. Such molecules are designated by the prefix D or L, abbreviations for dextro and levo, designating the direction of rotation of the polarized light. Except for certain special adaptations involving bacterial cell walls and biochemical defense mechanisms, all living organisms today incorporate only L amino acids. Various attempts have been made to explain why only one optical isomer of amino acids came to be favored, ranging from the asymmetric crystal structure of minerals that could act as surface catalysts to the natural polarization of cosmic rays and Coriolis forces arising from the rotation of the earth (which differ in the

two hemispheres). It seems likely that the primitive selection of the L isomers over the D isomers was a matter of chance. We do know that enzymes must bind molecules to their surface and that enzymes will be more efficient if they are designed to bind only one isomer or the other. There may at one time have been primitive life or precursors of life based on both D- and L-amino acids, with a 50 percent chance that the L amino acids would eventually prevail.

What are the detailed chemical steps by which amino acids are synthesized by a spark discharge or ultraviolet radiation? In following the appearance and disappearance of intermediates during week-long synthesis runs Miller and Urey observed that the concentration of ammonia fell steadily and that its nitrogen atoms appeared first in molecules of hydrogen cyanide (HCN) and cyanogen (C_2N_2), which along with aldehydes were the first substances formed. Amino acids were synthesized more slowly at the expense of hydrogen cyanide and aldehydes. This progression suggests that the amino acids are formed from aldehydes by a mechanism well known to organic chemists, the Strecker synthesis (see equations 1, 2 and 3 in Figure 12.8).

The aldehyde first adds ammonia and loses water to form an imine; the imine then adds hydrogen cyanide to form an aminonitrile. These two steps are freely reversible. The

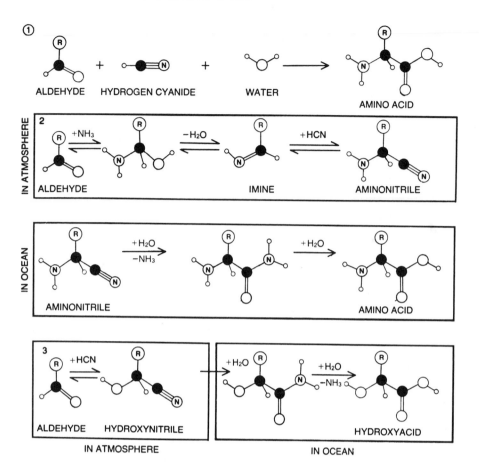

Figure 12.8. Sources of amino acids in experiments simulating the primitive atmosphere are aldehydes, hydrogen cyanide (HCN) and water (*1*). In the reaction *R*, a side chain of the aldehyde, appears as the side group of the amino acid. The reaction probably follows the steps of what is known as the Strecker synthesis (*2*). The first three steps, in which water is removed, could have taken place in the primitive atmosphere. If the aminonitrile formed entered ocean, the final hydrolysis steps could proceed. Hydroxyacids, such as lactic acid and glycolic acid, which have an additional hydroxyl group (OH), can also be formed by a Strecker synthesis (*3*).

amino acid is formed by the irreversible hydrolysis of the aminonitrile, with the addition of two molecules of water and the loss of ammonia. On the primitive earth the aminonitriles could have been synthesized in the atmosphere and dissolved and hydrolyzed in the ocean. In laboratory applications of the Strecker synthesis hydrolysis is carried out in solutions that are either acidic or basic because the rate of hydrolysis in a neutral solution is low. On the primitive earth, however, the hydrolysis could have been stretched over tens of thousands of years without penalty since there would have been no free oxygen to degrade the aminonitriles. Hydroxyacids also are formed by the Strecker synthesis. Formaldehyde (CH_2O) is converted into glycolic acid ($C_2H_4O_3$) and the amino acid glycine ($C_2H_5NO_2$), and acetaldehyde (C_2H_4O) is converted into lactic acid ($C_3H_6O_3$) and the

amino acid alanine ($C_3H_7NO_2$). Some of the more complex amino acids require more complex aldehydes as starting materials. On the other hand, serine ($C_3H_7NO_3$), an amino acid with a hydroxyl group (OH), can be made by the condensation of two molecules of formaldehyde followed by a Strecker synthesis (see equations 4 and 5 in Figure 12.9). Other special pathways have been proposed for the synthesis of most of the naturally occurring amino acids.

Electric discharges and ultraviolet radiation are not the only conceivable sources of energy for prebiological synthesis. Other sources include the emanations from radioactive elements in surface rocks and shock waves from lightning and meteors. Although by far the greatest amount of energy comes from the sun, much of it is in the visible and infrared regions of the spectrum, where the

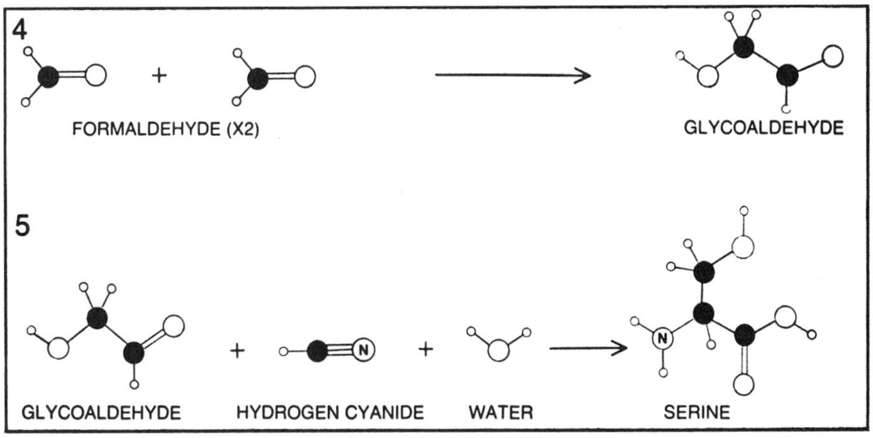

Figure 12.9. Synthesis of amino acid serine in primitive atmosphere could have begun with condensation of two molecules of formaldehyde to form glycoaldehyde (*4*). Strecker synthesis could have converted glycoaldehyde into serine, which has hydroxyl group in its side chain (*5*).

photons do not have enough energy to make and break chemical bonds. Moreover, most of the ultraviolet radiation would have been ineffective in triggering chemical synthesis because methane and other small hydrocarbon molecules, water, carbon monoxide and carbon dioxide can absorb only wavelengths shorter than about 200 nanometers, or less than 1.2 percent of the available ultraviolet radiation (about 1,720 kilojoules out of a total of 143,000 kilojoules per square meter of the earth's surface per year). Of the gases presumed to have been present in the primitive atmosphere only ammonia and hydrogen sulfide can absorb longer wavelengths: ammonia up to 220 nanometers and hydrogen sulfide up to 240. As a result the two gases may have been important in the primitive atmosphere as collectors of solar energy.

If the weather on the primitive earth was no more violent than today's, lightning and the corona discharges of atmospheric electricity would have provided about 170 kilojoules per square meter per year. The amount of energy from natural radioactivity, calculated by extrapolating backward to the early years of the earth's history, would have been about 117 kilojoules. Shock waves in the atmosphere would have provided on the order of 46 kilojoules, again assuming that the primitive weather resembled our own. The "wind" of energetic particles from the sun and volcanism together may have contributed another 14 kilojoules, perhaps more if the primitive earth were more active tectonically than it is today. Of all the energy sources for prebiological synthesis electric discharges probably were the most significant, both because of the amount of energy involved and because this energy would be released close to the surface of the ocean where the products could easily be dissolved in the water.

Twenty amino acids are enough building blocks for proteins. For nucleic acids one must have two kinds of sugar (ribose for RNA and deoxyribose for DNA), phosphate and two kinds of nitrogen-containing bases: purines and pyrimidines. The sugars can be built up as condensation products of formaldehyde by the synthesis known as the formose reaction. The mechanism involves several steps, but the overall reaction is simple: five molecules of formaldehyde combine to form one molecule of ribose (see equation 6 in Figure 12.10). There are problems with the formose reaction (the ribose produced is unstable in aqueous solution and the experimental conditions are not realistic simulations of primitive earth conditions), but some such reaction could have yielded the necessary ribose molecules.

Among the organic bases the purine molecule adenine is the most readily synthesized. It is simply a pentamer of hydrogen cyanide: 5HCN yields $C_5H_5N_5$. It seems likely that four molecules of hydrogen cyanide combine initially to form a tetramer of HCN, diaminomaleonitrile. The diaminomaleonitrile molecule is an important intermediate in many reactions leading to base synthesis. In the presence of light the molecule can rearrange itself and add one more hydrogen cyanide to form adenine (see equations 7, 8 and 9 in Figure 12.11). The synthesis proceeds under conditions that would have been reasonable on the primitive earth. Guanine, the other purine needed in nucleic acids, can be obtained from diaminomaleonitrile by hydrolysis reactions involving cyanogen. Other less convincing syntheses have been proposed for the pyrimidine bases: thymine, uracil and cytosine.

When adenine is joined to a molecule of ribose, the product is adenosine, a nucleoside. With the simple addition of a triphosphate tail adenosine becomes adenosine triphosphate (ATP), the molecule that serves as the pri-

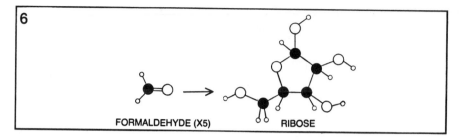

Figure 12.10. Synthesis of sugar ribose, a constituent of nucleic acids, can be achieved by a multistep reaction in which five molecules of formaldehyde combine to form one molecule of ribose.

mary currency of energy exchange in all living organisms (see equation 10 in Figure 12.12). It is noteworthy that the nucleoside selected for coupling to the triphosphate is adenosine and not guanosine, cytidine or uridine. There is no obvious reason why ATP is better suited for energy storage than GTP, CTP or UTP. It may be that the relative simplicity of the synthesis of adenine led to its being present in greater concentrations than the other bases in the primitive Haldane soup. The use of ATP may represent nothing more than another cosmic throw of the dice.

It is not difficult to account for the appearance of the bases and sugars of nucleic acids on the primitive earth. An unexpected stumbling block arises, however, when one tries to account for the particular way in which the bases and sugars are joined to make nucleosides, such as the coupling of adenine and ribose to form the adenosine molecule (see equation 11 in Figure 12.12).

The problem is that ribose has four hydroxyl groups, any one of which could serve as the linkage point to adenine. Moreover, three of the hydroxyl groups are joined to "asymmetric" carbon atoms, so that adenine could be linked to each of the carbon atoms (designated 1', 2' and 3') in two structurally different ways, yielding either an alpha or a beta configuration. No one has yet proposed a convincing method of getting good yields of the beta-1' connection between adenine and ribose that is universally found in DNA and RNA.

In spite of all these qualifying statements, current knowledge of the chemistry by which amino acids, bases, sugars and other monomers of life could have been synthesized on the primitive earth is really rather impressive. Although the answers are not yet known, the problems can at least be defined, and they seem to be chemical problems that are likely to yield to further effort and experimentation. There are no fundamental or philosophical difficulties to be encountered with the synthesis of the monomers. As will now be apparent, one can say nearly as much for the formation of the biological polymers.

The central problem in understanding how the polymers were formed on the primitive earth is understanding how reactions requiring both the input of energy and the

removal of water could take place in the ocean. Each joining of subunits to lengthen a polymer chain calls for the removal of the elements of water from the ends being joined (see equation 12 in Figure 12.13). Since such reactions are reversible, an excess of water will drive them toward the left, in the direction of hydrolysis rather than polymerization. Furthermore, if all the reactants and products are present in comparable concentrations, the reaction to the left gives off free energy and hence is spontaneous, whereas the desired reaction requires free energy and hence must be driven "uphill" to the right. There are two ways to drive the polymerization reaction to the right: concentrate the reactants and remove water from the products or couple the process to some energy-releasing reaction that will drive polymerization toward completion. Both approaches have been investigated.

The energy to drive polymerization reactions in living organisms today is supplied by molecules of ATP. The coupling of reactions that require energy with those that release energy is accomplished by enzymes. On the pre-biological earth, before enzymes existed, the two functions could have been carried out by certain compounds that possess large amounts of free energy and provide their own coupling to the reactant molecules. Such coupling reagents are familiar to organic chemists. The molecules of a typical class, the carbodiimides, have a carbon atom linked by energetic double bonds to two nitrogen atoms (N=C=N). If the carbodiimide is brought in contact successively with two monomers or polymers, A and B, one of which has a terminal hydroxyl group and the other a terminal hydrogen atom, it removes water and joins the two monomers or polymers end to end. The uptake of water by the carbodiimide releases enough energy to make the combined reaction go (see equations 13 and 14 in Figure 12.13).

Carbodiimides are only illustrative of the principle of coupling. Potential coupling agents that have actually been made in prebiological synthesis experiments include cyanogen (N≡C−C≡N), cyanamide (N≡C−NH₂), cyanoacetylene (N≡C−C≡C−H) and diaminomaleonitrile, all of which incorporate carbon and nitrogen atoms joined

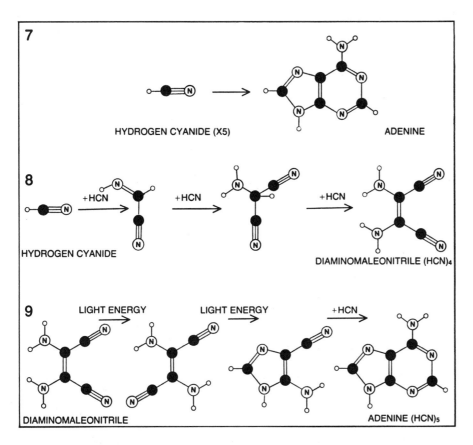

Figure 12.11. Adenine, a pentamer of hydrogen cyanide (*7*), is the most easily synthesized of the four organic bases that serve as the coding units of DNA. Presumably four molecules of hydrogen cyanide combine to form a tetramer, diaminomaleonitrile (*8*). The tetramer rearranges itself so that it forms a five-member ring. A fifth molecule of hydrogen cyanide closes a second ring (*9*).

by high-energy triple bonds. Cyanoacetylene can be produced by an electric discharge through a mixture of hydrogen cyanide; cyanogen can be produced from hydrogen cyanide both by electric discharge and by ultraviolet radiation. The energy of the electric spark or of the photons is stored as free chemical energy in the triple bonds of the product molecules for release later in the coupling reaction. In this way prebiological polymerizations could be driven indirectly by ultraviolet radiation or lightning in much the same way that animals living on plant starches today are driven indirectly by solar energy.

A major problem presented by such coupling mechanisms under prebiological conditions is to explain how the coupling agent can be prevented from combining directly with the ubiquitous water molecules and thereby short-circuiting the desired polymerization reactions. In the laboratory carbodiimide coupling reactions are carried out in nonaqueous solutions, but that is clearly not a reasonable model for the primitive earth. For example, if the primitive coupling agent were cyanogen, water could abort polymerization in either the first or the second step of the two-step reaction (see equation 15 in Figure 12.13).

One suggestion is that prebiological coupling reactions could have succeeded in aqueous solution if the molecules to be polymerized had been previously coupled to negatively charged ions, such as the phosphate ion (HPO_4^{--}). Organophosphates can compete quite successfully with water for the energetic bonds of coupling reagents. Phosphate condensations have been used successfully to make dipeptides from amino acids, to make adenosine monophosphate from adenosine and phosphate, to connect the ribose-phosphate backbone of nucleic acids and to build up polyphosphates from phosphate ions.

This nonbiological synthesis of polyphosphates, long-chain polymers of phosphate, may have been of great significance in the evolution of life. ATP is useful to living cells as an energy-storage molecule precisely because the hydrolysis of one bond to produce adenosine diphosphate (ADP) and one unit of inorganic phosphate releases a large amount of chemical free energy. The energy comes

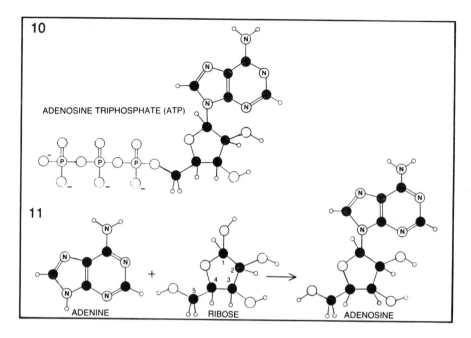

Figure 12.12. Adenosine triphosphate (ATP), the principal medium for the storage and exchange of energy in all living organisms, is created from adenine, ribose and a triphosphate tail (*10*). The nonbiological synthesis of the adenosine presents a special difficulty because the adenine might be coupled to any one of the four carbons in the ribose (1′, 2′, 3′ or 5′) that carries a hydroxyl group (*11*). Moreover, three of the four hydroxyl carbons (1′, 2′ and 3′) are asymmetric, so that alpha and beta forms of the molecule can be synthesized at each of them. In organisms today adenine and ribose are coupled at the 1′ carbon in the molecule's beta configuration.

in part from the repulsion between the negative charges on the molecular fragments of the reaction. As might be expected, comparably large amounts of energy are released when polyphosphates are hydrolyzed to phosphate (see equations 16 and 17 in Figure 12.14). ATP can be regarded simply as a small polyphosphate with an adenine "label" attached so that it can be recognized by enzymes. Some present-day bacteria store energy in polyphosphate bodies within their cytoplasm. It may be that polyphosphates produced by condensing agents were the first energy source tapped by living organisms or their immediate precursors.

Glycolysis, or anaerobic fermentation, is the most universal and probably the oldest energy-extracting pathway found in life on the earth today. Its function is to break down glucose or similar molecules and store the energy in the form of "labeled polyphosphate": ATP. The glycolytic pathway may have arisen only in response to a shortage of natural polyphosphates, as the energy needs of a growing population of primitive organisms exceeded the natural rate of production of polyphosphates by condensation. If one assumes that coupling reagents were synthesized with the energy from ultraviolet radiation or electric discharge, if the coupling reagents were responsible for the buildup of polyphosphates and if hydrolysis of the polyphosphates provided the energy source for primitive

life, then the first organisms would have lived on the energy of lightning and ultraviolet radiation at third hand.

The difficulty of preventing competition by water molecules in a coupling reaction has encouraged biochemists to think of ways in which the amount of water in the vicinity of the polymerizing species might be reduced. An obvious possibility is evaporation, and one can imagine a small portion of the Haldane soup being concentrated by the solar evaporation of a tide pool near some Archean beach. A freshwater pond would be even better, since salt would not crystallize as the water dried up. One objection to this proposal is that several of the important precursors of biological molecules, such as hydrogen cyanide, cyanogen, formaldehyde, acetaldehyde and ammonia, are themselves volatile. The evaporation of pools might be more effective in concentrating monomers for polymerization than in the synthesis of monomers themselves.

A possibly more attractive mechanism for concentrating prebiological molecules, first pointed out by Bernal, involves the adsorption of molecules on the surface of common minerals. Micas and clays, for example, consist of stacked silicate sheets held together by positive ions, with layers of water molecules between the sheets. The water layers make both sides of the silicate sheets accessible to molecules that diffuse into the clays, so that the total

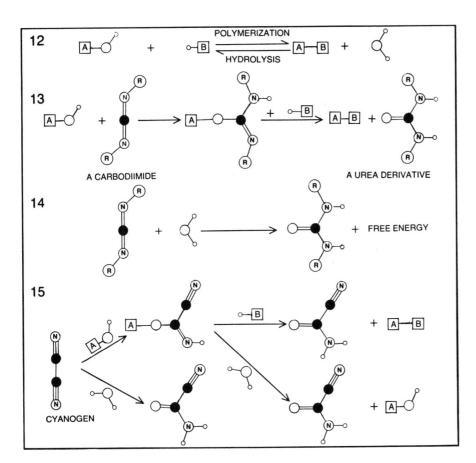

Figure 12.13. Formation or lengthening of a polymer chain requires the removal of the constituents of water from the ends of the subunits being joined. Here the subunits, which can be either monomers or polymers, are labeled *A* and *B* (*12*). In the presence of water the reaction is driven to the left, thereby uncoupling the subunits. To drive the reaction to the right requires not only the removal of water but also an input of energy. On the prebiological earth a coupling agent such as a carbodiimide could have brought about such polymerizations (*13*). The uptake of water by the carbodiimide releases the free energy needed for the reaction (*14*). Polymerization can be thwarted, however, if water is present (*15*). In this example the coupling agent is cyanogen. At each step the reactions with water are thermodynamically favored.

surface area available for adsorbing molecules is enormous. In kaolinite clay the silicate layers are separated by only .71 nanometer, which means that a cube of kaolinite one centimeter on a side provides a total surface area of about 2,800 square meters, or two-thirds the area of a football field. Furthermore, the silicate sheets themselves are negatively charged and the aluminum ions are bonded to the sheets by triple positive charges. The abundance of negative and positive charges not only can serve to bind charged molecules to the sheets but also can act as primitive catalytic centers for reactions.

Aharon Katchalsky of the Weizmann Institute of Science in Israel demonstrated that the montmorillonite clays will promote the polymerization of protein-like polypeptide chains from amino acid adenylates, which are esters formed from amino acids and adenosine monophosphate. Because they are rich in free energy and incorporate phosphate ions the adenylates polymerize efficiently even in aqueous solution (see equation 18 in Figure 12.15). When adenylates are adsorbed on clay surfaces, they form polypeptide chains of 50 or more amino acids with nearly 100 percent efficiency. Amino acid adenylates are the precursors of protein synthesis in all living organisms, and so it is tempting to imagine that this clay-surface polymerization with the same precursors might be an early step in the evolution of biological protein synthesis. Once the polymers had been formed they could be leached back into solution to accumulate slowly over the aeons, ready for further reactions.

Two other means of concentration and polymerization of prebiological substances have been proposed: freezing and heating to dryness. Miller and Orgel have pointed out

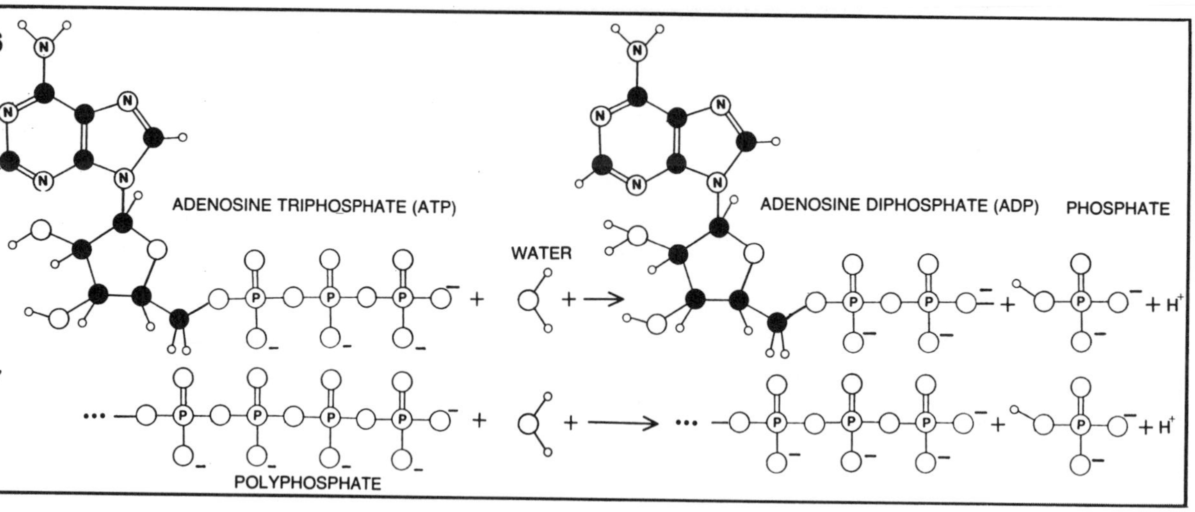

Figure 12.14. ATP is an effective storehouse of free energy because a large amount of energy is released when it is hydrolyzed by water to adenosine diphosphate (ADP) and a phosphate ion (16). This energy arises in part because the repulsion between the three bound phosphates in ATP can be relieved by breaking the connecting bonds and letting the phosphates move apart. For the same reason polyphosphate chains can also store energy (17). Early organisms might have relied on nonbiologically formed polyphosphates for their energy.

that solutions can be concentrated by freezing out the water as ice crystals, a procedure familiar to many as a means of making applejack from hard cider. On the prebiological earth the freezing of ice crystals out of a dilute solution of hydrogen cyanide could finally yield a solution containing 75 percent hydrogen cyanide by weight, freezing at minus 21 degrees Celsius.

At the other end of the temperature range, as Sidney W. Fox of the University of Miami has shown, dry mixtures of pure amino acids will polymerize spontaneously in a few hours at temperatures as low as 130 degrees C. to produce what Fox calls thermal proteinoids. If polyphosphates are present, similar results can be obtained by merely warming the amino acid mixture to 60 degrees for a day or so. Provided that the amino acids in the mixture are predominantly either acidic or basic and have side chains that are electrically charged, Fox's method will build polymers consisting of 200 or more amino acid units. Although most of the peptide bonds formed are of the normal type, a small fraction exhibit "wrong" connections involving the side chains. This is hardly surprising. One would not expect a prebiological polymer to show the degree of perfection found in a product of living metabolism. Fox speculates that amino acids formed in the ocean could have been washed up on volcanic cinder cones, evaporated to dryness and polymerized by heat. The resulting proteinoids, on being washed back into the sea, would have been available for further prebiological processing.

The problem of ensuring that only the right connections are made in nonbiological polymerization is rather more acute for nucleic acids than it is for proteins. As we have seen, each ribose molecule has four hydroxyl groups that can be involved in binding a purine or pyrimidine base and in polymerizing with bridging phosphates. Assuming that an efficient nonenzymatic method could be found for making nucleotides (a base plus a ribose plus a phosphate) with all the correct linkages, there would still be the problem of joining the nucleotides correctly to make polymers of nucleic acid. Although nucleotides can be polymerized nonbiologically with mild heat (about 55 degrees C.) in the presence of polyphosphates, the most readily formed connection is from the 5' hydroxyl of one sugar to the 2' hydroxyl of the next sugar rather than to the 3' hydroxyl, the connection that is found in all DNA's and RNA's today. The 5',3' linkage must have had a significant advantage over the 5',2' one to have been adopted for storage of genetic information even though it is less favored chemically.

Studies with molecular models show that it is possible to construct a double-strand DNA helix with paired bases and a 5',2' connection, but the helix appears to be less stable than one with a 5',3' structure. Hence a genetic message stored in a 5',2' helix may have been less secure than a message stored in a 5',3' helix. One way of ensuring that a 5',2' helix does not form is to remove the 2' hydroxyl group, and that is exactly what makes DNA different from RNA. DNA may be the more primitive of the two information-storing polymers, with RNA appearing only after enzymes had been developed that would avoid making the connection to the 2' hydroxyl group.

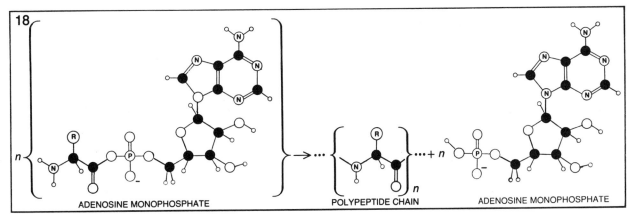

Figure 12.15. Formation of polypeptide chains has been shown to take place on the surface of particles of clay. In a typical example montmorillonite clay promotes the polymerization of polypeptides from amino acid adenylates, esters formed by the reaction of amino acids with adenosine monophosphate (AMP). Energy released by phosphate ions enables the polymer to form in the presence of water.

Living organisms that share an environment with other organisms must be clearly set off from that environment by a boundary surface to avoid being diluted out of existence. The segregation of matter in solution into droplets that were possible precursors of life has been studied mainly by two men and their co-workers: Oparin and Fox. Oparin has focused for many years on the tendency of aqueous solutions of polymers to separate spontaneously into coacervates: polymer-rich colloidal droplets suspended in a water-rich surrounding medium. Various combinations of biological polymers will give rise to coacervates: protein-carbohydrate (histone and gum arabic), protein-protein (histone and albumin) and protein-nucleic acid (histone and clupein with DNA or RNA). Such coacervates are to be regarded not as ancestors of living cells, since the polymers employed by Oparin in his experiments are definitely not primitive, but rather as analogues of the kinds of complex chemical behavior that can arise under the influence of natural forces.

The coacervate droplets range in diameter from one micrometer to 500 micrometers. Many seem to be set off from the surrounding medium by a kind of membrane, a thickening around the outside of the droplet of the polymer that causes it to separate from the bulk medium in the first place. Some coacervate systems are unstable: the droplets settle to the bottom of the liquid within minutes and coalesce into a nonaqueous layer. Oparin and his co-workers have sought conditions that will stabilize the suspensions of coacervate droplets for hours or weeks. Interestingly enough, they have found that one way to stabilize the droplets is to give them a primitive kind of metabolism.

One important property of coacervates, or of any two-phase system, is that substances whose solubility differs in the two phases will be preferentially concentrated in one phase or the other. Oparin found that when he added the enzyme phosphorylase to a solution containing histone and gum arabic, the enzyme was concentrated within the coacervate droplets. If glucose-1-phosphate was then added to the surrounding water, it diffused into the droplets and was polymerized to starch by the enzyme. Since gum arabic itself is a sugar polymer, the starch adds to the bulk of the droplet, causing it to grow in size. Energy for polymerization comes from the phosphate bond in glucose-1-phosphate. The inorganic phosphate that is released diffuses back out of the droplet into solution as a waste product.

When the coacervate droplets get too big, they tend to break up spontaneously into several daughter droplets. Those that happen to receive molecules of phosphorylase enzyme can continue to grow, although they do so at a lower rate because the original supply of enzyme molecules is dispersed among many droplets. If there were some way for the droplets to make more phosphorylase molecules (and it is a very big if), such coacervates would be examples of self-perpetuating protoorganisms with a one-step energy metabolism. They would be able to survive, grow and multiply on a restricted diet of glucose-1-phosphate.

If both phosphorylase and amylase are added to the coacervate preparation, both enzymes accumulate within the droplets and a two-step reaction ensues. Glucose-1-phosphate diffuses into the droplets and is polymerized to starch by phosphorylase. Amylase then cuts the starch polymer down to maltose, a dimer of glucose. The maltose diffuses back into the bulk solution along with the inorganic phosphate. The coacervates are thus small factories, driven by the energy of the glucose-phosphate bond, for dimerizing glucose-1-phosphate to maltose.

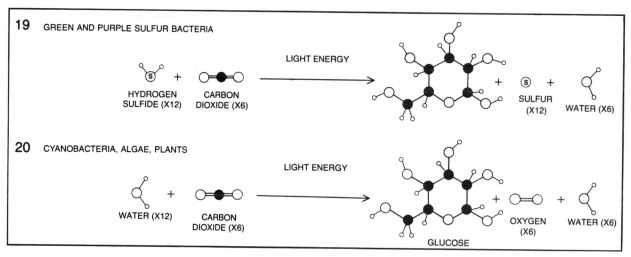

Figure 12.16. Invention of photosynthesis enabled living organisms to become the primary producers of energy-rich molecules instead of mere consumers of those provided by prebiological processes. The first organisms capable of photosynthesis probably made use of hydrogen sulfide (H₂S) as a source of hydrogen atoms for converting carbon dioxide (CO₂) into glucose (19). Later the predecessors of the cyanobacteria (blue-green algae), green algae and higher plants mastered the technique of obtaining hydrogen from water, a more elaborate two-step process (20) that released oxygen and transformed the atmosphere into the one in which all subsequent life developed.

Oparin has reported another self-growing system in which the coacervate droplets are made from histone and RNA. The enzyme RNA polymerase is introduced into the droplets, and ADP is added to the surrounding medium as "food." When the ADP enters the droplet, it encounters the RNA polymerase and is polymerized into polyadenylic acid. The energy for polymerization is contained within the ADP itself. The new polyadenylic acid adds to the total RNA in the coacervates. The droplets grow with time and break up into daughter droplets. Such systems eventually wind down because the supply of enzyme molecules for polymerizing ADP does not increase with the total mass of the coacervate droplets. As we saw earlier, however, nucleic acids can be polymerized nonenzymatically with small, energy-rich coupling-agent molecules such as cyanogen. It should be possible to construct coacervate droplets from protein and RNA, to provide them with ADP and the appropriate coupling reagents, and to see them grow and multiply without limit as long as their "nutrients" continue to be supplied.

Oparin has also set up coacervate-droplet experiments that mimic electron transport. The droplets contain a dehydrogenase enzyme from bacteria: nicotinamide adenine dinucleotide dehydrogenase (NADH). NADH and the dye methyl red are added to the medium and diffuse into the droplets. At the active site on the enzyme the NADH gives up hydrogen, which reduces the dye. The reduced dye and the oxidized NAD+ diffuse out of the droplets.

In another dye-reduction experiment, which mimics photosynthesis, chlorophyll is incorporated into the droplets; methyl red and ascorbic acid are added to the surroundings as nutrients. Ascorbic acid by itself is not a strong enough reducing agent to reduce methyl red. If, however, the droplets are illuminated with visible light, excited electrons from the chlorophyll can reduce the methyl red, and the electrons can be replaced on the chlorophyll by taking them away from ascorbic acid. In this way ascorbic acid, assisted by the energy of the photons of light, can reduce methyl red in a process that is analogous to the way water molecules, when they are assisted by photon energy, can reduce NADP+ to NADPH in the photosynthesis conducted by green plants.

Fox's interest in coacervate-like droplets has developed from his work with thermal proteinoids. The proteinoids have a remarkable property: when they are heated in a concentrated aqueous solution at 130 to 180 degrees C., they aggregate spontaneously into microspheres one or two micrometers in diameter. Although no lipids are present, many of the microspheres develop an outer boundary that resembles the double lipid layer of a cell membrane. Under the proper conditions the microspheres will grow at the expense of the dissolved proteinoid and will bud and fission in a most bacteriumlike manner.

Whereas Oparin has constructed artificial systems with catalysts incorporated, Fox has looked for catalytic activity inherent in the microspheres themselves. For example, he has found that microsphere preparations can catalyze the decomposition of glucose and can function as esterases and peroxidases. It would be surprising indeed if a polypeptide chain with positive and negative charges on its side groups did not exhibit some kind of generalized acid-base catalytic activity. Perhaps specific enzymes

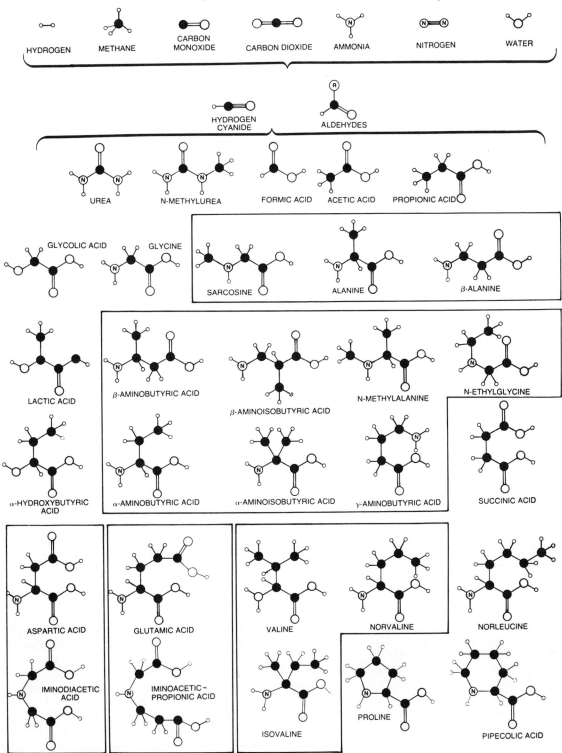

Figure 12.17. Products of spark-discharge experiments include many compounds found in living organisms and others that, although they are closely related, are not found in living matter today. The raw materials for the experiment are at the top. The first intermediates formed are hydrogen cyanide and a variety of aldehydes. The products are arranged so that the number of carbon atoms increases from left to right and from top to bottom. Isomers (molecules with the same atoms but different configurations) are enclosed by lines. These 30 typical products include 20 compounds listed in Figure 12.6. Six compounds are amino acids found in proteins.

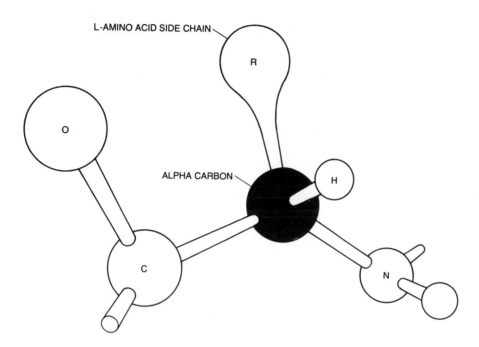

Figure 12.18. Stereoisomers are molecules that have two configurations, one the mirror image of the other. Among organic molecules stereoisomers can be formed when a carbon atom in the molecule has four different atoms or groups of atoms attached to it. Such a carbon atom is termed asymmetric. The central, or alpha-carbon, atom in an amino acid is asymmetric. An amino acid whose side chain (R) projects to the left as one crosses an imaginary bridge from the carbonyl group (CO) to alpha carbon to nitrogen atom (N) is an L-amino acid. In a D-amino acid side chain R projects to right. Living organisms make proteins only from L-amino acids.

evolved from such randomly ordered polymers by a gradual improvement in the positioning of electron-donating and electron-accepting side chains at active sites that were tailored to favor one reaction over another.

The Oparin and Fox experiments are only analogies to life, but they are suggestive ones. They demonstrate the extent to which lifelike behavior is grounded in physical chemistry, and they illustrate the concept of chemical selection for survival. This is the only kind of natural selection and evolution that could have existed prior to the development of information-storing molecules and genetic selection. Such experiments demonstrate that separation into coacervate suspensions or microspheres is a common behavior of polymers in solution, that all such microsystems are not equally stable and that the probability of their survival is enhanced if they have within themselves the ability to carry out simple reactions increasing their bulk or strengthening their barrier against the outside world.

One can visualize that before living cells evolved the primitive ocean was teeming with droplets possessing special chemistries that survived for a time and then were dispersed again. Those droplets that by sheer chance contained catalysts able to induce "useful" polymerizations would survive longer than others; the probability of survi-

val would be directly linked to the complexity and effectiveness of their "metabolism." Over the aeons there would be a strong chemical selection for the types of droplets that contained within themselves the ability to take molecules and energy from their surroundings and incorporate them into substances that would promote the survival not only of the parent droplets but also of the daughter droplets into which the parents were dispersed when they became too big. This is not life, but it is getting close to it. The missing ingredient is an orderly mechanism for ensuring that all the daughter droplets receive the catalysts they need for all the reactions important to their survival. This is the pragmatic definition of a genetic apparatus, the subject to which I shall now turn.

The evolution of the genetic machinery is the step for which there are no laboratory models; hence one can speculate endlessly, unfettered by inconvenient facts. The complex genetic apparatus in present-day organisms is so universal that one has few clues as to what the apparatus may have looked like in its most primitive form.

Some 30 years ago Norman H. Horowitz of the California Institute of Technology made the provocative suggestion that metabolic systems evolved, so to speak, from back to front. If today a series of metabolic steps goes

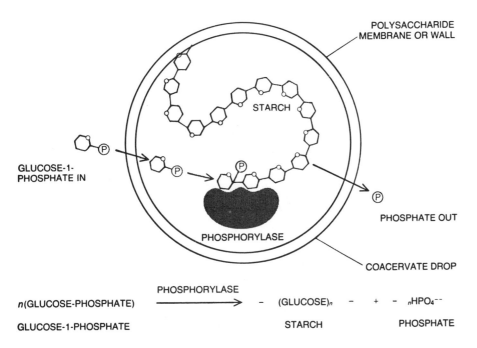

$$n(\text{GLUCOSE-PHOSPHATE}) \xrightarrow{\text{PHOSPHORYLASE}} - (\text{GLUCOSE})_n - + - {}_n\text{HPO}_4^{--}$$

GLUCOSE-1-PHOSPHATE   STARCH   PHOSPHATE

Figure 12.19. Polymerization inside a coacervate droplet causes the wall of the droplet to thicken and the droplet to grow. The droplet, consisting of protein and polysaccharide, contains the enzyme phosphorylase.

Glucose-1-phosphate diffuses into the droplet and is polymerized to starch by the enzyme. The starch migrates to the wall and increases volume of droplet.

from substance A to substance B and then to substances C, D and E, the oldest need was probably for substance E, and the oldest reaction was the one that made E from D, which then was a raw material obtained from the surroundings. Only when the supply of D began to run low would there have been a strong selection pressure for the ability to make D from another raw material, C. An eventual shortage of C would have led to competition in finding ways of making it from some other precursor, B, and in this manner an entire metabolic chain could have evolved slowly in reverse order.

In this way of thinking photosynthesis evolved as a means of providing an alternate source of glucose for organisms that depended on anaerobic fermentation of compounds rich in free energy, at a time when competition had depleted the natural supply of such compounds. Anaerobic fermentation, or glycolysis, itself may have evolved as an alternate means of providing a supply of ATP for more primitive organisms that previously had been dependent on an external source of nucleotides and polyphosphates for driving energy. Hence the familiar metabolic series of reactions (1) photosynthesis of glucose, (2) glycolysis with energy storage through ATP and (3) the utilization of ATP as the energy source in cell activities may be the result of the kind of back-to-front evolution Horowitz postulated.

At a still earlier stage enzymes themselves may not

have been essential if a plentiful supply of activated monomers and condensing reagents was available. Enzymes not only are catalysts but also have a directing or coupling function, ensuring that the chemical free energy released by one reaction is utilized productively by another reaction rather than being dissipated as heat. As soon as it became important for a limited supply of free energy to be channeled into one or a few of the many possible reactions, directed catalysis by enzymes would have become essential.

The first protoenzymes may have been the polymer chains that were themselves being formed. Some polymerizations tend to be autocatalytic: the presence of a particular polymer favors the formation of more of the same polymer. The double-strand helix of DNA is an outstanding example of autocatalytic polymerization, and for this reason alone DNA may have been the natural candidate for a central role in living organisms. As soon as the reaction ceases to be strictly autocatalytic, that is, when the catalyst for the reaction is no longer just the product of the reaction, the problem arises of ensuring that the supply of catalyst increases and is passed on to the descendants of the protobiont.

The first successfully stabilized protobionts may have been autocatalytic coacervates of nucleic acids similar to those constructed by Oparin but dependent on activated monomers and coupling reagents rather than on polymer-

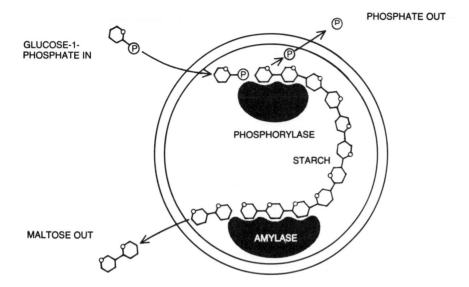

GLUCOSE-1-
PHOSPHATE IN

PHOSPHATE OUT

PHOSPHORYLASE

STARCH

MALTOSE OUT

AMYLASE

Figure 12.20. Two-step reaction takes place inside a protein-carbohydrate droplet provided with two enzymes. One enzyme, phosphorylase, polymerizes glucose-1-phosphate to starch. The second enzyme, amylase, degrades the starch to maltose. Droplets in this instance do not grow because the starch disappears as fast as it is made. The maltose diffuses back into surrounding medium.

ase enzymes. If the nucleic acid could have served as a template for the polymerization of protein chains, even of a random sequence, then this protein might have been useful as a skin to protect the nucleic acid coacervate. In this way a cooperative interaction between nucleic acid and protein would have existed from the beginning, with the nucleic acid playing the autocatalytic and template role, and the protein playing a structural and protective role. If a particular pattern of positive and negative charges along the polypeptide chain proved to be helpful in polymerizing either the nucleic acid or the polypeptide, then the first protein catalyst or enzyme function would have arisen. There would then have been strong selection pressures for those nucleic acid sequences that favored the continued formation of just that pattern of positive and negative amino acid side chains. In this way template replication in the nucleic acid and enzymatic catalysis in the polypeptide could have evolved in tandem, and there may never have been an era either of "life without DNA" or of "naked genes."

Such speculations all require the existence of some kind of mutual recognition or complementarity at the molecular level between amino acid sequences in proteins and base sequences in nucleic acids. Many attempts have been made to find a natural fit between protein sequences and nucleic acid sequences that could have existed before the appearance of the present-day elaborate machinery involving transfer-RNA molecules, ribosomes and charging enzymes. None of these attempts has been fully convinc-

ing. In all present-day life a charging enzyme attaches a specific amino acid to a transfer-RNA molecule that has at its other end an anticodon for that amino acid. (An anticodon is a triplet of bases that are complementary to a codon: a triplet that codes for a particular amino acid.) The specificity of matching amino acids to codons lies neither in the codon nor in the transfer RNA but in the charging enzyme. How did the matching arise before charging enzymes existed? This looks like another chicken-and-egg paradox, since the charging enzymes themselves are synthesized by the translation machinery they help to operate. The answer to the original chicken-and-egg paradox was that neither the chicken nor the egg came first; they evolved together from lower forms of life. The same must be true of the genetic machinery; the entire apparatus evolved in concert from simpler systems now driven out of existence by competition. Although we can examine fossil remains of chicken ancestors, we have no fossil enzymes to study. We can only imagine what probably existed, and our imagination so far has not been very helpful.

The system today that is most likely to shed light on a primitive association between nucleic acid and the replication of protein is the repressor-operator system of genetic control. Although the direct recognition of nucleic acid sequences by amino acid side chains is no longer a part of the readout of the genetic message, when certain genes are shut down in bacteria, a protein molecule of definite amino acid sequence (the repressor) must recognize and bind to a particular sequence of base pairs (the operator

a

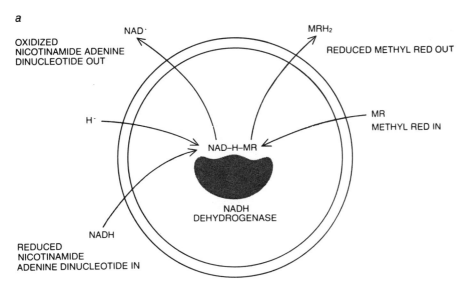

OXIDIZED
NICOTINAMIDE ADENINE
DINUCLEOTIDE OUT

NAD·

MRH₂

REDUCED METHYL RED OUT

H·

MR
METHYL RED IN

NAD–H–MR

NADH
DEHYDROGENASE

NADH

REDUCED
NICOTINAMIDE
ADENINE DINUCLEOTIDE IN

Figure 12.21. Electron transport is mimicked in an Oparin experiment in which coacervate droplets are supplied with NADH dehydrogenase, an enzyme from bacteria. The medium contains methyl red, a dye, and nicotinamide adenine dinucleotide in its reduced form (NADH). When the two substances diffuse into the droplet, the enzyme effectuates the transfer of hydrogen from NADH to the dye, thus reducing it. Products of reaction then diffuse out of droplet.

DNA). The nature of this sequence recognition is being studied in many laboratories. When it is finally understood exactly how the protein repressor recognizes the base sequence of the DNA operator, we may begin to hypothesize intelligently how a given sequence of bases could have produced a specific polypeptide chain sequence in the days before transfer RNA, ribosomes and charging enzymes.

Through some gradual means about which we can now only speculate, an association of nucleic acid as the archival material with protein as the working catalysts evolved into the complex genetic transcription and translation machinery that all forms of life exhibit today. This made it possible to preserve all the biochemical abilities of a parent protocell in its offspring. But since the genetic message was subject to alteration by the slow accumulation of errors and by direct mutation brought about by ionizing radiation and other agents, the environment could now serve as a screen, selecting for or against the possessors of the altered messages. Evolution by natural selection in its Darwinian sense could begin.

The steps I have outlined so far or something similar to them are probably responsible for the appearance of the first living organisms on the earth. They were presumably one-celled entities resembling modern fermenting bacteria such as *Clostridium,* which had a complete genetic apparatus but were totally dependent on the breakdown of nonbiologically formed energy-rich molecules for their survival. They would have been scavengers of the organic matter produced by electric discharges and ultraviolet radiation. Hence the total amount of life the earth could have sustained would have been limited by the rate of production of such compounds by nonbiological means. Living organisms in that era would have been strictly consumers of organic matter, not producers.

The capacity of the earth to support life was enormously enhanced by the invention of photosynthesis, which enabled living organisms to capture solar energy for the synthesis of organic molecules. The first photosynthesizers removed themselves from the competition for a dwindling supply of natural energy-rich molecules and set themselves up as primary producers. Photosynthesis using hydrogen sulfide as the source of hydrogen atoms for reducing carbon dioxide, which is the process conducted today by the green and purple sulfur bacteria, undoubtedly preceded the more elaborate two-step form of photosynthesis wherein water supplies the hydrogen, which is the process conducted today by the cyanobacteria, or blue-green algae, and by green plants (see equations 19 and 20 in Figure 12.16). On the primitive earth hydrogen sulfide would have been sufficiently abundant for it to have served as a practical reductant. Water is even more abundant, however, and organisms that found ways of taking hydrogen atoms for synthesis from water rather than from hydrogen sulfide would have had a great advantage over their sulfur-using cousins.

This brings the story of life on the earth up to the cyanobacteria, whose fossilized ancestors seem to be present in sediments at least 3.2 billion years old. It is apparent that not only life but also photosynthetic life evolved within a billion years of the formation of the

b

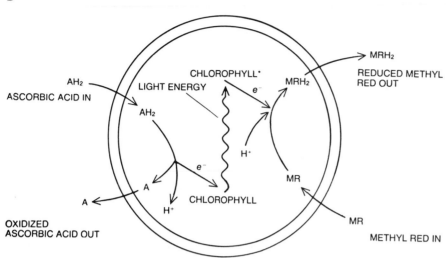

Figure 12.22. Photosynthesis is mimicked in another Oparin experiment with coacervates containing chlorophyll. Here again the dye methyl red is the substance to be reduced. Ascorbic acid, which diffuses into the droplets, is not in itself a strong enough reducing agent to serve the purpose. If the droplets are exposed to light, however, excited electrons from the chlorophyll are capable of reducing the methyl red. The chlorophyll then regains the expended electrons from the ascorbic acid, which is oxidized in the process and diffuses out of the droplets.

planet. It is not absolutely clear that those ancient organisms split water by photosynthesis and released free oxygen into the atmosphere, but it seems likely.

The next two billion years saw a revolution in the nature of the atmosphere of the planet: from a reducing atmosphere with little or no free oxygen to an oxidizing atmosphere in which one out of every five molecules is oxygen. One consequence was the formation of an ozone layer in the upper atmosphere that sharply reduced the ultraviolet radiation at the earth's surface. Although this effectively ended the nonbiological synthesis of organic matter, biological photosynthesis working with the energy in the visible wavelengths more than made up for it. The pattern of life driven by solar energy was fixed for all time on our planet, and the stage was set for true biological evolution.

13

Fred Hoyle and N. C. Wickramasinghe
Prebiotic Molecules and Interstellar Grain Clumps

Interstellar molecules detected by radioastronomical techniques in clouds such as OMC 1 and 2 in the Orion Nebula and Sgr B2 in the galactic centre span a wide range of types and complexity. Among the heaviest of the molecules recently discovered is cyanodiacetylene ($H-C\equiv C-C\equiv C-C\equiv N$)(Avery et al., 1976). There have been earlier detections of precursors to the simplest known amino acid glycine (formic acid and methanimine), and probable detections of polyoxymethylene polymers and copolymers (Wickramasinghe, 1974 and 1975; Mendis and Wickramasinghe, 1975; Cooke and Wickramasinghe, 1977) in interstellar clouds. We discuss here a possible identification of organic molecules of even greater complexity, and its implications for the start of biological activity.

Large departures from thermodynamic equilibrium in the interstellar medium and the co-existence of solid grains, molecules, radicals, ions and ultraviolet photons provide conditions which are ideal for the assembly of 'exotic' molecular species. If clumping of 100 Å-sized dust grains occurred by a process which we have discussed (Hoyle and Wickramasinghe, 1976), highly complex organic molecules could be formed and become securely trapped during the 'welding' process of smaller grains. Such grain clumps could be widely dispersed amongst the particulate material in the Galaxy, being responsible for most of extinction and polarisation of starlight.

The spectral identification of highly complex organic species in the interstellar medium might seem hopeless. A property common to a wide class of such molecules, however, is an absorption band centred at $\lambda \simeq 2,200$ Å with a half-width $\simeq 250$ Å, which arises mainly due to $\pi-\pi^*$ or $n-\pi^*$ electronic transitions in conjugated multiple bonds such as:

$$C=C-C=C \qquad C=C-C=O$$
$$C=C-C=C \qquad C=C-C=N$$
$$C=C-C=C \qquad C=C-C\equiv N.$$

The absorption occurs due to transitions between electrons in n and π-orbital states to antibonding π^* states (*Organic Electronic Spectral Data,* 1946–61; *Ultraviolet Reference Spectra*). Specific examples of such transitions and relevant spectroscopic data have been cited elsewhere (Wickramasinghe et al., 1977). In general, this absorption band occurs in most highly complex organic species, with a typical value of $\sim 200-300$ Å for the half-width, and with a molar extinction coefficient $\epsilon \sim 10^4$ cm^{-1} (g mol 1^{-1})$^{-1}$ which implies an absorption cross-section of $\sim 4 \times 10^{-17}$ cm^2. A typical case is the curve for mesityl oxide (Fig. 13.1).

It is tempting to identify this feature with the well-known 2,200-Å band of the interstellar extinction curve. The smeared-out mass density ρ of molecules with molecular weight M required to produce the observed extinction coefficient at 2,200 Å of ~ 1.5 mag kpc^{-1} (Bless and Savage, 1972; Nandy et al., 1975) in the interstellar medium is

$$\rho \simeq 2 \times 10^{-27} \, (M/100) \text{ g cm}^{-3}.$$

For $M \sim 100$ we obtain a value of ρ which is $\sim 10\%$ of the total available CNO density in the interstellar medium. If this material is assumed to be homogeneously mixed with silicate-type grains of radius a, the condition for optical depth less than unity within a single grain at the band centre is

$$a < \sim 3 \times 10^{-5} \text{ cm}.$$

It may not be fortuitous that this upper limit is about twice the 'typical radius' required for interstellar silicate grains.

Several amino acids of biological importance have been discovered in carbonaceous chondrites (Cronin and Moore, 1976) and there has also been an identification which is strongly suggestive of the existence of polyoxymethylene polymers or copolymers in the Allende carbonaceous chondrite (Breger et al., 1972). Such polymers could have served as glueing agents for smaller grains in prestellar molecular clouds. It is also of interest that an extract of organic materials from the Murchison chondrite using hexafluoroisopropanol as a solvent showed a 2,200 Å absorption feature, typical of conjugated double bonds of the type considered above, and similar to the interstellar extinction hump at this wavelength (Sakata et al., 1977).

We tentatively conclude that these data could be interpreted as independent new chemical evidence of the existence of composite grain clumps in the interstellar medium and in carbonaceous chondrites. Moreover, such grain clumps probably include a significant mass fraction of highly complex organic, prebiotic molecules which could have led to the start and dispersal of biological activity on the Earth and elsewhere in the Galaxy.

Even if all the CNO elements on the Earth's surface were initially in the form of the biologically important 20 amino acids, their linkage into feasible peptide chains could not have occurred by chance. Some process of natural selection is required, and this is usually held to have occurred under terrestrial conditions. If interstellar grain assembly occurred in prestellar molecular clouds already containing these amino acids, pre-Darwinian selection leading to the production of self-replicable peptide chains, could have depended upon the sticking properties of the various molecules on to the polymeric coatings of 100 Å-sized grains. A welded composite grain with a 'cellular' structure on the scale of $2\pi \times 100$ Å could have served as the host system for the earliest primitive gene in a manner similar to that discussed by Cairns-Smith (Cairns-Smith, 1965).

We now consider the selective mechanisms which could have operated in the growth and evolution of grain clumps, and their possible bearing on the development of a primitive biological system. Not all polymer coatings have the same sticking probabilities α. Some combinations are more probable than others. Suppose $\langle \alpha \rangle$ denotes the average sticking probability. What would be the optimum situation for clump growth? A reproducible situation with sticking probability only slightly above $\langle \alpha \rangle$ would be better than a non-reproducible situation with probability much greater than $\langle \alpha \rangle$. This is because the latter system, if it occurred, would soon be covered up by further growth of clumps. Once the possibility of a reproducible system is admitted, what reproducible system, among possibly many such systems, would become most widespread? The answer to this question is: a system which also promoted clump division. This leads not only to more clumps but also to a greater total mass of clump material, because of the advantageous surface/volume ratio.

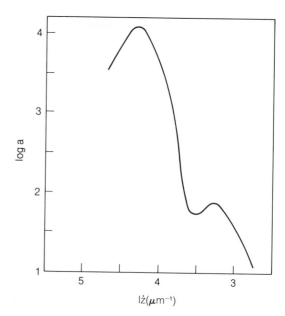

Figure 13.1. Molar extinction coefficient ϵ (cm⁻¹(g mol l⁻¹)⁻¹) of mesityl oxide

as a function of inverse wavelength (Ultraviolet Reference Spectra).

Lastly, we turn to the problem of protection of prebiotic material against external disruptive agencies, for example, ultraviolet light. Whilst this protection could be provided by an outer shield of solid grains, such a system would be biologically inert. A clump stripped of its outer polymeric coating would not be suitable for sticking, even if it was able to preserve itself. A more refined protection system in the nature of a biological cell wall would be much better, since such an outer surface could permit organic molecules to stick, as well as to diffuse through it.

Thus there seem to be two somewhat different types of selective process which could be operative under interstellar conditions. First, a competition for clump growth in the absence of disruptive agencies, for example, within a pre-stellar cloud; and second, a competition for organic materials (or even for sticking) in a more hostile environment in the presence of disruptive agencies. With the development of a cell wall, the last step could presumably be to 'split out' the inorganic grains which started off the whole process.

III

The Search for Life in the Solar System

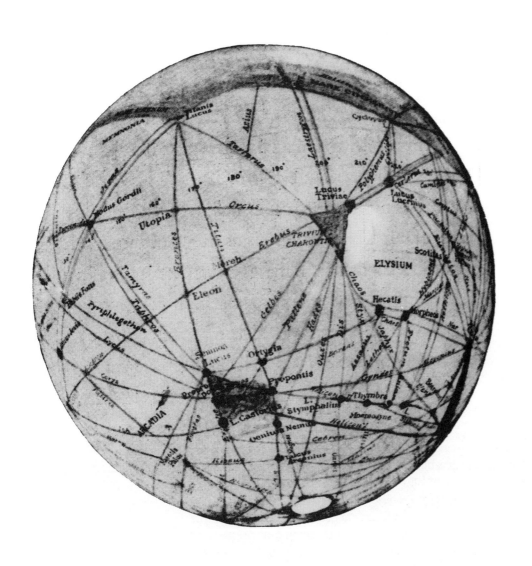

Our sun's family of nine planets contains just one on which life is known to exist, leaving eight candidates for speculation. Of these eight, by far the most intriguing to mankind has been Mars, the next planet out from the sun, whose rust-red color has suggested the god of war to many civilizations that admired it from Earth.

The possibility of life on Mars seemed to receive a tremendous increase, if not downright confirmation, from the announcement of *canali,* an Italian word meaning "channels" or "canals," made by Giovanni Schiaparelli more than a century ago. Schiaparelli made no statements about the nature of these apparent straight lines on the Martian surface, but the American astronomer Percival Lowell, possessed of a fortune large enough to build his own observatory in Flagstaff, Arizona, and a self-confidence to match, proclaimed that his observations showed an indisputable network of water-carrying channels, covering the planet except for the polar caps, from which they presumably carried irrigation water as the caps melted.

Lowell was wrong, and Schiaparelli was wrong: the Martian canals are optical illusions, straight lines produced by the human eye and brain, which links together small, faint markings in straight patterns. But Lowell's impact was tremendous, coming at a time when human interest in life on other planets was growing. Lowell's "social Darwinism,' his conclusions about how Martian society must have evolved under the harsh conditions of that planet's environment, fed straight into the mainstream of his era's political thought. Lowell's ability as a popular writer (he also wrote interesting books about his travels in China) helped his "discoveries" attain worldwide recognition. In England, H. G. Wells responded by writing one of his most famous science-fiction novels, *War of the Worlds* (published in 1897), which described an invasion of Earth by intelligent Martians, who proved unable in the end to survive terrestrial microbes.

Radio searches for signals from Mars were made as early as 1922, but in the absence of success, and with the onset of the Great Depression and the Second World War, support for such efforts faded. But the popular belief in life on other worlds showed itself vividly on Halloween night, 1938, when Orson Welles led a radio dramatization of H. G. Wells's novel, telling in a quasi-realistic fashion of a Martian invasion of New Jersey. Thousands of listeners throughout the East panicked; many of them fled their homes for supposedly safer regions of the country. Reassuring news soon greeted them, and many were later reluctant to admit the full extent of their participation in the evacuation.

By the time that the eminent British astronomer Sir James Jeans reviewed the search for extraterrestrial life in 1942, Lowell's influence had been replaced by a realization that the canals probably did not exist. Jeans concluded that none of the inner planets, Mercury, Venus, and Mars, offered a likely site for our sort of life, and he dismissed the possibility of life on the giant planets, Jupiter, Saturn, Uranus, and Neptune, because their outer layers were known to be tremendously cold.

By sending space probes to five of Earth's eight sister planets, and by studying them with better equipment from Earth's surface, we have found during the past decade that the generally negative outlook for life elsewhere in the solar system appears correct. However, we have discovered that the giant planets—Jupiter in particular—grow warmer as one descends into their gaseous envelopes, so that it would be premature to conclude, as Jeans did, that life cannot exist there simply because of the temperature problem. Temperature does indeed rule out life on Mercury (which alternately bakes and freezes), on the moon (for the same reason), and on Venus, which stifles under a blanket of carbon-dioxide atmosphere that keeps the planet's surface above 700° C. But the possibility of life on the giant planets seems more threatened—though by no means eliminated—through the lack of oxygen or a solid surface on which molecules can collect.

To search for life on Mars, we spent a billion dollars to send the two Viking orbiters and landers to the red planet in 1976. As Norman Horowitz describes, we found tantalizing indications, but on balance no clear evidence of life. When we manage to send probes that will enter the atmospheres of Jupiter and the other giant planets, we may find equally fascinating data, perhaps equally indecisive, on the outer members of the solar system.

14

Percival Lowell
Mars as the Abode of Life

From what has taken place on earth, we see that cooling and complexity of organism have advanced together. Life originated here as soon as the temperature fell below the boiling-point, and it started in water, the liquefying of which out of steam gave it at once an essential factor of its substance and an environment of the most easily satisfying kind.

An upward step in evolution occurred when life stepped out upon the land. While less directly favorable to life, the land was fraught with more possibilities for organisms capable of turning them to account. Brain was needed, and brain evolved.

Brain, indeed, now became the chief concern of nature. The character of the habitat undoubtedly brought this about through the prizes it offered the clever, and the snuffing out to which it consigned the crass.

For long the animal remained thus the creature of its environment, its view restricted in both time and space. Greater possibilities came in with man. Doubtless his was no very dignified entry, though something better than on all fours. Brain now finally distanced brawn, and even in his savage state man became a being that others feared. From this standing *primus inter pares*, he soon developed into first, "with the rest nowhere." Fire and clothes raised him to some independence of his surroundings, and slowly he began to take possession of the earth. His breeching, the putting on by the race of the *toga virilis*, was both an incident of his rise and part cause of it as well, for it made him superior to climate. But the fertility of brain, however humble in its beginning, which suggested the means of protecting the body, devised the methods by which he was to subjugate the earth.

For some centuries now this has been his goal, unconscious or confessed. The true history of man has consisted not in his squabbles with his kind, but in his steady conquest of all earth's animals except himself. He has enslaved all that he could; he is busy in exterminating the

rest. From this he has gone on to turn the very forces of nature to his own ends. This task is recent and is yet in its infancy, but it is destined to great things. As brain develops, it must take possession of its world.

Subjugation carries its telltale in its train; for it alters the face of its habitat to its own ends. Already man has begun to leave his mark on this his globe in deforestation, in canalization, in communication. So far his towns and his tillage are more partial than complete. But the time is coming when the earth will bear his imprint, and his alone. What he chooses, will survive; what he pleases, will lapse, and the landscape itself become the carved object of his handiwork.

Equally applicable is this deduction to planets other than the earth. Instead of its being true, as a recent writer remarked, that "we cannot expect to see any signs of the works of inhabitants of Mars if such exist," precisely the opposite is the case. Until the animal attain to dominance of his world, his presence on it would not be seen. Too small in body himself to show, it would be only when his doings had stamped themselves there that his existence could with certainty be known. Then and not till then would he stand disclosed. It would not be by what he was, but through what he had brought about. His mind would reveal him by its works—the signs left upon the world he had fashioned to his will. And this is what I mean by saying that through mind and mind alone we on earth should first be cognizant of beings on Mars. . . .

The evidence of observation thus bears out what we might suspect from the planet's smaller size: that it is much farther along in its planetary career than is our earth. This aging in its own condition must have its effect upon any life it may previously have brought forth. That life at the present moment would be likely to be of a high order. For whatever its actual age, any life now existent on Mars must be in the land stage of its development, on the whole a much higher one than the marine. But, more than this, it should probably have gone much farther if it exist at all, for in its evolving of terra firma, Mars has far outstripped the earth. Mars's surface is now all land. Its forms of life

From *Mars as the Abode of Life* (New York: The Macmillan Company, 1908). By permission of Lowell Observatory.

must be not only terrestrial as against aquatic, but even as opposed to terraqueous ones. They must have reached not simply the stage of land-dwelling, where the possibilities are greater for those able to embrace them, but that further point of pinching poverty where brain is needed to survive at all.

The struggle for existence in their planet's decrepitude and decay would tend to evolve intelligence to cope with circumstances growing momentarily more and more adverse. But, furthermore, the solidarity that the conditions prescribed would conduce to a breadth of understanding sufficient to utilize it. Intercommunication over the whole globe is made not only possible, but obligatory. This would lead to the easier spreading over it of some dominant creature,—especially were this being of an advanced order of intellect,—able to rise above its bodily limitations to amelioration of the conditions through exercise of mind. What absence of seas would thus entail, absence of mountains would further. These two obstacles to distribution removed, life there would tend the quicker to reach a highly organized stage. Thus Martian conditions themselves make for intelligence.

Our knowledge of it would likewise have its likelihood increased. Not only could any beings there disclose their presence only through their works, but from the physical features the planet presents, we are led to believe that such disclosure would be distinctly more probable than in the case of the earth. Any markings made by mind should there be more definite, more uniform, and more widespread than those human ones with which we are familiar. More dominant of its domicile, it should so have impressed itself upon its habitat as to impress us across intervening space.

What the character of such markings might be, we shall best conceive by letting the pitiless forbiddingness of the Martian surface take hold upon our thought. Between the two polar husbandings of the only water left, stands the pathless desert—pathless even to the water semiannually set free. Only overhead does the moisture find natural passage to its winter sojourn at the other pole. Untraversable without water to organic life, and uninhabitable, the Sahara cuts off completely the planet's hemispheres from each other, barring surface commerce by sundering its supplies. Thirst—the thirst of the desert—comes to us as we realize the situation, parching our throat as we think of a thirst impossible of quenching except in the far-off and by nature unattainable polar snows. . . .

Thirty years ago what were taken for the continents of Mars seemed, as one would expect continents seen at such a distance to appear, virtually featureless.

In 1877, however, a remarkable observer made a still more remarkable discovery; for in that year Schiaparelli, in scanning these continents, chanced upon long, narrow markings in them which have since become famous as the canals of Mars. Surprising as they seemed when first imperfectly made out, they have grown only more wonderful with study. It is certainly no exaggeration to say that they are the most astounding objects to be viewed in the heavens. There are celestial sights more dazzling, spectacles that inspire more awe, but to the thoughtful observer who is privileged to see them well there is nothing in the sky so profoundly impressive as these canals of Mars. Fine lines and little gossamer filaments only, cobwebbing the face of the Martian disk, but threads to draw one's mind after them across the millions of miles of intervening void.

Although to the observer practised in their detection they are at certain times not only perfectly distinct, but are not even difficult objects—being by no means at the limit of vision, as is often stated from ignorance,—to one not used to the subject, and observing under the average conditions of our troublesome air, they are not at first so easy to descry. Had they been so very facile, they had not escaped detection so long, nor needed Schiaparelli, the best observer of his day, to discover them. But in good air they stand out at times with startling abruptness. I say this after having had twelve years' experience in the subject—almost entitling one to an opinion equal to that of critics who have had none at all.

How beside the mark it is to credit them to illusion may at once be appreciated from the fact that experiment shows the main ones to appear through the telescope of the same size as a telegraph wire seen with the naked eye at a distance of a hundred and fifty feet. But if the air be not steady, they are blurred almost out of recognition.

With our air at its best, the first thing to strike one in these strange phenomena is their geometric look. It has impressed every observer who has seen them well. It would be hard to determine to which of their peculiar characteristics this effect was specially due. Indeed, it is probably attributable to their combination; for distinctive as each trait is alone, their summation is multiplicitly telling. That the lines run quite straight from point to point—that is, on arcs of great circles, or else curve in an equally determinate manner; that they are of uniform width throughout; that their tenuity is extreme and that they are of enormous length, are attributes each of which is geometrically startling and which, taken together, enhance this in geometric ratio.

That the lines are absolutely straight—which means that on a sphere like Mars they follow arcs of great circles—is shown by two facts which fay into one another. One of these is that they look straight to the observer when central enough not to have foreshortening tell. This could not happen unless they were the shortest possible lines between their termini. The other proof consists in their fitting together to form a self-agreeing whole when the result of all the drawings—hundreds in number at each opposition—are plotted on a globe.

In regard to their width, it would be nearest the mark to say that they had none at all. For they have been found narrower and narrower as the conditions of scanning have improved. By careful experiments at Flagstaff it has been shown that the smallest appear as they should were they but a mile across. The reason so slender a filament is visible is due to its length, and this probably because of the number of retinal cones that are struck. Were only one affected, as would be the case were the object a point, it certainly could not be detected.

So much for the smallest canal now visible with our present means. The larger are much more conspicuous. These look not like gossamers, as the little ones do, but like strong pencil-lines. Comparison with the thread of the micrometer gives for the average canal a breadth of about ten miles. The canals, hoever, are by no means of a uniform width. Indeed, they are of all sizes, from lines it would seem impossible to miss to others it taxes attention to descry.

All the more surprising for their relative diversity is the remarkably uniform size of each throughout its course. So far as it is possible to make out, there is no perceptible difference in width of a canal, when fully developed, from one end of it to the other. Certainly it takes a well-ruled line on paper to look its peer for regularity and deportment.

True thus to itself, each canal differs from its neighbor not only in width, but in extension. For the canals are of very various length. Some are not above 250 miles long, while others stretch 2500 miles from end to end. Nor is this span by any means the limit. The Eumenides-Orcus runs 3450 miles from where it leaves the Phoenix Lake to where it enters the Trivium Charontis. Enormous as these distances are for lines which remain straight throughout, they become the more surprising when we consider the size of the planet on which they are found. For Mars is only 4220 miles through, while the earth is 7919. So that a canal 3450 miles long, for all its unswervingness to right or left, actually curves in its own plane through an arc of some 90° round the planet. It is much as if a straight line joined London to Denver, or Boston to Bering Strait.

It should be remembered, however, that it is the actual, not the relative, length we have really to consider. But this is surprising enough—more than sufficient in the Eumenides-Orcus to span the United States.

Odd as is the look of the individual canal, it is nothing to the impression forced upon the observer by their number and still more by their articulation. When Schiaparelli finished his life-work, he had detected 113 canals; this figure has now been increased to 437 by those since added at Flagstaff. As with the discovery of the asteroids, the later found are as a rule smaller and in consequence less evident than the earlier. But not always; and, unlike asteroid hunting, it is not because of easy missing in the vast field of sky. The cause is intrinsic to the canal.

This great number of lines forms an articulate whole. Each stands jointed to the next (to the many next, in fact) in the most direct and simple manner—that of meeting at their ends. But as each has its own peculiar length and its special direction, the result is a sort of irregular regularity. It resembles lace-tracery of an elaborate and elegant pattern, woven as a whole over the disk, veiling the planet's face. By this means the surface of the planet is divided into a great number of polygons, the areolas of Mars.

Schiaparelli detected the existence of the canals when engaged in a triangulation of the planet's surface for topographic purpose. What he found was a triangulation already made. In his own words, the thing "looked to have been laid down by rule and compass." Indeed, no lines could be more precisely drawn, or more meticulously adjusted. Not only do none of them break off in mid-career, to vanish, as rivers in the desert, in the great void of ochre ground, but they contrive always in a most gregarious way to rendezvous at special points, running into the junctions with the space punctuality of a train on time. Nor do one or two only manage this precision; all without exception converge from far points accurately upon their centres. The meetings are as definite and direct as is possible to conceive. None of the large ochre areas escapes some filament of the mesh. No single secluded spot upon them could be found, were one inclined to desert isolation, distant more than three hundred miles from some great thoroughfare.

For many years—in fact, throughout the period of observation of the great Italian—the canals were supposed to be confined to the bright or reddish ochre regions of the disk. None had been seen by him elsewhere, and none was divined to exist. But in 1892, W. H. Pickering, at Arequipa, saw lines in the dark regions; and, in 1894, Douglass, at Flagstaff, definitely detected the presence of a system of canals criss-crossing the blue-green similar to that networking the ochre. Later work at Flagstaff has shown all the dark areas to be thus seamed with lines, and lastly has brought out with emphasis the pregnant fact that these are continued by others connecting with the polar snows. Thus the system is planet-wide in its application, while it ends by running up to the confines of the polar cap. The first gives it a generality that opened up new conceptions of its office, the second vouchsafes a hint as to its origin.

For many years the pioneers in this discovery of another world had their revelations strictly to themselves, decried as baseless views and visions by the telescopically blind. So easily are men the dupes of their own prejudice. But in 1901 attempts began to be made at Flagstaff to make them tell their own story to the world, writing it by self-registration on a photographic plate. It was long before they could be compelled to do so. The first attempts showed nothing; the next, two years later, did better, evoking faint forms to the initiate, but to them alone; but two years later still, success crowned the long endeavor. At last these strange geometricisms have stood successfully for their pictures. The photographic feat of making them keep still sufficiently long—or, what with heavenly objects is as near as man may come to his practice with human subjects, the catching of the air-waves still long enough to secure impression of them upon a photographic plate—has been accomplished by Mr. Lampland. After great study, patience, and skill he has succeeded in this remarkable performance, of which Schiaparelli wrote in wonder to the present writer: "I should never have believed it possible." . . .

Next in interest to the canals come the oases. Many years after the detection of the canals, scrutiny revealed another class of detail upon the planet of an equally surprising order. This was the presence there of small, round, dark spots dotted over the surface of the disk. Seen in any number, first by W. H. Pickering in 1892, they lay at the meeting-places of the canals. He called them lakes. Some few had been caught earlier, but were not well recognized. We now know of 186 of them, and we are very certain they are not lakes. In the case of one of them, the Ascraeus Lucus, no less than seventeen canals converge to it.

It thus appears that the spots make, as it were, the knots of the canal network. They emphasize the junctions in look and at the same time indicate their importance in the system. For just as no spot but stands at a junction, so reversely, few prominent junctions are without a spot, and the better the surface is seen, the more of these junctions prove to be provided with them.

Their form is equally demonstrative of their function. They are apparently self-contained and self-centred, being small, dark, and, as near as can be made out, round. It is certain that they are not mere reënforcements of the canals due to crossing, for crossings do occur where none are seen, while the lines themselves are perfectly visible, and of the same strength at the crossing as before and after.

We now come to yet a more surprising detail. The existence of the single canals had scarcely been launched upon a world quite unprepared for their reception, and duly distant in their welcome in consequence, before that world was asked to admit something more astounding still; namely, that at certain times some of these canals appeared mysteriously paired, the second line being an exact replica of the first, running by its side the whole of its course, however long this might be, and keeping equidistant from it throughout. The two looked like the twin rails of a railway track.

To begin by giving an idea of the phenomenon, I will select a typical example, which happened also to be one of the very first observed by me—that of the great Phison. The Phison is a canal that runs for 2250 miles between two important points upon the planet's surface, the Portus Sigaeus, halfway along the Mare Icarium, and the Pseboas Lucus, just off the Protonilus. In this long journey it traverses some six degrees of the southern hemisphere and about forty degrees of the northern. In 1894 the canal was first seen as a single, well-defined line—not a line that admitted of haziness or doubt, but which was as strictly self-contained and slenderly distinguished as any other single canal on the planet. A Martian month or more after it thus expressed itself, it suddenly stood forth an equally self-confessed double, two parallel lines replacing the solitary line of some months before. Not the slightest difference in the character, direction, or end served was to be detected between the two constituents. Just as certainly as a single line had shown before, a double line now showed in its stead.

15

No Mars Message Yet, Marconi Radios; Ends Yacht Trip "Listening In" On Planet Today

Guglielmo Marconi, Italian radio wizard, has apparently failed to pick up any messages from Mars in his trip across the Atlantic on his floating laboratory, the yacht Electra.

Replying today to a radiogram sent yesterday by The Associated Press asking for a statement on the results of his experiment, Senator Marconi said:

"Have no sensational announcement to make."

He added that he would arrive in New York today.

Marconi, who sailed from Southampton on May 23 to lecture here before the Institute of Radio Engineers and the Institute of Electrical Engineers on "Radio Telegraphy," is not alone in his theory that communication with Mars by radio is possible. Director Baillaud of the Paris Observatory and other authorities have expressed faith in the idea.

Marconi has spent the time crossing the Atlantic performing many electrical experiments, principally by listening for signals from Mars. Several months ago he picked up what he took to be signals of 150,000-meter wave length. As these were much longer than any ever sent from a known earthly source, he believed they might have come from Mars. He admitted, however, that they might have come from any region in the universe where electrons are in vibration.

As Mars is now at its nearest approach to the earth he was especially hopeful that he might on this trip gain some definite knowledge regarding the signals which might at least indicate whether communication with Mars or other planets is possible.

16

Sir James Jeans
Is There Life on the Other Worlds?

So long as the earth was believed to be the center of the universe the question of life on other worlds could hardly arise; there were no other worlds in the astronomical sense, although a heaven above and a hell beneath might form adjuncts to this world. The cosmology of the *Divina Commedia* is typical of its period. In 1440 we find Nicholas of Cusa comparing our earth, as Pythagoras had done before him, to the other stars, although without expressing any opinion as to whether these other stars were inhabited or not. At the end of the next century Giordano Bruno wrote that "there are endless particular worlds similar to this of the earth." He plainly supposed these other worlds—"the moon, planets and other stars, which are infinite in number"—to be inhabited, since he regarded their creation as evidence of the Divine goodness. He was burned at the stake in 1600; had he lived only ten years longer, his convictions would have been strengthened by Galileo's discovery of mountains and supposed seas on the moon.

The arguments of Kepler and Newton led to a general recognition that the stars were not other worlds like our earth but other suns like our sun. When once this was accepted it became natural to imagine that they also were surrounded by planets and to picture each sun as showering life-sustaining light and heat on inhabitants more or less like ourselves. In 1829 a New York newspaper scored a great journalistic hit by giving a vivid, but wholly fictitious, account of the activities of the inhabitants of the moon as seen through the telescope recently erected by His Majesty's Government at the Cape.

It will be a long time before we could see what the New York paper claimed to see on the moon—bat-like men flying through the air and inhabiting houses in trees— even if it were there to see. To see an object of human size on the moon in detail we should need a telescope of from 10,000 to a 100,000 inches aperture, and even then we should have to wait years, or more probably centuries,

Reprinted from *Science*, 95 (1942), 589. Copyright 1942 by the American Association for the Advancement of Science.

before the air was still and clear enough for us to see details of human size.

To detect general evidence of life on even the nearest of the planets would demand far larger telescopes than anything at present in existence, unless this evidence occupied an appreciable fraction of the planet's surface. The French astronomer Flammarion once suggested that if chains of light were placed on the Sahara on a sufficiently generous scale, they might be visible to Martian astronomers if any such there be. If this light were placed so as to form a mathematical pattern, intelligent Martians might conjecture that there was intelligent life on earth. Flammarion thought that the lights might suitably be arranged to illustrate the theorem of Phythagoras (Euclid, I. 47). Possibly a better scheme would be a group of searchlights which could emit successive flashes to represent a series of numbers. If, for instance, the numbers 3, 5, 7, 11, 13, 17, 19, 23 . . . (the sequence of primes) were transmitted, the Martians might surely infer the existence of intelligent Tellurians. But any visual communication between planets would need a combination of high telescopic power at one end and of engineering works on a colossal, although not impossible, scale at the other.

Some astronomers—mainly in the past—have thought that the so-called "canals" on Mars provide evidence of just this kind, although of course unintentionally on the part of the Martians. Two white patches which surround the two poles of Mars are observed to increase and decrease with the seasons, like our terrestrial polar ice. Over the surface of Mars some astronomers have claimed to see a geometrical network of straight lines, which they have interpreted as an irrigation system of canals, designed to bring melted ice from these polar caps to parched equatorial regions. Percival Lowell calculated that this could be done by a pumping system of 4,000 times the power of Niagara. It is fairly certain now that the polar caps are not of ice, but even if they were, the radiation of the summer sun on Mars is so feeble that it could not melt more than a very thin layer of ice before the winter cold came to freeze it solid again. Actually the caps are observed to change

81

very rapidly and are most probably clouds consisting of some kind of solid particles.

The alleged canals can not be seen at all in the largest telescopes nor can they be photographed, but there are technical reasons why neither of these considerations is conclusive against the existence of the canals. A variety of evidence suggests, however, that the canals are mere subjective illusions—the result of overstraining the eyes in trying to see every detail of a never very brightly illuminated surface. Experiments with school-children have shown that under such circumstances the strained eye tends to connect patches of color by straight lines. This will at least explain why various astronomers have claimed to see straight lines not only on Mars, where it is just conceivable that there might be canals, but also on Mercury and the largest satellite of Jupiter, where it seems beyond the bounds of possibility that canals could have been constructed, as well as on Venus, on which real canals could not possibly be seen since its solid surface is entirely hidden under clouds. It may be significant that E. E. Barnard, perhaps the most skilled observer that astronomy has ever known, was never able to see the canals at all, although he studied Mars for years through the largest telescopes.

A more promising line of approach to our problem is to examine which, if any, of the planets is physically suitable for life. But we are at once confronted with the difficulty that we do not know what precise conditions are necessary for life. A human being transferred to the surface of any one of the planets or of their satellites, would die at once and this for several different reasons on each. On Jupiter he would be simultaneously frozen, asphyxiated and poisoned, as well as doubly pressed to death by his own weight and by an atmospheric pressure of about a million terrestrial atmospheres. On Mercury he would be burned to death by the sun's heat, killed by its ultra-violet radiation, asphyxiated from want of oxygen and desiccated from want of water. But this does not touch the question of whether other planets may not have developed species of life suited to their own physical conditions. When we think of the vast variety of conditions under which terrestrial life exists on earth—plankton, soil-bacteria, stone-bacteria and the great variety of bacteria which are parasitic on the higher forms of life, it would seem rash to suggest that there are any physical conditions whatever to which life can not adapt itself. Yet, as the physical states of other planets are so different from that of our own, it seems safe to say that any life there may be on any of them must be very different from the life on earth.

The visible surface of Jupiter has a temperature of about −138° C., which represents about 248 degrees of frost on the Fahrenheit scale. The planet probably comprises an inner core of rock, with a surrounding layer of ice some 16,000 miles in thickness, and an atmosphere which again is several thousands of miles thick and exerts the pressure

of a million terrestrial atmospheres which we have already mentioned. The only known constituents of this atmosphere are the poisonous gases methane and ammonia. It is certainly hard to imagine such a planet providing a home for life of any kind whatever. The planets Saturn, Uranus, Neptune and Pluto, being further from the sun, are almost certainly even colder than Jupiter and in all probability suffer from at least equal disabilities as abodes of life.

Turning sunwards from these dismal planets, we come first to Mars, where we find conditions much more like those of our own planet. The average temperature is about −40° C., which is also −40° on the Fahrenheit scale, but the temperature rises above the freezing point on summer afternoons in the equatorial regions. The atmosphere contains at most only small amounts of oxygen and carbon dioxide, perhaps none at all, so that there can be no vegetation comparable with that of the earth. The surface, in so far as it can be tested by a study of its powers of reflection and polarization, appears to consist of lava and volcanic ash. To us it may not seem a promising or comfortable home for life, but life of some kind or other may be there nevertheless.

Being at the same average distance from the sun as the earth, the moon has about the same average temperature, but the variations around this average temperature are enormous, the equatorial temperature varying roughly from 120° C. to −80° C. The telescope shows high ranges of mountains, apparently volcanic, interspersed with flat plains of volcanic ash. The moon has no atmosphere and consequently no water; it shows no signs of life or change of any kind, unless perhaps for rare falls of rock such as might result from the impact of meteors falling in from outer space. A small town on the moon, perhaps even a large building, ought to be visible in our largest telescopes, but, needless to say, we see nothing of the kind.

Venus, the planet next to the earth, presents an interesting problem. It is similar to the earth in size but being nearer the sun is somewhat warmer. As it is blanketed in cloud we can only guess as to the nature of its surface. But its atmosphere can be studied and is found to contain little or no oxygen, so that the planet's surface can hardly be covered with vegetation as the surface of the earth is. Indeed, its surface is probably so hot that water would boil away. Yet no trace of water-vapor is found in the atmosphere, so that the planet may well be devoid of water. There are reasons for thinking that its shroud of clouds may consist of solid particles, possibly hydrates of formaldehyde. Clearly any life that this planet may harbor must be very different from that of the earth.

The only planet that remains is Mercury. This always turns the same face to the sun and its temperature ranges from about 420° C. at the center of this face to unimaginable depths of cold in the eternal night of the face which never sees the sun. The planet is too feeble gravitationally to retain much of an atmosphere and its surface, in so far as this can be tested, appears to consist mainly of volcanic ash like the moon and Mars. Once again we have a planet which does not appear promising as an abode of life and any life that there may be must be very different from our own.

Thus our survey of the solar system forces us to the conclusion that it contains no place other than our earth which is at all suitable for life at all resembling that existing on earth. The other planets are ruled out largely by unsuitable temperatures. It used to be thought that Mars might have had a temperature more suited to life in some past epoch when the sun's radiation was more energetic than it now is, and that similarly Venus can perhaps look forward to a more temperate climate in some future age. But these possibilities hardly accord with modern views of stellar evolution. The sun is now thought to be a comparatively unchanging structure, which has radiated much as now through the greater part of its past life and will continue to do the same until it changes cataclysmically into a minute "white-dwarf" star. When this happens there will be a fall in temperature too rapid for life to survive anywhere in the solar system and too great for new life ever to get a foothold. As regards suitability for life, the earth seems permanently to hold a unique position among the bodies surrounding our sun.

Our sun is, however, only one of myriads of stars in space. Our own galaxy alone contains about 100,000 million stars, and there are perhaps 10,000 million similar galaxies in space. Stars are about as numerous in space as grains of sand in the Sahara. What can we say about the possibilities of life on planets surrounding these other suns?

We want first to know whether these planets exist. Observational astronomy can tell us nothing; if every star in the sky were surrounded by a planetary system like that of our sun, no telescope on earth could reveal a single one of these planets. Theory can tell us a little more. While there is some doubt as to the exact manner in which the sun acquired its family of planets, all modern theories are at one in supposing that it was the result of the close approach of another star. Other stars in the sky must also experience similar approaches, although calculation shows that such events must be excessively rare. Under conditions like those which now prevail in the neighborhood of the sun, a star will only experience an approach close enough to generate planets about once in every million, million, million years. If we suppose the star to have lived under these conditions for about 2,000 million years, only one star in 500 million will have experienced the necessary close encounter, so that at most one star in 500 million will be surrounded by planets. This looks an absurdly minute fraction of the whole, yet when the whole consists of a thousand million million million stars, this minute fraction represents two million million stars. On this calculation, then, two million million stars must already be surrounded by planets and a new solar system is born every few hours. The calculation probably needs many adjustments; for instance, conditions near our sun are not at all typical of conditions throughout space and the conditions of to-day are probably not typical of conditions in past ages. But even so the calculation suggests, with a large margin to spare, that although planetary systems may be rare in space, their total number is far from insignificant. Out of the thousands or millions of millions of planets that there must surely be in space, a very great number must have physical conditions very similar to those prevailing on earth.

We can not even guess whether these are inhabited by life like our own or by life of any kind whatever. The same chemical atoms exist there as exist here and must have the same properties, so that it is likely that the same inorganic compounds have formed there as have formed here. If so, we would like to know how far the chain of life has progressed but present-day science can give no help. We can only wonder whether any life there may be elsewhere in the universe has succeeded in managing its affairs better than we have done in recent years.

17

Soviet Scientist on Radio Denies Martian Invasion

LONDON, April 30 (Reuters)—A prominent Russian scientist solemnly went before a microphone today to allay Russian jitters that a space ship invasion from Mars had taken place.

The "Martian invasion" rumors apparently buzzing around Russia would meet sympathetic understanding in the United States, which still ruefully recalls localized panic in 1938 when Orson Welles broadcast a fictitious account of men from Mars landing in New Jersey.

The Moscow radio talk today said that news of supposed Maritan landings on earth and Martian space ship stories found a gullible audience among young Russians. It disclosed that space ships and Martians are a favorite topic with Soviet youth.

E. Krinov, secretary of the Meteorite Committee of the Soviet Academy of Science, today denied "fantastic accounts" of a "Martian space ship which was supposed to have crash-landed on earth." Mr. Krinov said the story had been published by a popular Soviet magazine.

18

Other-World Bid to Earth Doubted

Signal From Strange Planet Amusing, but Unsound Idea, Shapley Informs Rabbis

George Dugan
Special to The New York Times.

DETROIT, June 23—Dr. Harlow Shapley, Professor Emeritus of Astronomy at Harvard University, questioned here tonight the suggestion that beings on other planets might be seeking contact with earth.

In a speech at the annual convention of the Central Conference of American Rabbis, the astronomer agreed that "organisms" probably exist on some planet "not too far away." They could be more "advanced" than people on earth, he also remarked.

But, Dr. Shapley told the rabbis, suggestions that these "organisms" are trying to communicate with this planet are "amusing and constructive, if not yet very practical."

He noted that "effective" signaling would be difficult, mainly because no suitable planetary station is close enough to the earth. To ask a question and get an answer across the emptiness of space, he observed, would take about twenty years.

Earlier today the 500 Reform rabbis adopted a series of resolutions that endorsed student sit-in demonstrations against segregation, urged more medical care for the aged, commended efforts to reduce armaments, and called on candidates for public office to "refrain directly or indirectly from utterance or action designed to arouse religious or racial prejudice."

The rabbis also approved a "platform on justice and peace" covering the conference's stand on social problems of the day.

Highlights of the platform were:

¶Civil Rights—Condemned racial segregation and discrimination and pledged to work for desegregated schools, transportation and recreational facilities.

¶Capital Punishment—Renewed opposition to the death penalty, calling it vengeful and no deterrent to crime.

¶Housing—Advocated greatly expanded Federal programs to aid municipalities in slum clearance and to provide low-cost housing projects.

¶Labor—Recommended that management and labor write provisions for arbitration into labor contracts.

¶Birth Control—Urged that the United Nations through the World Health Organization, be enabled, where requested, to provide education and materials.

The convention is being held at the Sheraton-Cadillac Hotel.

19

Norman H. Horowitz
The Search for Life on Mars

Is there life on Mars? The question is an interesting and legitimate scientific one, quite unrelated to the fact that generations of science-fiction writers have populated Mars with creatures of their imagination. Of all the extraterrestrial bodies in the solar system Mars is the one most like the earth, and it is by far the most plausible habitat for extraterrestrial life in the solar system. For that reason a major objective of the Viking mission to Mars was to search for evidences of life.

The two Viking spacecraft were launched from Cape Canaveral in the summer of 1975. Each spacecraft consisted of an orbiter and an attached lander. When the spacecraft arrived at Mars in July and August of 1976, each was put in a predetermined orbit around the planet, and the search for a landing place began. Cameras aboard the orbiters were the principal source of information on which the choice of the landing sites was based; important data also came from infrared sensors on the orbiters and from radar observatories on the earth. The sole consideration in the final selection of the sites was the safety of the spacecraft. It would be a mistake to suppose, however, that the sites were therefore without biological interest. Biological criteria dominated the initial decisions as to the latitude at which each spacecraft would land. Once the latitudes had been chosen there was relatively little difference between sites at different longitudes.

On command from the earth each lander separated from its orbiter. With the help of its retroengines and parachute it dropped to the surface of Mars. Both orbiters continued to circle the planet, operating their own scientific instruments and relaying to the earth data transmitted from the landers. Both landings were in the northern hemisphere of Mars, and the Martian season was summer. (Mars has seasons like those on the earth, but each season lasts approximately twice as long. The Martian year is 687 Martian days; each Martian day, named a sol by the Viking team to distinguish it from a terrestrial day, is 24

hours, 39 minutes long.) On July 20, 1976, the *Viking I* lander came to rest in the Chryse Planitia region of Mars, some 23 degrees north of the equator. Six weeks later the *Viking 2* lander settled down in the Utopia Planitia region, some 48 degrees north of the equator. In longitude the two landers are separated by almost exactly 180 degrees, thus placing them on opposite sides of the planet. Since the instrumentation of the two landers is identical, the difference in their landing sites is the only distinction between them.

The first biologically significant task carried out by each lander was the analysis of the Martian atmosphere. Life is based on the chemistry of light elements, notably carbon, hydrogen, oxygen and nitrogen. To be suitable as an abode of life a planet must have those elements in its atmosphere. Spectroscopic observations from the earth and from spacecraft that had flown past Mars in previous years had already shown that carbon dioxide was the principal component of the Martian atmosphere. Small quantities of carbon monoxide, oxygen and water vapor had also been detected. Nitrogen had not been detected in any form, however, and atmospheric theory suggested that Mars had lost most of its nitrogen in the past.

Each Viking lander analyzed the atmosphere by means of two mass spectrometers. One spectrometer, operating during the descent to the surface, sampled and analyzed the atmospheric gases every five seconds. The second spectrometer operated on the ground. The results showed that the atmosphere near the ground was approximately 95 percent carbon dioxide, 2.5 percent nitrogen and 1.5 percent argon, and that it also held traces of oxygen, carbon monoxide, neon, krypton and xenon. At both landing sites the atmospheric pressure was 7.5 millibars (The atmospheric pressure at sea level on the earth is 1,013 millibars.)

Since the Viking spacecraft revealed that nitrogen is indeed present in the Martian atmosphere, we can say that the elements necessary for life are available on Mars. Missing from the list of gases, however, is one critically

Reprinted from *Scientific American*, 237 (November 1977), 52–61. Reprinted with permission.

Figure 19.1. The landing site of Viking 1 lander on Mars shows a wasteland of sand and rock. Soil samples from the lander vicinity were delivered to instruments in the spacecraft to be tested for their chemical composition and for signs of life. The meteorology sensor appears at the center.

important compound: water vapor. Although earlier measurements had shown that traces of water vapor are present in the Martian atmosphere, the quantity varies with season and place. The Viking orbiters carried out a survey of water vapor over the entire planet with infrared spectrometers. The results showed that the highest concentration of atmospheric water vapor was at the edge of the north polar cap (the summertime hemisphere), and that the concentration fell off toward the south (the opposite of what is found on the earth). In the polar region the amount of water vapor in the atmosphere would form a film only a tenth of a millimeter thick if all of it were to be condensed on the planet's surface. At the landing sites the concentration of water vapor ranged between 10 and 30 percent of the concentration at the pole.

These numbers put into quantitative terms a long-known fact about Mars: It is a very dry place. Mars has ice at its poles, but nowhere on its surface are there oceans or lakes or any other bodies of liquid water. The absence of liquid water is related to the dryness of the atmosphere through a fundamental law of physical chemistry: the phase rule. The phase rules states that for liquid water to exist on the surface of a planet the pressure of the water vapor in the atmosphere must at some times and in some places be at least 6.1 millibars. The Viking measurements imply that the vapor pressure of water at the surface of Mars in the northern hemisphere is at most .05 millibar, even if all the water vapor is concentrated in the lower atmosphere. At that low pressure liquid water cannot remain in the liquid phase; depending on the temperature, it must either freeze or evaporate. By the same token raindrops cannot form in the Martian atmosphere and ice cannot melt on the Martian surface.

The extreme dryness presents a difficult problem for

Figure 19.2. The landing site of Viking 2 lander is a field of boulders superficially similar to the landing site of the Viking 1 lander. The horizon is tilted because the lander is not quite level.

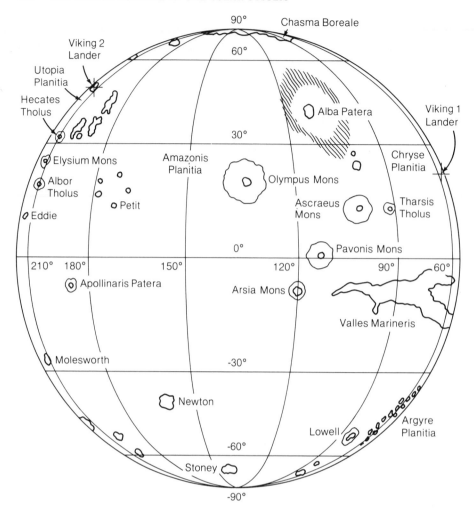

Figure 19.3. Locations of the two Viking landers are indicated on this map of Mars, which shows some of the major geological features of the planet. The two spacecraft are on opposite sides of the planet, some 4,600 miles apart. Both are in the northern hemisphere at sites selected partly for their possible biological interest. *Viking 1* lander is in Chryse Planitia region at a latitude of 23 degrees; *Viking 2* lander is in Utopia Planitia region at a latitude of 48 degrees.

any Martian biology. Liquid water is essential for life on the earth. All terrestrial species have high and apparently irreducible requirements for water; none could live on Mars. If there is life on Mars, it must operate on a different principle as far as water is concerned. If Mars had a more favorable environment in the past, however, and if the planet did not dry up too fast, species may have had time to evolve and adapt to present conditions. Pictures made by the *Mariner 9* spacecraft, which went into orbit around Mars in 1971, suggested that Mars may indeed have had running water on its surface in the past. The pictures from the Viking orbiters have confirmed that impression. The evidence consists of channels in the Martian desert that resemble dry riverbeds. There seems to be little doubt that

the channels were carved by rapidly flowing liquid, and there is widespread agreement that the most probable liquid is water.

If liquid water once existed on Mars, could life have arisen on the planet? If the life evolved to meet changing conditions, could it exist there still? There is no way to settle these questions by deductive reasoning or even by experimentation in laboratories on the earth. They can be answered only by the direct exploration of Mars, and that is what the Viking spacecraft did.

Five different types of instruments on each Viking lander were involved in the search for evidences of life: two cameras for photographing the landscape, a combined gas chromatograph and mass spectrometer for analyzing the surface for organic material and three instruments designed to detect the metabolic activities of any microorganisms that might be present in the soil. In this brief account I shall not be able to mention the names of the many scientists, engineers and managers whose joint efforts made all the Viking projects possible. They work in universities, industrial laboratories and the National Aeronautics and Space Administration and its field centers. Their names are recorded in the growing technical literature dealing with this historic mission.

Each of the Viking landers carried two cameras of the facsimile type, which built up a picture of the scene by scanning it in a series of narrow strips. Such cameras make pictures slowly, but they are rugged and versatile. Their resolution was moderately high: a few millimeters at a distance of 1.5 meters. They produced pictures in black and white, in color and in stereo. The two cameras on each lander could between them survey the entire horizon around the spacecraft.

As life-seeking tools cameras have inherent advantages and disadvantages. Their chief advantage lies in the fact that a picture contains a large amount of information. In principle it would be possible to prove unequivocally the existence of life on Mars with a single photograph. For example, if a line of trees were visible on the horizon or if footprints appeared on the ground in front of the spacecraft one morning, there would be no room for doubt that there is life on Mars. Another advantage lies in the fact that pictorial evidence is independent of all assumptions about the chemistry and physiology of Martian organisms. The organisms need not respond in certain ways to certain substances or treatments in order to be recognized. The cameras could identify, say, a mushroom made of titanium as a form of life if one were to sprout up from under a rock in the course of the mission. Of course, reliance on pictorial evidence rests on its own set of assumptions about the morphology of living things. The most obvious disadvantage of the camera as a life-seeking instrument is the fact that an entire world of life can exist below the camera's limit of resolution.

Of all the results of the Viking mission the wonderful photographs of the Martian desert at the two landing sites are the most impressive. The photographs have been eagerly scanned by alert and hopeful eyes, but no investigator has yet seen anything suggesting a living form.

The next step was to analyze the soil for any organic constituents. Among the elements carbon is unique in the number, variety and complexity of the compounds it can form. The special properties of carbon that enable it to form large and complex molecules arise from the basic structure of the carbon atom. That structure enables the carbon atom to form four strong bonds with other atoms, including other carbon atoms. The molecules thus formed are very stable at ordinary temperatures, so stable, in fact,

that there seems to be no limit to the size they can attain. The connection between life and organic chemistry (that is, the chemistry of carbon) rests on the fact that the attributes by which we identify living things—their capacity to replicate themselves, to repair themselves, to evolve and to adapt—originate in properties that are unique to large organic molecules. It is the highly complex information-rich proteins and nucleic acids that endow all the living things we know, even "simple" ones such as bacteria and viruses, with their essential nature. No other element, including that favorite of science-fiction writers, silicon, has the capacity to form large and complex structures that are so stable. It is no accident that even though silicon is far more abundant than carbon on the earth, it has only minor and nonessential roles in biochemistry. Biochemistry is largely a chemistry of carbon.

Such fundamental facts lead to the conclusion that wherever life arises in the universe it will most likely be based on carbon chemistry. That view has been strengthened by the discovery of organic compounds of biological interest in meteorites and in clouds of dust in interstellar space. Although these compounds are non-biological in origin, they are closely related to the amino acids and the nucleotides that are the respective building blocks of proteins and of nucleic acids. The fact that they are formed in settings remote from the earth implies that carbon chemistry gives rise to familiar organic compounds throughout the universe. This fact in turn suggests that life elsewhere in the universe will be based on an organic chemistry similar to our own, although not necessarily identical with it.

Such considerations led to the decision to include an organic-analysis experiment aboard the Viking landers. The instrument used in the experiment was the mass spectrometer that had analyzed the atmosphere combined with a gas chromatograph and a pyrolysis furnace. A sample of the Martian soil was first heated in the furnace through a series of steps up to a temperature of 500 degrees Celsius. Any volatile materials released were passed through the gas chromatograph. Since each of the different compounds has a different molecular weight, composition and polarity, among other properties, it passed through the columns of the gas chromatograph at a unique rate, and so the compounds were separated from one another. As each compound emerged from the chromatograph column it was directed into the mass spectrometer for identification. Since essentially all organic matter is cracked, or decomposed, into smaller fragments at 500 degrees C., the method is capable of detecting organic compounds that have a wide range of molecular weights.

Two soil samples were analyzed at each landing site. The only organic compounds detected were traces of cleaning solvents known to have been present in the apparatus. The fact that the solvents were detected shows the

90

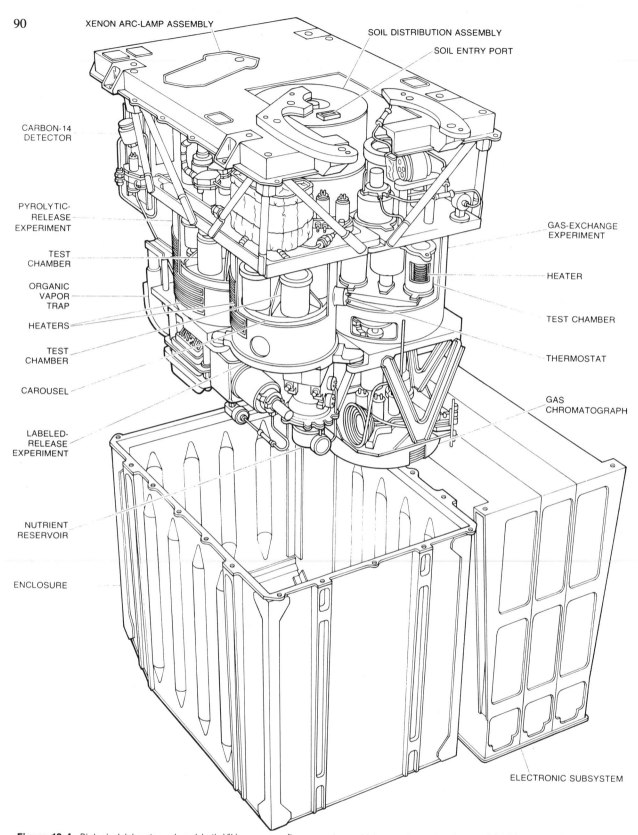

XENON ARC-LAMP ASSEMBLY

SOIL DISTRIBUTION ASSEMBLY

SOIL ENTRY PORT

CARBON-14 DETECTOR

PYROLYTIC-RELEASE EXPERIMENT

TEST CHAMBER

ORGANIC VAPOR TRAP

HEATERS

TEST CHAMBER

CAROUSEL

LABELED-RELEASE EXPERIMENT

NUTRIENT RESERVOIR

ENCLOSURE

GAS-EXCHANGE EXPERIMENT

HEATER

TEST CHAMBER

THERMOSTAT

GAS CHROMATOGRAPH

ELECTRONIC SUBSYSTEM

Figure 19.4. Biological laboratory aboard both Viking spacecraft occupies a volume of only one cubic foot. The three biological experiments were the gas-exchange experiment (*right*), the labeled-release experiment (*bottom left*) and the pyrolytic-release experiment (*top left*). Each experiment, shown cut away, had several test chambers on a carousel so that the experiment could test several samples of Martian soil. The soil was dumped into an entry port at the top of the laboratory, where it fell into a hopper. For each sample of soil one test chamber of each experiment was rotated under the hopper in order to receive a portion of the sample. All together approximately half a dozen samples of soil were tested at each landing site. The experiments were completed in April. Results are given in following figures.

instruments were functioning properly. The heated samples gave off carbon dioxide and a small amount of water vapor; nothing else was found.

This result is surprising and weighs heavily against the existence of biological processes on Mars. The combined gas chromatograph and mass spectrometer aboard each Viking lander is a sensitive instrument, capable of detecting organic compounds at a concentration of a few parts per billion, a level that is between 100 and 1,000 times below their concentration in desert soils on the earth. Even if there is no life on Mars, it has been supposed the fall of meteorites onto the Martian surface would have brought enough organic matter to the planet to have been detected. Because Mars is near the asteroid belt, from which meteorites originate, it is believed to receive a much larger number of meteorite impacts than either the earth or the moon. Indeed, a question that was frequently discussed before the Viking spacecraft were launched was whether or not it would be possible to distinguish biological organic matter on Mars from the meteoritic organic matter that was expected to be present. The absence of organic matter at the parts-per-billion level, however, suggests that on Mars organic compounds are actively destroyed, probably by the strong ultraviolet radiation from the sun.

The other experiments aboard the Viking landers searched not just for organic matter in the soil but for living organisms. On the earth microorganisms such as bacteria, yeasts and molds are the hardiest of all species. There are few places on the earth where microbial forms do not live; they are the last survivors in environments of extreme temperature and aridity. The reasons for their hardiness are interesting but need not detain us here. Suffice it to say that if there is life on Mars, the chance of detecting it would be maximized by searching for microorganisms in the Martian soil.

Each Viking lander carried three instruments designed to detect the metabolic activities of soil microorganisms. First, the gas-exchange experiment was designed to detect changes in the composition of the atmosphere caused by microbial metabolism. Second, the labeled-release experiment was designed to detect decomposition of organic compounds by soil microbes when they were fed with a nutrient. Third, the pyrolytic-release experiment was designed to detect the synthesis of organic matter in Martian soil from gases in the atmosphere by either photosynthetic or nonphotosynthetic processes. All three experiments analyzed portions of each sample of Martian soil.

All the experiments detected chemical changes of one kind or another in the soil. All the experiments are now completed, and some of the changes they observed suggest biological processes. There has been much discussion both within the team of Viking investigators and outside it as to the best way to interpret the findings. Are the changes due to biological responses or are they just chemical reactions we would like to believe are biological? Indeed, since life is a form of chemistry, how can the two be told apart?

One way to decide whether or not a process is biological is to test its sensitivity to heat. Living structures are highly organized and fragile, and they are destroyed by temperatures that leave many chemical reactions unaffected. A process that is insensitive to heat is thus likely to be a nonliving chemical reaction, but a process that is sensitive to it could be either living or nonliving.

The decision as to whether a heat-sensitive process is biological or not must be based on additional evidence. In the end, however, the judgment is based on Occam's razor: the traditional principle that the hypothesis most likely to be correct is the one that accounts for the maximum number of observations with the minimum number of assumptions.

The gas-exchange experiment and the labeled-release experiment were frankly terrestrial in orientation. In both experiments a nutrient medium composed of an aqueous solution of organic compounds was mixed with a sample of Martian soil. Since liquid water cannot exist on Mars, the experiments could not be conducted under Martian conditions; the test chambers had to be heated to prevent the water from freezing and pressurized to prevent it from boiling. Both experiments were based on the universal property of terrestrial organisms to evolve gas as they metabolize food. If a sample of soil from the earth is moistened with a nutrient solution, the microorganisms in the soil take up the nutrients and convert them partly into more microorganisms (that is, the population of microorganisms grows) and partly into various by-products, including gases. Among the gases given off in microbial metabolism are carbon dioxide, methane, nitrogen, hydrogen and hydrogen sulfide. On the earth gases evolved by one species of organisms are eventually consumed by other species of organisms. In that way the light elements at the earth's surface are continually cycled through the biosphere and the atmosphere.

In the gas-exchange experiment a complex nutrient solution was added to a sample of Martian soil in a closed chamber, and the gases were analyzed periodically by means of a gas chromatograph. The experiment proceeded in two stages. In the first stage a small volume of the nutrient solution was introduced into the soil chamber in such a way that it humidified the chamber without actually wetting the soil, and the resulting gases were analyzed several times. In the second stage a large volume of the nutrient was poured into the chamber, saturating the soil. With the soil now in direct contact with the medium the main part of the experiment began. The soil was incubated for nearly seven months, so that whatever microorganisms might be in the sample had enough time to

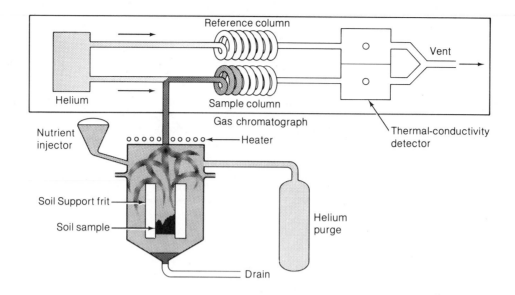

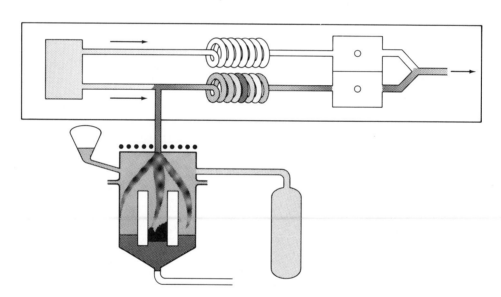

Figure 19.5. Gas-exchange experiment tested the Martian soil to see if there were any microorganisms in it that took in atmospheric gases and nutrients and gave off gaseous by-products. The experiment proceeded in two stages. In the first stage a small volume of a complex nutrient solution was injected into the test chamber in such a way that it humidified the chamber without wetting the soil (*top*). The gases evolved were flushed into a gas chromatograph with a stream of helium where they were analyzed for organic compounds and compared with results of reference analysis run as a standard. In second stage of experiment a large volume of nutrient was poured into chamber to wet the soil (*bottom*).

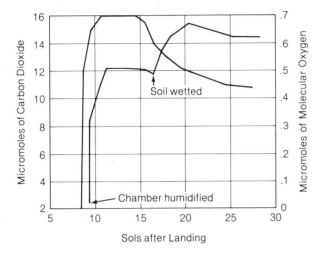

Figure 19.6. Results of the gas-exchange experiment, according to data of Vance I. Oyama of the Ames Research Center of the National Aeronautics and Space Administration, showed that in the first humid stage of the experiment a large amount of carbon dioxide and molecular oxygen surged into the test chamber. In the second wet stage the amount of carbon dioxide continued to rise at a decreasing rate and then declined. The amount of oxygen, however, quickly fell. It is believed the gases were released by physical and chemical processes, not by biological ones. One micromole is a millionth of a mole, where one mole is the amount of a substance that has a weight in grams equal to its molecular weight. Oxygen curve is displaced to the left by one sol so that the curves do not overlap. A sol is one Martian day.

signal their presence by producing or consuming gases. During the period of incubation the atmosphere in the chamber was periodically analyzed.

The findings of the first stage of the experiment were both surprising and simple. Immediately after the soil sample was humidified carbon dioxide and oxygen were rapidly released. The release of the gases ceased soon after it had begun but not before the pressure in the chamber had risen measurably. At the Chryse site in a period of little more than one sol the quantity of carbon dioxide in the incubation chamber of the *Viking I* lander increased by a factor of five and the quantity of oxygen increased by a factor of 200. At the Utopia site the increases were less, but they were still considerable.

The rapidity and the brevity of the response recorded by both landers clearly suggested that the process observed was a chemical reaction, not a biological one. The appearance of the carbon dioxide is readily explained. Carbon dioxide gas would be expected to be adsorbed on the surface of the dry Martian soil; if the soil was exposed to a very humid atmosphere, the gas would be displaced by water vapor. The appearance of the oxygen is more complex. The production of so much oxygen seems to require an oxygen-generating chemical reaction, not just a physi-

cal liberation of preexisting gas. It is likely that the oxygen was released when the water vapor decomposed an oxygen-rich compound such as a peroxide. Peroxides are known to decompose if they are exposed to water in the presence of iron compounds, and according to the X-ray fluorescence spectrometer aboard each Viking lander, the Martian soil is 13 percent iron.

At both landing sites the second phase of the gas-exchange experiment was anticlimactic. When the soil sample was saturated with the nutrient medium and incubated, carbon dioxide continued to be released. The production of the carbon dioxide gradually tapered off, however, and the oxygen gradually disappeared. The slow increase in the amount of carbon dioxide was probably a continuation of the reaction in the humid stage of the experiment. The disappearance of the oxygen also can be easily explained: one of the ingredients of the nutrient medium was ascorbic acid, which combines readily with oxygen. And so after seven months it became clear that everything of interest had happened in the humid stage of the experiment, before the soil came in contact with the nutrient! What the gas-exchange experiment detected was not metabolism but the chemical interaction of the Martian surface material with water vapor at a pressure that has not been reached on Mars for many millions of years.

The labeled-release experiment differed from the gas-exchange experiment in several ways. The nutrient medium employed was a simpler one containing only a few cosmically abundant organic compounds such as formic acid ($HCOOH$) and the amino acid glycine (NH_2CH_2COOH). All the compounds were labeled with atoms of the radioactive isotope carbon 14. The labeled-release instrument was designed to detect radioactive gases, principally carbon dioxide, released when the nutrient medium was added to a sample of soil. The number of radioactive disintegrations in gases can be counted quite efficiently, so that the labeled-release experiment is faster and more sensitive than the gas-exchange experiment in detecting microbial activity in terrestrial soil. The labeled-release experiment's sequence of operations did not include a humid stage as such, but it attempted to accomplish the same end by injecting a volume of nutrient medium that was insufficient to wet the entire soil sample but sufficient to humidify the chamber. If the experiment worked on Mars as planned, subsequent injections of the medium, which were controlled by commands sent from the earth, brought the medium into contact with some soil that had previously been wetted and with other soil that had been humidified but not wetted.

As in the gas-exchange experiment, immediately after the nutrient medium was added to the soil in the labeled-release experiment, gas surged into the chamber. The release of gas tapered off soon after the first sol. The gas, undoubtedly carbon dioxide, was radioactive, showing

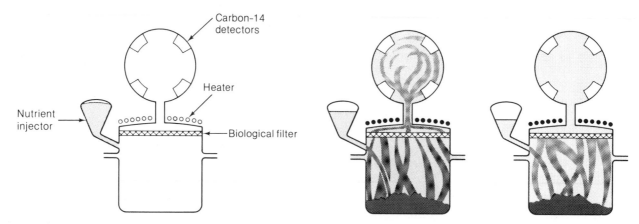

Figure 19.7. Labeled-release experiment on the Viking landers tested the Martian soil for microorganisms that could metabolize simple organic, or carbon, compounds. The nutrient medium was composed of several organic compounds that are widely abundant in the universe. The compounds were labeled with radioactive carbon. If microorganisms exist in the Martian soil, they might consume the labeled nutrient and give off radioactive gases (particularly carbon dioxide), which would be detected by the carbon-14 counters. Before the soil was tested the background level of radiation was measured (*left*). The soil was dumped into the test chamber, injected with a small amount of medium (*middle*) and incubated for up to 11 sols. The amount of nutrient in this first injection was planned to wet only part of the soil but to humidify the entire chamber. A subsequent injection of the nutrient (*right*), controlled by signals from the earth, thus brought the medium into contact with soil that had already been wetted and with other soil that had been humidified but not wetted. If the labeled-release experiment worked on Mars as planned, its results would serve as a check on the results of the gas-exchange experiment.

that it had been formed from the radioactive compounds of the medium and not from compounds in the Martian soil. Nonradioactive gases, which also must have formed when the aqueous medium came in contact with the soil, were not detectable in the experiment.

The production of radioactive carbon dioxide in the labeled-release experiment is understandable in the light of the evidence from the gas-exchange experiment suggesting that the surface material of Mars contains peroxides. Formic acid, one of the compounds of the labeled-release nutrient medium, is oxidized with particular ease: if a molecule of formic acid ($HCOOH$) reacts with one of hydrogen peroxide (H_2O_2), it will form a molecule of carbon dioxide (CO_2) and two molecules of water ($2H_2O$). The amount of radioactive carbon dioxide given off in the labeled-release experiment was only slightly less than what would have been expected if all the formic acid in the medium had been oxidized in this way.

If the source of the oxygen released in the humid stage of the gas-exchange experiment was indeed peroxides in the soil decomposed by water vapor, then in the labeled-release experiment all the peroxides should also have been decomposed by the first injection of nutrient. Thus the next injection should have evolved no additional radioactive gas in spite of the fact that part of the sample presumably had not yet been wetted by the medium. That proved to be the case. When a second volume of medium was injected into the chamber, the amount of the gas in the chamber was not increased; indeed, it decreased. The decrease is explained by the fact that carbon dioxide is quite soluble in water; when fresh nutrient medium was added to the chamber, it absorbed some of the carbon dioxide in the head space above the sample.

This result was obtained with all the samples tested by the labeled-release experiment at both Viking sites. In that respect the results of the labeled-release experiment did not parallel the results of the gas-exchange experiment. At both sites both experiments tested soil gathered from the ground's exposed surface; at the Utopia site the experiments also tested soil gathered from under a rock. Although the labeled-release experiment found essentially no difference in the amount of gas released by any of the samples, the gas-exchange experiment recorded about three-fourths as much carbon dioxide from the surface samples at the Utopia site as it had from the surface samples at the Chryse site, and it recorded even less carbon dioxide from the sample from under the rock.

The gas-exchange experiment also recorded less oxygen from the samples from the Utopia region, but the interference of the ascorbic acid in the complex nutrient medium of that experiment makes it difficult to quantify the difference. In every case, however, the gas-exchange experiment detected considerably more gas than the labeled-release experiment did with portions of the same sample. Those results, however, do not contradict the thesis that the production of oxygen detected by the gas-exchange experiment and the production of radioactive carbon dioxide detected by the labeled-release experiment

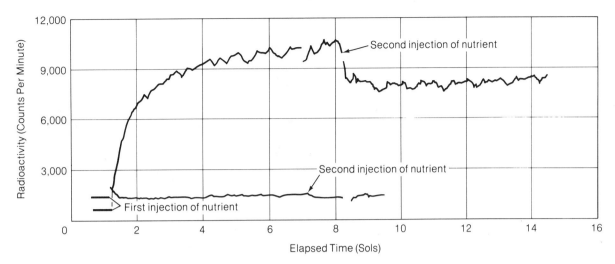

Figure 19.8. Results of the labeled-release experiment for the first sample of soil analyzed at the Chryse site were indeed consistent with the results of the gas-exchange experiment, according to data from Gilbert V. Levin and Patricia A. Straat of Biospherics Inc. Immediately after the first injection of the nutrient, radioactive gases surged into the chamber. The radioactivity was measured at 16-minute intervals throughout the experiment except for the first two hours after the first injection, when measurements were made every four minutes. After the second injection the amount of gas in the chamber dropped, then remained at a nearly constant level until the end of the experiment. In order to test the sensitivity of reaction to heat a second portion of the soil sample was sterilized at a temperature of 160 degrees Celsius for three hours and the experiment was repeated. The reaction was abolished. Although such behavior is consistent with a biological process, it is more likely that experiment again detected only a chemical reaction.

are simply different measurements of the same surface chemistry. The gas-exchange experiment measures the total amount of oxidant in the surface; the labeled-release experiment measures only a fraction of it.

The labeled-release experiment also tested the stability of the reaction to heat. When the soil was preheated to 160 degrees C., for three hours before incubation, the reaction was abolished. When it was heated to 46 degrees for the same length of time, the magnitude of the reaction was reduced by about half. These results have been regarded by some as evidence in favor of the hypothesis that the reaction is biological. The results are of course consistent with such a hypothesis, but they are also consistent with a chemical oxidation in which the oxidizing agent is destroyed or evaporated at relatively low temperatures. A variety of both inorganic peroxides and organic peroxides could probably have produced the same results.

The third microbiological experiment, the pyrolytic-release experiment, differed from the gas-exchange and labeled-release experiments in two respects. First, it attempted to measure the synthesis of organic matter from atmospheric gases rather than its decomposition. Second, it was designed to operate under the conditions of pressure, temperature and atmospheric composition that actually obtain on Mars, since those are the conditions under which any form of Martian life must exist. In practice the conditions in the chamber were a reasonably good approximation of Martian conditions except for the temperature, which stayed warmer than the outside temperature because of heat sources within the spacecraft.

A sample of Martian soil was sealed in a chamber along with some Martian atmosphere. A quartz window in the chamber admitted simulated Martian sunlight from a xenon arc lamp. Into this Martian microcosm small amounts of radioactive carbon dioxide and radioactive carbon monoxide were introduced. Both gases are present in the Martian atmosphere but not in radioactive form. After five days the lamp was turned off, the atmosphere was removed from the chamber and the soil was analyzed for the presence of radioactive organic matter.

First the soil was heated in the pyrolysis furnace to a temperature high enough to crack any organic compounds into small volatile fragments. The fragments were swept out of the chamber by a stream of helium and passed through a column that was designed to trap organic molecules but allow carbon dioxide and carbon monoxide to pass through. The radioactive organic molecules were thus transferred from the soil to the column and at the same time were separated from any remaining gases of the incubation atmosphere. The organic molecules were released from the column by raising the column's temperature. Simultaneously the radioactive organic molecules were decomposed into radioactive carbon dioxide by copper oxide in the column. The carbon dioxide was then carried by the stream of helium into a radiation counter. If organic compounds had been synthesized in the soil, they

96

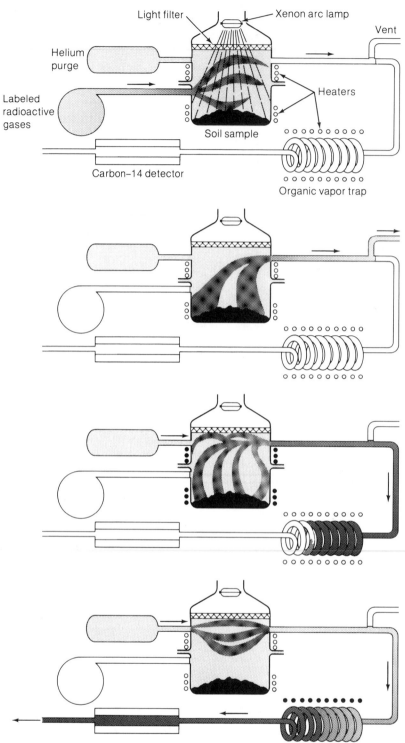

Figure 19.9. Pyrolytic-release experiment tested the Martian soil to see if there were microorganisms in it that would create organic compounds out of atmospheric gases by either a photosynthetic process or a non-photosynthetic process. A sample of soil was sealed into a chamber along with some Martian atmosphere and a small amount of radioactive carbon dioxide and carbon monoxide. A xenon arc lamp irradiated the soil with simulated Martian sunlight (1). After five days lamp was turned off and the atmosphere was removed from the chamber (2). Soil was heated to a temperature high enough to pyrolyze (decompose) into small volatile fragments any radioactive organic compounds produced. Fragments were swept out of the chamber (3) by a stream of helium into a column designed to trap organic molecules but pass carbon dioxide and carbon monoxide. In column trapped radioactive organic molecules were released by raising column's temperature; the molecules were oxidized to form carbon dioxide, which was carried into a radiation counter (4).

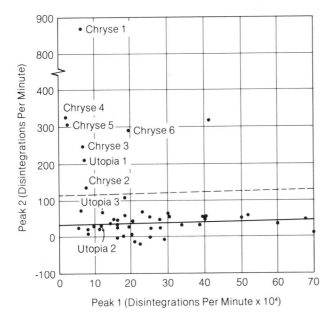

Figure 19.10. Results of pyrolytic-release experiment are shown for all the samples tested on Mars. The axis labeled Peak 1 shows how much radioactivity in the form of carbon dioxide and carbon monoxide passed through the column during the pyrolysis of each sample. The axis labeled Peak 2 shows how much radioactivity, representing newly synthesized organic matter, remained attached to the column in each case. Labeled and numbered dots indicate the site and experiment. (For example, "Chryse 1" means the result of the first experiment at Chryse, "Utopia 1" means the result of the first experiment at Utopia, and so on.) Unlabeled dots are the data obtained from tests of sterilized soil samples in a duplicate of the Viking pyrolytic-release instrument on the earth. The line drawn through those points represents the best fit to the points. The dashed line above the black line is a statistically significant dividing line; any point lying above this line is a positive result. The single black point above the dashed line is believed to be due to a technical error in performing that particular test. Seven of the nine pyrolytic-release experiments performed on Mars, however, yielded firmly positive results.

would be detected as radioactive carbon dioxide; if no organic compounds had been synthesized, no radioactive carbon dioxide would have been formed.

Surprisingly, seven of the nine pyrolytic-release tests executed on Mars gave positive results. The two negative results were obtained at the Utopia site, but a third sample tested at Utopia was positive. This third sample was actually incubated in the dark, implying that light may not be required for the reaction. The amount of carbon fixed in the soil by the experiment was small: enough to furnish organic matter for between 100 and 1,000 bacterial cells. The quantity is so small, in fact, that it could not have been detected by the organic-analysis experiment. The quantity is nonetheless significant; it was surprising that in such a strongly oxidizing environment even a small amount of organic material could be fixed in the soil.

Even more significant, the pyrolytic-release instrument had been rigorously designed to eliminate nonbiological sources of organic compounds. During the development of the experiment it had been found that in the presence of short-wavelength ultraviolet radiation, carbon monoxide spontaneously combined with water vapor to form organic molecules on glass, quartz and soil surfaces in the experimental chamber. In order to avoid those reactions and the confusion they would have caused, the short-wavelength ultraviolet was filtered out of the radiation allowed to enter the incubation chamber. To receive positive results from the soil on Mars in spite of that precaution was startling.

Nevertheless, it appears that the findings of the pyrolytic-release experiment must also be interpreted nonbiologically. The reason is that the reaction detected was less sensitive to heat than one would expect of a biological process. In two of the nine pyrolytic-release experiments performed on Mars the soil sample was heated before the radioactive gases were injected and the incubation was begun. In once case the sample was held at 175 degrees C. for three hours and in the other it was held at 90 degrees for nearly two hours. The effect of the higher temperature was to reduce the reaction by almost 90 percent but not to abolish it. The effect of the lower temperature was nil. When it is recalled that the temperature at the surface of Mars at the two landing sites does not rise above zero degrees C. at any time, and that the temperature below the surface is even lower, it becomes difficult to reconcile the results with a biological source. Any organisms living in the Martian soil should have been killed by those temperatures.

On the other hand, it is not easy to point to a nonbiological explanation for the positive results. Investigations into the problem are now under way in terrestrial laboratories with synthetic Martian soils formulated on the basis of the data from the inorganic analyses carried out by the Viking landers. The solution to the puzzle will probably also explain why the organic-analysis experiment detected no organic material in the Martian surface. Until the mystery of the results from the pyrolytic-release experiment is solved, a biological explanation will continue to be a remote possibility.

Even though some ambiguities remain, there is little doubt about the meaning of the observations of the Viking landers: At least those areas on Mars examined by the two spacecraft are not habitats of life. Possibly the same conclusion applies to the entire planet, but that is an intricate problem that cannot yet be addressed. The most surprising finding of the life-seeking experiments is the extraordinary chemical reactivity of the Martian soil: its oxidizing capacity, its lack of organic matter down to the level of several parts per billion and its capacity to fix atmospheric carbon (presumably into organic molecules) at a still lower level. It seems Mars has a photochemically activated surface that, due to the low temperature and the absence of water, is maintained in a state far from chemical equilibrium.

These conclusions drawn from the results of the life-seeking experiments on the Viking landers are undeniably disappointing. The discovery of life would have been much more interesting, to say the least. There are doubtless some who, unwilling to accept the notion of a lifeless Mars, will maintain that the interpretation I have given is unproved. They are right. It is impossible to prove that any of the reactions detected by the Viking instruments were not biological in origin. It is equally impossible to prove from any result of the Viking experiments that the rocks seen at the landing sites are not living organisms that happen to look like rocks. Once one abandons Occam's razor the field is open to every fantasy. Centuries of human experience warn us, however, that such an approach is not the way to discover the truth.

IV

Intelligent Life Outside the Solar System?

Among the classic scientific papers of the twentieth century, "Searching for Interstellar Communications" by Giuseppe Cocconi and Philip Morrison seems likely to rank high—especially if we eventually establish interstellar communication using the techniques they outlined. In this work, published in 1959, Cocconi and Morrison reopened the question of searching for extraterrestrial life as a scientific endeavor that can begin immediately.

Cocconi and Morrison sought to answer a basic question. Since interstellar distances are so vast, communications ought to proceed by radio waves traveling at the speed of light. But *which frequencies* carry the messages? How should we set our receivers (and possibly our transmitters as well) to participate in the flow of interstellar communication? Cocconi and Morrison pointed out that our galaxy, and indeed the universe, show a single frequency of tremendous importance: the 1,420-MHz frequency of radio emission by hydrogen atoms. Hydrogen, by far the most abundant element in the universe, radiates at this one frequency only, unless it is "excited" by local sources of energy such as young, hot stars. This natural radio emission provides a beacon in the frequency spectrum, close to one that we might expect civilizations to use to exchange messages.

But do other civilizations exist that seek, or enjoy, mutual intercommunication? Ronald Bracewell asked a logical next question: What would a civilization far more advanced than ours do to find its neighbors? Bracewell suggested that message probes moving slowly (in comparison to radio) might be highly efficient *if* the time delay of many millenia made little difference to a long-lived civilization. Each probe could begin to broadcast only upon receiving radio signals from a target area, evidence that another civilization had achieved radio capability.

We may face a difficult challenge in discriminating between an advanced civilization and an apparently natural, though mysterious, celestial object. Freeman Dyson proposed that some civilizations may capture all the visible-light output from their parent stars and radiate only the "waste heat" of infrared photons. Such "Dyson spheres" would, oddly enough, resemble stars in the process of formation, which also emit copious infrared radiation. R. N. Schwartz and C. H. Townes considered the possibility that high-powered masers, operating at visible-light ("optical") frequencies, may be efficient message carriers, provided that those using them can distinguish them from natural sources of visible light. Frank Drake, Philip Morrison, A. G. W. Cameron, and N. S. Kardashev, among others, have written about the more "conventional" problem of interstellar radio communication. Kardashev's work draws attention for his introduction of a classification scheme for civilizations: Type I for those which, like our own, have the ability to communicate by radio over interstellar distances; Type II for the "Dyson-sphere" sort, which use trillions of times more energy; and Type III for civilizations that use the energy of an entire *galaxy*, thus capturing billions of times more energy per second than the Type II civilizations.

Carl Sagan has used the Kardashev terminology in his analysis of the number of civilizations and the chances of detecting them with our present capabilities. He concludes that we must face the possibility that civilizations far in advance of our own may have little interest in meeting those like ourselves, but that nonetheless we may hope to discover their existence. Gerrit Verschuur's search for radio emission from other civilizations, like Frank Drake's Project Ozma, aimed only at nearby stars similar to our sun, and at radio emission roughly similar to our own. Robert Machol's article gives a good overview of the SETI (Search for Extraterrestrial Intelligence) problem, which returns to the difficulty of searching in many directions and at many different frequencies.

If we cannot find other civilizations immediately, perhaps we can at least find evidence that other stars besides our sun have planets, which are considered the likeliest places to find life. Our best chance for this consists in observations of nearby stars' motions through space. Deviations from straight-line trajectories may indicate the gravitational effect of the star and its planets orbiting the center of mass of another solar system. The article by David Black and Graham Suffolk tends to confirm the observations of "Barnard's star," the fourth closest star to the sun, which were made by Peter van de Kamp and his associates, that suggest at least two planets with near-Jupiter masses near the star. These observations remain controversial, and will be vastly improved by the use of the Space Telescope, high above the blurring effects of our atmosphere. Ronald Bracewell and Robert MacPhie suggest other methods besides direct observation in visible light as the means of detecting extrasolar planets.

Thomas Kuiper and Mark Morris attempt to confront the question of why, if life is prevalent in our galaxy, no civilization has yet colonized at least our region of the Milky Way (at least so far as we can tell!), and what this implies about our SETI strategy. Bernard Oliver, in many ways the great planner of SETI, analyzes the average separation of civilizations and the chance that most civilizations discover at least one other during their lifetimes. This chance remains undetermined.

20

Giuseppe Cocconi and Philip Morrison
Searching for Interstellar Communications

No theories yet exist which enable a reliable estimate of the probabilities of (1) planet formation; (2) origin of life; (3) evolution of societies possessing advanced scientific capabilities. In the absence of such theories, our environment suggests that stars of the main sequence with a lifetime of many billions of years can possess planets, that of a small set of such planets two (Earth and very probably Mars) support life, that life on one such planet includes a society recently capable of considerable scientific investigation. The lifetime of such societies is not known; but it seems unwarranted to deny that among such societies some might maintain themselves for times very long compared to the time of human history, perhaps for times comparable with geological time. It follows, then, that near some star rather like the Sun there are civilizations with scientific interests and with technical possibilities much greater than those now available to us.

To the beings of such a society, our Sun must appear as a likely site for the evolution of a new society. It is highly probable that for a long time they will have been expecting the development of science near the Sun. We shall assume that long ago they established a channel of communication that would one day become known to us, and that they look forward patiently to the answering signals from the Sun which would make known to them that a new society has entered the community of intelligence. What sort of channel would it be?

The Optimum Channel

Interstellar communication across the galactic plasma without dispersion in direction and flight-time is practical, so far as we know, only with electromagnetic waves.

Since the object of those who operate the source is to find a newly evolved society, we may presume that the channel used will be one that places a minimum burden of frequency and angular discrimination on the detector.

Moreover, the channel must not be highly attenuated in space or in the Earth's atmosphere. Radio frequencies below ~ 1 Mc./s., and all frequencies higher than molecular absorption lines near 30,000 Mc./s., up to cosmic-ray gamma energies, are suspect of absorption in planetary atmospheres. The band-widths which seem physically possible in the near-visible or gamma-ray domains demand either very great power at the source or very complicated techniques. The wide radio-band from, say, 1Mc. to 10^4 Mc./s., remains as the rational choice.

In the radio region, the source must compete with two backgrounds: (1) the emission of its own local star (we assume that the detector's angular resolution is unable to separate source from star since the source is likely to lie within a second of arc of its nearby star); (2) the galactic emission along the line of sight.

Let us examine the frequency dependence of these backgrounds. A star similar to the quiet Sun would emit a power which produces at a distance R (in metres) a flux of:

$$10^{-15} f^2/R^2 \quad \text{W.m.}^{-2}(\text{c./s.})^{-1}$$

If this flux is detected by a mirror of diameter l_d, the received power is the above flux multiplied by l_d^2.

The more or less isotropic part of the galactic background yields a received power equal to:

$$\frac{10^{-12.5}}{f}\left(\frac{\lambda}{l_d}\right)^2 (l_d)^2 \quad \text{W.(c./s.)}^{-1}$$

where the first factor arises from the spectrum of the galactic continuum, the second from the angular resolution, and the third from the area of the detector. Thus a minimum in spurious background is defined by equating these two terms. The minimum lies at:

$$f_{\text{min.}} \approx 10^4 \left(\frac{R}{l_d}\right)^{0.4} \quad \text{c./s.}$$

With $R = 10$ light years $= 10^{17}$ m. and $l_d = 10^2$ m., $f_{\text{min.}} \approx 10^{10}$ c./s.

Reprinted by permission from *Nature*, 184 (1959), 844. Copyright © 1959 Macmillan Journals Limited.

The source is likely to emit in the region of this broad minimum.

At what frequency shall we look? A long spectrum search for a weak signal of unknown frequency is difficult. But, just in the most favoured radio region there lies a unique, objective standard of frequency, which must be known to every observer in the universe: the outstanding radio emission line at 1,420 Mc./s. ($\lambda = 21$ cm.) of neutral hydrogen. It is reasonable to expect that sensitive receivers for this frequency will be made at an early stage of the development of radio-astronomy. That would be the expectation of the operators of the assumed source, and the present state of terrestrial instruments indeed justifies the expectation. Therefore we think it most promising to search in the neighborhood of 1,420 Mc./s.

Power Demands of the Source

The galactic background around the 21-cm. line amounts to:

$$\frac{dW_b}{dS\ d\Omega\ df} \approx 10^{-21.5}\quad \text{W.m.}^{-2}\ \text{ster.}^{-1}\ (\text{c./s.})^{-1}$$

for about two-thirds of the directions in the sky. In the directions near the plane of the galaxy there is a background up to forty times higher. It is thus economical to examine first those nearby stars which are in directions far from the galactic plane.

If at the source a mirror is used l_s metres in diameter, then the power required for it to generate in our detector a signal as large as the galactic background is:

$$\frac{dW_s}{df} = \frac{dW_b}{dS\ d\Omega\ df}\left(\frac{\lambda}{l_s}\right)^2 \frac{\lambda}{l_d} R^2$$
$$= 10^{-24.2}\ R^2/l_s{}^2 l_d{}^2\quad \text{W.(c./s.)}^{-1}$$

For source and receiver with mirrors like those at Jodrell Bank ($l = 80$ m.), and for a distance $R \simeq 10$ light years, the power at the source required is $10^{2.2}$ W.(c./s.)$^{-1}$, which would tax our present technical possibilities. However, if the size of the two mirrors is that of the telescope already planned by the U.S. Naval Research Laboratory ($l = 200$ m.), the power needed is a factor of 40 lower, which would fall within even our limited capabilities.

We have assumed that the source is beaming towards all the sun-like stars in its galactic neighbourhood. The support of, say, 100 different beams of the kind we have described does not seem an impossible burden on a society more advanced than our own. (Upon detecting one signal, even we would quickly establish many search beams.) We can then hope to see a beam toward us from any suitable star within some tens of light years.

Signal Location and Band-Width

In all directions outside the plane of the galaxy the 21-cm. emission line does not emerge from the general background. For stars in directions far from the galactic plane search should then be made around that wavelength. However, the unknown Doppler shifts which arise from the motion of unseen planets suggest that the observed emission might be shifted up or down from the natural co-moving atomic frequency by $\pm\sim300$ kc./s. (±100 km.s.$^{-1}$). Closer to the galactic plane, where the 21-cm. line is strong, the source frequency would presumably move off to the wing of the natural line background as observed from the direction of the Sun.

So far as the duration of the scanning is concerned, the receiver band-width appears to be unimportant. The usual radiometer relation for fluctuations in the background applies here, that is:

$$\frac{\Delta B}{B} \propto \sqrt{\frac{1}{\Delta f_d\ \tau}}$$

where Δf_d is the band-width of the detector and τ the time constant of the post-detection recording equipment. On the other hand, the background accepted by the receiver is:

$$B = \frac{dW_b}{df}\Delta f_d \quad\text{and}\quad \tau \propto \frac{\Delta f_d}{(\Delta B)^2}$$

If we set ΔB equal to some fixed value, then the search time T required to examine the band F within which we postulated the signal to lie is given by:

$$T = \frac{F\tau}{\Delta f_d} \propto \frac{F}{(\Delta B)^2}$$

independent of receiver band-width Δf_d.

Of course, the smaller the band-width chosen, the weaker the signal which can be detected, provided $\Delta f_d \geqslant \Delta f_s$. It looks reasonable for a first effort to choose a band-width Δf_d normal in 21 cm. practice, but an integration time τ longer than usual. A few settings should cover the frequency range F using an integration time of minutes or hours.

Nature of the Signal and Possible Sources

No guesswork here is as good as finding the signal. We expect that the signal will be pulse-modulated with a speed not very fast or very slow compared to a second, on grounds of band-width and of rotations. A message is likely to continue for a time measured in years, since no answer can return in any event for some ten years. It will then repeat, from the beginning. Possibly it will contain different types of signals alternating throughout the years. For indisputable identification as an artificial signal, one signal might contain, for example, a sequence of small prime numbers of pulses, or simple arithmetical sums.

The first effort should be devoted to examining the closest likely stars. Among the stars within 15 light years, seven have luminosity and lifetime similar to those of our Sun. Four of these lie in the directions of low background. They are τ Ceti, o$_2$ Eridani, ϵ Eridani, and ϵ Indi. All these happen to have southern declinations. Three others, α Centauri, 70 Ophiuchi and 61 Cygni, lie near the galactic plane and therefore stand against higher backgrounds. There are about a hundred stars of the appropriate luminosity among the stars of known spectral type within some fifty light years. All main-sequence dwarfs between perhaps $G0$ and $K2$ with visual magnitudes less than about $+6$ are candidates.

The reader may seek to consign these speculations wholly to the domain of science-fiction. We submit, rather, that the foregoing line of argument demonstrates that the presence of interstellar signals is entirely consistent with all we now know, and that if signals are present the means of detecting them is now at hand. Few will deny the profound importance, practical and philosophical, which the detection of interstellar communications would have. We therefore feel that a discriminating search for signals deserves a considerable effort. The probability of success is difficult to estimate; but if we never search, the chance of success is zero.

21

Ronald N. Bracewell
Communications from Superior Galactic Communities

Since Morrison and Cocconi (Reading 20) published the suggestion that there might be advanced societies elsewhere in the Galaxy, superior to ourselves in technological development, who are beaming transmissions at us on a frequency of 1,420 Mc./s., Drake (Reading 24) has described equipment under construction to look for such transmissions. The confidence necessary to commence actual observations is based on an opinion that planets are a common by-product of the formation of stars. One argument among others is that stars of spectral type later than $F5$ have low angular momenta, just as the Sun has; and in the case of the Sun we know that it is because the momentum (98 per cent of it) resides in planets (Struve, 1960). Of the thousands of millions of planets in the Galaxy likely to be situated similarly to the Earth in relation to their star, it is hard to dismiss the possibility that some have more advanced civilizations than ours. In view of the acceleration with which technology develops, advanced societies could be incredibly more advanced.

Any simple test of this possibility would be well worth while. Drake plans to look at τ Ceti and ϵ Eridani. Of the list of likely neighbouring stars given by Morrison and Cocconi, these two, and ϵ Indi, are the only ones left when we eliminate double stars. Because of orbital perturbations, the planets of double stars are, with some exceptions, not expected to possess equable climates over the geological periods deemed necessary for evolution (Huang, 1959a).

But do we really expect a superior community to be on the nearest of those stars which we cannot at the moment positively rule out? Unless superior communities are extremely abundant, is it not more likely that the nearest is situated at least ten times farther off, say, beyond 100 light years? Let us assume that there are one thousand likely stars within the same range as the nearest superior community. This makes it hard for us to select the right one. Furthermore, if this advanced society is looking for us, we

can only expect to find them expending such effort as they could afford to expend on the thousand likely stars within the same range of them. It does not seem likely that they would maintain a thousand transmitters at powers well above the megawatt estimated by Drake as a minimum for spanning only 10 light years, and run them for many years, and we could scarcely count on them paying special attention to us. Remember that throughout most of the thousands of millions of years of the Earth's existence such attention would have been fruitless.

Would not this other more advanced society, on the contrary, be doing what we ourselves are now discussing and are on the point of doing, probably during this century, namely, sending probes to nearby stars. Their exploration and other activity would be intense in their immediately neighbouring planetary systems. Beyond their immediate neighbourhood, it might be feasible for them to spray some number of suitable stars, say, one thousand, with modest probes. Each probe would be sent into a circular orbit about one of the thousand stars, at a distance within the habitable zone of temperature. Armoured against meteorites and radiation damage, and stellar powered, the probes could contain durable radio transmitters for the purpose of attracting the attention of technologies such as ours.

Using this plan, our hypothetical advanced neighbours could lay down a stronger signal here than they could with a home-based transmitter handicapped by inverse-square attenuation over interstellar distances. They would also eliminate their dependence on our ingenuity in selecting the right star and the right wave-length.

For this reason we might better devote our efforts to scrutinizing our solar system for signs of probes sent here by our more advanced neighbours. In this way we would be effectively paying attention to all stars capable of reaching us. We need not expect, however, that any community other than the nearest is trying to reach us, because the superior communities throughout the galaxy are probably already linked together into an existing galaxy-wide chain of communication. They will act in concert and avoid duplication in searching. Our impending contact cannot be expected to be the first of its kind; rather it will be our induction into the chain of superior communities, who have had long experience in effecting contacts with emerging communities like ours.

For suggestions as to how the superior communities may detect us, consider what we might do to detect them. A very good first project for us, when we come to probe outside the solar system, would be to seek the presence of technological development on τ Ceti and ϵ Eridani by means of a probe that would listen for the existence of monochromatic radio-communication, and report back by star-to-star relay. We would see whether there is in those solar systems a radio-frequency line emission spectrum such as the Earth now emits. It is possible, in fact, that the hypothetical feelers sent out in large numbers by our nearest superior community did no more than listen for this radiation. If so, a positive answer could have been on the way back to the home star several decades ago, and we may look forward in due course to the arrival of a more sophisticated mission.

However, since interstellar transfer of material things is time-consuming, and transfer of information is in any event more important, it would be commensurate with the effort of delivering a material probe into our solar system if the very first probe sent here contained a quite elaborate store of information and a complex computer, so that it could not only detect our presence, but could also converse with us. Such a probe may be here now, in our solar system, trying to make its presence known to us. For this purpose a radio transmitter would seem essential. On what wave-length would it transmit, and how should we decode its signal? To ensure use of a wave-length that could both penetrate our ionosphere and be in a band certain to be in use, the probe could first listen for our signals and then repeat them back. To us, its signals would have the appearance of echoes having delays of seconds or minutes, such as were reported thirty years ago by Størmer and van der Pol (Størmer, 1928; van der Pol, 1928; see also Budden and Yates, 1951) and never explained.

To notify the probe that we had heard it, we would repeat back to it once again. It would then know that it was in touch with us. After some routine tests to guard against accident, and to test our sensitivity and band-width, it would begin its message, with further occasional interrogation to ensure that it had not set below our horizon. Should we be surprised if the beginning of its message were a television image of a constellation?

These details, and the matter of teaching the probe our language (by transmitting a pictorial dictionary?), are fascinating but present no problems once contact has been made with the probe. The latter is the main problem. The important thing for us is to be alert to the possible interstellar origin of unexpected signals. We must avoid relegating them, if they are there, to the fate of the very strong

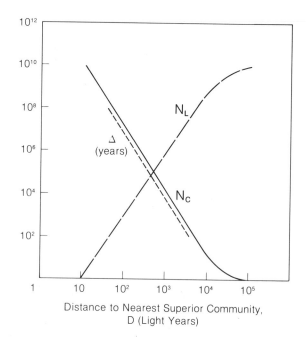

Figure 21.1. N_C, total number of communities in the galaxy the technology of which is superior to ours; N_L, total number of likely stars out to a distance d; Δ, average life-time of a superior community.

emissions from Jupiter (of the order of 1,000 megawatts per Mc./s.) which were heard and ignored for decades (Burke and Franklin, 1955; Shain, 1956).

If after a few years of careful attention we find no signs, radio or other, of such probe, we shall have to admit the possibility that our nearest superior community is beyond the range where attempts at contact with us would be assured of much certainty of success.

To survey the possibilities of there being a superior community within reaching distance of us, consider Fig. 21.1, which shows the number of superior communities in the Galaxy, N_C, plotted against the distance to the nearest superior community, d, in light years. This graph is obtained from the broken curve which shows a quantity N_L, the number of likely stars at a range less than d. By likely stars I mean those 5 per cent (Huang, 1959a), here taken as 10^{10} in number, that cannot be ruled out at present as unsuitable to support life. The curve is based on a galactic mass distribution model and cannot be considered accurate to better than an order of magnitude.

Now consider the consequences if $Nc = 10^7$. Then $d = 100$ light years, and the number of likely stars within this range is 10^3. The frequency of occurence, p, of superior communities among likely stars, is 10^{-3}. Although we have no evidence for intelligent life elsewhere, yet if we consider than on the average it takes 5×10^9 years for a likely star to produce one superior community, when then endures for an average life-time Δ measured in years, then $p = 10^{-3}$ implies a Δ of 5×10^6 years, assuming that we are in a state of secular equilibrium. This would seem to offer ample time to explore the 10^3 stars out to 100 light years and establish a chain of communication.

Consider the consequences, however, if technology in our galaxy is less abundant, for example, take $Nc = 10^3$. Then $p = 10^{-7}$, $N_L = 10^7$, $d = 2,000$ light years, and $\Delta = 500$ years. The duration of communities which can maintain a frequency of occurrence of only 10^{-7} is thus, on the average, too short to permit interstellar traffic.

If intelligent life does develop on other likely systems at the same tempo as ours has developed, and if some superior community has not made contact with us, it may simply be that the mortality rate for advanced civilizations is too high for them to become abundant in the Galaxy. Even so, it is rather striking that there would be a thousand superior communities present in the Galaxy at any time even though it takes as long as five thousand million years to produce a technological community that is viable, on the average, for only 500 years beyond the point we have reached.

Even in the event of technology being rare, there is, however, the possibility of a chain existing. Thus, in a Galaxy supporting only 10^3 superior communities with brief expectation of life, there may be some communities that have achieved durability, even quasi-permanence, perhaps by gaining control of the circumstances that lead to short average life-times. Aided by accidental proximity due to random spacing, some of these could be in contact. Presumably such an ancient association would be very able indeed technically, and might seek us out by special means that we cannot guess. Whether they would be interested in rudimentary societies which, in their experience, would usually have burnt themselves out before they could be located and reached, is hard to say. Such communities would be collapsing at the rate of two a year (10^3 in 500 years), and they might already have satisfied their curiosity by archaeological inspection made at leisure on sites nearer home. On the other hand, the prospect of catching a technology near its peak might be a strong incentive for them to reach us.

22

Freeman J. Dyson
Search for Artificial Stellar Sources of Infrared Radiation

Cocconi and Morrison (Reading 20) have called attention to the importance and feasibility of listening for radio signals transmitted by extraterrestrial intelligent beings. They propose that listening aerials be directed toward nearby stars which might be accompanied by planets carrying such beings. Their proposal is now being implemented (*Science,* April 29, 1960, p. 1303).

The purpose of this report is to point out other possibilities which ought to be considered in planning any serious search for evidence of extraterrestrial beings. We start from the notion that the time scale for industrial and technical development of these beings is likely to be very short in comparison with the time scale of stellar evolution. It is therefore overwhelmingly probable that any such beings observed by us will have been in existence for millions of years, and will have already reached a technological level surpassing ours by many orders of magnitude. It is then a reasonable working hypothesis that their habitat will have been expanded to the limits set by Malthusian principles.

We have no direct knowledge of the material conditions which these beings would encounter in their search for *lebensraum*. We therefore consider what would be the likely course of events if these beings had originated in a solar system identical with ours. Taking our own solar system as the model, we shall reach at least a possible picture of what may be expected to happen elsewhere. I do not argue that this is what *will* happen in our system; I only say that this is what *may have* happened in other systems.

The material factors which ultimately limit the expansion of a technically advanced species are the supply of matter and the supply of energy. At present the material resources being exploited by the human species are roughly limited to the biosphere of the earth, a mass of the order of 5×10^{19} grams. Our present energy supply may be generously estimated at 10^{20} ergs per second. The quantities of matter and energy which might conceivably become accessible to us within the solar system are $2 \times$ 10^{30} grams (the mass of Jupiter) and 4×10^{33} ergs per second (the total energy output of the sun).

The reader may well ask in what sense can anyone speak of the mass of Jupiter or the total radiation from the sun as being accessible to exploitation. The following argument is intended to show that an exploitation of this magnitude is not absurd. First of all, the time required for an expansion of population and industry by a factor of 10^{12} is quite short, say 3000 years if an average growth rate of 1 percent per year is maintained. Second, the energy required to disassemble and rearrange a planet the size of Jupiter is about 10^{44} ergs, equal to the energy radiated by the sun in 800 years. Third, the mass of Jupiter, if distributed in a spherical shell revolving around the sun at twice the Earth's distance from it, would have a thickness such that the mass is 200 grams per square centimeter of surface area (2 to 3 meters, depending on the density). A shell of this thickness could be made comfortably habitable, and could contain all the machinery required for exploiting the solar radiation falling onto it from the inside.

It is remarkable that the time scale of industrial expansion, the mass of Jupiter, the energy output of the sun, and the thickness of a habitable biosphere all have consistent orders of magnitude. It seems, then, a reasonable expectation that, barring accidents, Malthusian pressures will ultimately drive an intelligent species to adopt some such efficient exploitation of its available resources. One should expect that, within a few thousand years of its entering the stage of industrial development, any intelligent species should be found occupying an artificial biosphere which completely surrounds its parent star.

If the foregoing argument is accepted, then the search for extraterrestrial intelligent beings should not be confined to the neighborhood of visible stars. The most likely habitat for such beings would be a dark object, having a size comparable with the Earth's orbit, and a surface temperature of 200° to 300°K. Such a dark object would be radiating as copiously as the star which is hidden inside it, but the radiation would be in the far infrared, around 10 microns wavelength.

Reprinted from *Science,* 131 (1960), 1667. Copyright 1960 by the American Association for the Advancement of Science.

It happens that the earth's atmosphere is transparent to radiation with wavelength in the range from 8 to 12 microns. It is therefore feasible to search for "infrared stars" in this range of wavelengths, using existing telescopes on the earth's surface. Radiation in this range from Mars and Venus has not only been detected but has been spectroscopically analyzed in some detail (Sinton and Strong, 1960).

I propose, then, that a search for point sources of infrared radiation be attempted, either independently or in conjunction with the search for artificial radio emissions. A scan of the entire sky for objects down to the 5th or 6th magnitude would be desirable, but is probably beyond the capability of existing techniques of detection. If an undirected scan is impossible, it would be worthwhile as a preliminary measure to look for anomalously intense radiation in the 10-micron range associated with visible stars. Such radiation might be seen in the neighborhood of a visible star under either of two conditions. A race of intelligent beings might be unable to exploit fully the energy radiated by their star because of an insufficiency of accessible matter, or they might live in an artificial biosphere surrounding one star of a multiple system, in which one or more component stars are unsuitable for exploitation and would still be visible to us. It is impossible to guess the probability that either of these circumstances would arise for a particular race of extraterrestrial intelligent beings. But it is reasonable to begin the search for infrared radiation of artificial origin by looking in the direction of nearby visible stars, and especially in the direction of stars which are known to be binaries with invisible companions.

23

R. N. Schwartz and C. H. Townes
Interstellar and Interplanetary Communication by Optical Masers

Long-range communication by radio-waves is already well known, and the possibility of interstellar communication by radio-waves in the microwave region has been suggested in several interesting proposals (Reading 20; Purcell, 1959; Struve, 1959) to search for signals from intelligent beings on planets associated with nearby stars. The supposition is that curiosity such as our own would motivate advanced civilizations associated with stars other than our Sun to make determined efforts to communicate with whatever other intelligent life might exist on neighbouring planetary systems. Radio-waves have, because of our present state of technological development, dominated the field of very long-distance communication, and perhaps for this reason these proposals gave particular attention only to the radio region. It appears, however, that we are now not very far from the development of maser oscillators and other appropriate apparatus in or near the optical region which will also allow detectable light signals to be beamed between planets of two stars separated by a number of light years.

Our own maser techniques in the optical and nearby spectral regions are still in a rudimentary stage; no such operating device was known a year ago (Schawlow and Townes, 1958; Maiman, 1960b). Another ten years should bring very marked development. Further, only historical accident seems to have prevented discovery of optical masers thirty or more years ago, in which case they would probably already have been in an advanced stage of development. This implies that a separate civilization might have inverted our own history and become very sophisticated in the use of optical or infra-red masers rather than in the techniques of short radio-waves.

We propose to examine the possibility of broadcasting an optical beam from a planet associated with a star some few or some tens or light-years away at sufficient power-levels to establish communications with the Earth. There

is some chance that such broadcasts from another society approximately as advanced as we are could be adequately detected by present telescopes and spectrographs, and appropriate techniques now available for detection will be discussed. Communication between planets within our own stellar system by beams from optical masers appears *a fortiori* quite practical.

Optical Maser Characteristics

Present ruby masers have produced pulsed optical beams of 10-kW. power (Maiman, 1960a), or still shorter pulses of perhaps 100-kW. peak power. This radiation is concentrated in a band-width of about 0.02 cm.$^{-1}$. Another type of maser, operating continuously at about 0.02 W., emits a wave which has been shown to be in phase over the entire maser reflector surface (about ½ in. in diameter) and to be concentrated in a frequency-interval of about 10 kilocycles per sec (Javan, Bennett, and Herriot, 1961). The latter case is much closer to theoretical expectations (Schawlow and Townes, 1958) for an ideal maser in so far as coherence is concerned. There seems to be no general reason, other than the necessary dissipation of power, why solid-state optical masers cannot operate continuously at high power and with a short-term monochromaticity close to theoretical expectations, or hence with frequency-widths very much less than 1 megacycle/sec.

Now consider directivity of the beam from an optical maser. If a maser produces a wave of wave-length λ with constant phase over a surface of diameter d, the angular width of the radiating beam is approximately λ/d. However, it is obvious that its angular width can be still further reduced if it is operated in conjunction with an auxiliary optical system. Imagine that an ideal lens is put in the beam of focal length d and diameter d. Then, at the focal point the entire beam has a diameter approximately λ. If this focal point is made to coincide with that for a much larger ideal lens or reflecting mirror of diameter D and focal length D or larger, the beam emerges from the latter with an angular diffraction width determined by its aperture, or λ/D. To obtain this result, it is not necessary actually to focus the beam to a small spot; this was assumed only for heuristic purposes. The limiting factors in achieving an intense, directed beam probably reside, not in the source, but in the more familiar technical problems of producing large mirrors of the required accuracy and controlling the optical distortions due to heating by the beam.

For the sake of discussion, two optical maser systems will be assumed with the following characteristics:

System (*a*). Power-level, 10 kW. continuous; frequency, ~5000Å.; band-width, ~1 megacycle/sec. or 3×10^{-5} cm.$^{-1}$; diameter of reflector (D), 200 in. (maximum size of present telescopes); beam-width (λ/D), 10^{-7} radian.

System (*b*). A group of 25 masers, each of the same maser characteristics as system (*a*), but with an effective system aperture, D, of 4 in. (and therefore a beam width of 1 sec. of arc, or 5×10^{-6} radian). The entire group is to be pointed in the same direction within the accuracy of the beam-width.

The beam-width of system (*a*) could be attained in a system operating from a platform above (or in the absence of) a planetary atmosphere and of such accurate dimensions that diffraction is the limiting factor. (See following discussion.) When operating from within an Earth-like atmosphere, the atmospheric turbulence restricts the effective band-width to that which would be achievable in a 4-in. telescope system, suggesting system (*b*) in this case as a more reasonable choice.

Detectability of Maser Signals

Two primary criteria for detection of a maser beam emanating from near a star are: (1) it must produce enough photons per unit area at the receiver to be detectable with a lens of practical size and in a reasonable time; (2) it must be distinguishable from the background of stellar light.

The maser system operated from above an atmosphere (system (*a*)) would produce a beam of intensity, I, where: $I = \text{flux}/\lambda(D)^2 = 10^{18}$ W./sterad. at 5000 Å. or, per wave-number; $I_\nu = I/\Delta\nu = 3 \times 10^{22}$ W./sterad. cm.$^{-1}$ in a band-width $\Delta\nu = 3 \times 10^{-5}$ cm.$^{-1}$.

The intensity of radiation from system (*b*) would be less by a factor of 100. Under the most favourable conditions, a star of magnitude 8.3 can be seen by the naked eye; hence, the system (*a*) beam could be seen by eye at 0.1 light-year. With ordinary binoculars against the normal sky, the range would be about 0.4 light-year. Corresponding distances would be less by a factor of 10 for system (*b*).

From a distance of 10 light-years, the system (*a*) could be seen visually through the Hale 200-in. telescope or could be photographed with ordinary techniques with an exposure of about 1 min. This time of detection gives about 10^6 quanta of signal, and presumably could be much shortened by special effort. The distance of 10 light-years is significant because, as noted by G. Cocconi and P. Morrison (Reading 20), there are seven stars of nearly the same luminosity and spectral characteristics as our Sun within this distance. System (*b*) would require correspondingly about a 1½ hr. exposure in the Hale telescope for detection at 10 light-years. Thus criterion (1) is well satisfied.

The beam intensities indicated here are quite sufficient to communicate between planets of our solar system. They must, of course, compete with a background of light reflected from the planetary surface. The intensity of sunlight scattered by the Earth's surface is less than $1/\pi \times$ cross-sectional area $\times$ solar constant $= 6 \times 10^{16}$ W./sterad., so that the maser beams could be easily seen against the background of the Earth as viewed visually through an optical filter a few hundred angstroms wide, or as a bright spot superimposed on the image of the planet in a telescope of moderate size. This affords an alternate communications technique with some advantages for specific cases over radio communications. For example, because of directivity, light could be beamed from Mars or the Moon to a particular part only of the Earth. Furthermore, equipment for receiving simple signals need not be much more than the human eye.

Consider now the question of detection of such a signal from a distant planet against the background of light from its star. There is some, but little, hope of resolving the two spatially. The Earth and Sun subtend an angle of about ½ sec. of arc at a distance of 10 light-years. Hence light of equal intensity from the two might be resolved at this distance by a telescope of very high quality. However, one can perhaps more easily resort to high spectral resolution in order to discriminate between maser and stellar radiation.

We may take the spectrum of our own Sun as representative of the light to be discriminated against. The intensity per unit wave-number of the continuum has a maximum near 5000 Å. of about:

$$I_\nu(5000\text{Å.}) = 1.2 \times 10^{21} \text{ W./sterad. cm.}^{-1}$$

with a total intensity:

$$I = 3 \times 10^{25} \text{ W./sterad.}$$

Here it is seen that the maximum stellar intensity per wave-number is 1/25 that of maser system (*a*) or about four times that of maser system (*b*). The band-width of 1 mc./sec. chosen for the maser might possibly be reduced to increase the relative intensity of maser to star, but possible difficulty in obtaining high-power masers with this stability and some other considerations which will be mentioned later make use of such narrow band-widths unattractive.

For the Sun, I_ν is lower than the value of 5000 Å. by more than an order of magnitude below 2500 Å. and above 15,000 Å., two orders of magnitude below 2000 A. and above 4μ. Superimposed on this continuum is a discrete structure, the Fraunhofer absorption lines, and emission spectra in the far ultra-violet. The most pronounced absorption lines of Ca II, Sr II, and Na provide windows several angstroms wide in which the intensity is less by an order of magnitude than that of the continuum. Of course, it is to be remembered that the Earth's atmosphere limits terrestrial observation to wave-lengths between 2900 Å. and some few microns.

A logical selection in transmission frequency would be either to use a frequency as far in the violet as atmospheric absorption permits, or to pick a prominent Fraunhofer line (as the Ca II H or K line). In either case, a gain by a factor of about 10–20 in the ratio is to be had over transmission at 5000 Å., the choice probably being dictated by the availability of a suitable maser material to produce the desired frequency. Were it only a question of the solar intensity, another two or three orders of magnitude could be gained by recording above the Earth's atmosphere in the 1400–1800 Å. region. However, the loss of reflecting power in mirror systems and the anticipated difficulties in maser operation at these wave-lengths make the region unattractive—even if recording above the atmosphere were not already too awkward. The advantages of transmission in the infra-red over that at 5000 Å. would be offset by the widening of the beam were diffraction the limiting factor as in system (a).

Assuming a maser system (a) operating, say, within the Ca II H or K absorption lines, the ratio of maser to Sun intensity is about 300 in the narrow band-width of 3 × 10^{-5} cm.$^{-1}$. A spectrometer with resolution comparable to 0.01 cm.$^{-1}$ (that is, somewhat better than a resolution of 10^6 at this frequency) would give a maser signal just equal to the stellar background. Present grating spectrographs of high speed and resolving power give a resolution close to this figure (for example, the echelle spectrograph of G. Harrison) (Harrison, Archer and Camus, 1952). The use of interferometric techniques can provide a resolution of a few orders better than this. An instrument comparable to the coudé spectrographs developed for use in conjunction with the Hale 200-in. telescope and with the required resolution appears capable of producing an acceptable photographic image with an exposure of a little more than an hour. Much less resolution than 0.01 cm.$^{-1}$ would still yield signals detectable against background, although the maser signal would then not be so obviously narrower than some spectral emission lines.

Maser system (b) would give an intensity just three times that of stellar background, again in the band-width 3 × 10^{-5} cm.$^{-1}$. While convenient detectors operating within such a narrow band-width might be developed, none is yet available. A spectrometer of resolution 0.01 cm.$^{-1}$ would afford a maser signal 1 per cent as large as background, which might allow marginal discrimination.

One must consider the possibility that changes in Doppler shifts during the period of exposure will be larger than the spectral resolution used and thus weaken the image. The Earth's orbital or rotational motion, the orbital motion of a satellite around the Earth if such is used as an instrument platform, and corresponding motions of the transmitting source may all produce Doppler shifts which can change the frequency and smear out the maser signal. The maximum rate of variation of the fractional change of frequency due to a Doppler shift $\Delta\nu_D$ from a signal source following a circular path of radius r with angular speed ω is given by:

$$\frac{1}{\nu}\left|\frac{d(\Delta\nu)_D}{dt}\right|_{max.} = \frac{\omega^2 r}{c}$$

where ν is the frequency and c the velocity of light. Taking the motions within our own solar system as illustrative, one obtains maximum values of 2 × 10^{-10} per sec. for an Earth-based transmitter mainly due to rotation of the Earth and 10^{-11} per sec. for a Moon-based transmitter. For these Doppler changes on both the receiving and sending end of the transmission to be contained within the required limits of resolution of 0.01 cm.$^{-1}$, the times of photographic exposure in the visible spectrum, then, would be restricted to about an hour. This is, as already noted, the approximate time needed for a good spectrum if the transmitter were at a distance of 10 light-years. Alternatively, both sender and receiver might, of course, compensate for his own cyclical Doppler shifts, which would be well known to him.

The question of exposure, or integration time, has a different connotation depending on whether the observer is engaged in a search of the spectrum or has found a likely line. In the first situation, hundreds of angstroms of the spectrum can be examined in a single photograph and, consequently, the spectral search of a given star is not unduly hampered by fairly long times of exposure. The spectral line sought can be expected to be exceptionally narrow, at an abnormal frequency for the type of star in question, and varying in intensity. Observation of any of these characteristics should lead to closer examination. Once a likely line has been located, a photo-detection system of higher sensitivity could be used to reduce the required integration times to minutes for the purpose of recording modulations in the signal. Actually, with a 200-in. telescope and an efficient dispersing system, the modulation could be visible to the eye.

Discussion

Thus, it appears that if one postulates the existence of an optical maser system beamed towards us of the characteristics of system (a) at a distance of the order of 10 light-years, it is within the state-of-the-art for us to detect it. System (b) is marginally detectable, or detectable with good signal to background, if very narrow-band optical receivers are developed. One should hence consider the feasibility and expense of such systems as judged by our own experience, and of course in the light of any fundamental difficulties.

Advantages accruing from the coherence of radiation over a very large aperture, or from the theoretical possibility of obtaining coherence among several maser sources are nullified if the system is required to operate from within an Earth-like atmosphere. In this case, system (b) above, perhaps augmented in either power-level or numbers of system elements, appears most advantageous. An appreciable increase in the number of elements would not necessarily involve prohibitive expense. Accepting a criterion of minimum detectability as a signal of only 1 per cent of the background, system (b) as defined would be sufficient for the job. An added difficulty in working within an atmosphere should be noted. Local air turbulence caused by hot lens and mirror surfaces in contact with the atmosphere would widen the beam even more unless well controlled. The power density reflected from the mirror surface of system (a) is close to that of direct sunlight. Such problems, connected with handling a large amount of power in a carefully controlled optical system would probably be some of the most troublesome ones.

The unique advantages of maser sources can be better utilized in operation from a very high-altitude balloon, a space platform, or natural Moon—a possibility that may not have been seriously considered a few years ago but should be more acceptable to-day. The capability of the ideal system (a) could possibly be met by using a single large mirror, although the very large mirrors with which we have had experience have not attained the highest angular resolution possible (the Hale 200-in. has an angular resolution of about $1/3$ sec. of arc (Bowen, 1950). The task of obtaining the limit imposed by diffraction may be a difficult one. However, the beam intensities of system (a) might also be met by a bank of smaller mirrors of higher accuracy, perhaps operated in phase (the beam intensity varies in such case as the square of the number). In this connexion, it should be noted that the 36-in. mirror to be used in the current *Stratoscope II* high-altitude balloon programme (*Physics Today*, 14, 1961, 82) is a casting of fused quartz which is expected to achieve its theoretical angular resolution of $1/10$ sec. of arc.

It may be both an encouraging enlargement of possibilities, and at the same time an unwelcomed complication, to find that the frequency of the hydrogen line in the micro-wave region is not the only reasonable place at which to search for possible interstellar communications, and that the optical region also seems a logical one. What other methods are we overlooking which might appear natural to some other civilization? Far ultra-violet and infrared are absorbed by most imaginable atmospheres friendly to life, and hence would be avoided unless reception above the atmosphere is to be expected. Beams of charged particles would be subject to inconvenient bending in interstellar fields. We see no way of producing neutron beams or electromagnetic waves much shorter than the ultra-violet (Schawlow and Townes, 1958) with adequate intensity. On the other hand, the rapid progress of science implies that another civilization, more advanced than ourselves by only a few thousand years, might possess capabilities we now rule out—they may have already been able to send us an exploratory instrumented probe. Since none has yet been seen, perhaps it would be appropriate to examine high-resolution stellar spectra for lines which are unusually narrow, at peculiar frequencies, or varying in intensity.

Note added in proof. Dr. W. S. Boyle and Dr. R. J. Collins have independently arrived at similar conclusions (private communication).

24

Frank D. Drake

How Can We Detect Radio Transmissions from Distant Planetary Systems?

The question of whether there is intelligent life elsewhere in the universe, outside the bounds of the solar system, has long been fascinating, but apparently unanswerable. Optical telescopes offer no help with the problem. It would require a very large telescope outside the earth's atmosphere, say on the moon, merely to detect the existence of planets accompanying the closest stars. The presence of intelligent life would still be unrecognizable.

Could a radio telescope detect intelligent radio transmissions over interstellar distances? It is easy to show that the distance R over which a signal can be observed is given under most conditions by the accompanying formula. To see what ranges are obtainable with available equipment, let us consider an attempt by the 85-foot radio telescope of the National Radio Astronomy Observatory to detect a high-power radar signal.

Such a signal might be like that radiated by the Millstone Hill radar antenna at Westford, Massachusetts, when it achieved radar echoes from Venus (*Sky and Telescope,* May 1959, page 384). At such times, the Millstone radar has roughly an effective radiating power (P_e) of 10^{10} watts and a band-width (B) of 10 cycles per second. The 85-foot has an effective receiving area (A) of 370 square meters and, if we are using a maser, T might be only 10° Kelvin and t about 100 seconds.

Using these values in the formula, we find that the radar transmission would be detectable even if it originated 8.7 light-years away! This is as far as Sirius, and about twice as far as Alpha Centauri. Because of our large antennas and new sensitive receivers, we are already capable of detecting the radio transmissions of intelligent beings over interstellar distances.

When the receiving antenna is a parabolic reflector, this rule-of-thumb formula may be applied: The distance in light-years at which strong present-day transmitters can be detected is about equal to the diameter of the antenna in feet divided by 10. Thus the U.S. Navy's 600-foot an-

tenna, now under construction at Sugar Grove, West Virginia, could detect intelligent life transmissions from as far away as 60 light-years, while the Cornell University 1,000-foot bowl in Puerto Rico and NRAO's proposed 1,000-foot antenna will be able to see about 100 light-years. Within that distance of the earth, there are something like 10,000 stars.

It is very difficult to estimate how many of those stars may be supporting civilizations as advanced as ours. For many years, it was felt that the formation of the solar system was the result of a chance collision between two stars, or some other unlikely event. Furthermore, the development of life on planets was thought to be a very rare occurrence. This led to a pessimistic picture of the abundance of life in the galaxy.

In recent times, however, it has become clear that the formation of a second body or bodies is an essential part of the formation of a star. It is possible that most stars, not members of binary or multiple systems, are accompanied by families of planets or meteorites. As explained by Otto Struve (1960), probably at least a few per cent of all stars are accompanied by planets.

How many of these planets might have intelligent life? The pioneering experiments of Stanley Miller at the University of Chicago, as well as later studies, have shown that complicated organic molecules could well have been formed in great quantities during the early history of the earth or any similar planet. These molecules, which are the basic building blocks of life, should provide for its emergence on other planets just as they apparently did on earth.

We might expect life to be quite common on planets, then. However, recent papers by K. Kordylewski in Poland and Su-Shu Huang in the United States consider anew the fact that life will thrive only on planets that are at such distances from their star that the temperature is appropriate. This factor limits intelligent life to just a few planets in each planetary system, including our own. And the star must be long-lived, not changing its brightness

Reprinted by permission from *Sky and Telescope,* 19 (1960), 140.

appreciably during the billions of years that are required for the evolution of intelligent beings from a collection of organic molecules. The over-all result is to limit such life to planets of main-sequence stars of spectral types *F, G, K,* and possibly *M.* However, this still includes about half of all the stars.

The stars have quite varied ages. This, plus the good probability that biological evolution occurs at unequal rates on different planets, means that the present level of evolutionary development may be quite different from planet to planet. The age of the sun appears to be average. Thus, it is likely that throughout the galaxy there are scattered civilizations more advanced, at the same level, and less developed than ours.

From all the considerations above, it appears that strong intelligent radio transmission may emanate from the vicinities of, at best, 25 per cent of the stars, and at worst, perhaps one star in a million, which is the extremely conservative estimate recently suggested by Harlow Shapley in his book, *Of Stars and Men.* Obviously, these estimates are not very helpful, except in one respect: They make the possibility good that at least one of the 10,000 stars soon to be within our reach has a civilization using radio techniques. It is, then, worth while to apply our newly found technical prowess in an effort to detect interstellar radio transmission.

What search frequency would be best? Consider what might be called the principle of technical perfection. It is only about 50 years since radio communication was invented, yet we have already very nearly achieved technically perfect instruments, and within 50 more years we should have them. By *technical perfection* we mean that the limits on communication-system sensitivities are not set by deficiencies in the apparatus, such as receiver noise, but by natural phenomena over which man has no control. This is a state in which further improvements in apparatus will not improve the operational results.

A century is only about a hundred-millionth of the age of our galaxy. Thus, on the galactic time scale, a civilization passes abruptly from a state of no radio ability to one of perfect radio ability. If we could examine a large number of life-bearing planets, we might expect to find in virtually every case either complete ignorance of radio techniques, or complete mastery. This is the principle of technical perfection. Our civilization may be one of an extremely small minority in transition between the two possible states—this, in fact, may be the only major feature in which man is unique.

Therefore, it may be logical to assume that the civilizations we might detect possess complete mastery of radio already. The transmissions we seek will obviously be very powerful ones, in which large information transfer over long distances is being attempted. Frequencies will be chosen for which the natural limitations on performance are least. Two of these limitations are important: galactic

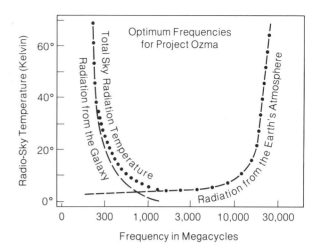

Figure 24.1. Only in a relatively narrow range of radio frequencies is there a reasonable chance of receiving transmissions from extraterrestrial beings. At low frequencies such signals would be smothered by radio noise of galactic origin; at high frequencies, radiation from the earth's atmosphere would overwhelm them. Both these types of interference are charted here, together with their resultant, the total sky radiation temperature. NRAO graph.

radio noise emission, and noise from the planetary atmosphere, if reception from beneath the atmosphere is being attempted.

Both these emissions insert noise into the receiver, and have the same effect as though the receiver itself were noisy. The graph shows for the earth the radio-sky temperature produced by each of these sources of noise, and their combined effect. This last would be the excess receiver noise temperature of an otherwise perfect receiver. Obviously, the best frequencies to use for our search are those where this total sky temperature is least.

For instance, from beneath the atmosphere of a planet like the earth, the band from 1,000 to 10,000 megacycles per second would be the optimum for reception of long-range transmissions. If, however, reception is being done from above the atmosphere, as the principle of technical perfection and our own success with satellites suggest, frequencies above 10,000 megacycles are also good candidates.

Giuseppe Cocconi and Philip Morrison, of Cornell University, have gone one step further in a paper that appeared in the British journal *Nature* (Reading 20). They speculate that civilizations in space may have produced strong radio beams directed toward their nearby neighbors, in an effort to establish two-way radio communication at the earliest possible time. Such an attempt would use the frequency at which high-sensitivity, narrow-band radio telescopes would be first operated extensively during radio development of civilizations in the Milky Way. Throughout the galaxy it could very well be

1,420 megacycles, the frequency of the 21-centimeter interstellar hydrogen line.

All these considerations suggest confining our search to the band between 1,000 and 10,000 megacycles, and that possibly frequencies around 1,420 megacycles offer the best chance for success.

What will be the characteristics of the sought-after signals? Communication theory states that the narrower the band-width of a transmitted signal of fixed total power, the greater the range of successful transmission. Since we are attempting, in any case, to intercept transmissions of great power, where great range is presumably the goal, we can expect the signals to be of narrow band-width. This is advantageous, as it will allow us to distinguish the signals from naturally occurring cosmic noise, which is extremely broad in band-width.

We should expect an appreciable Doppler shift in the transmitted frequency over short periods, because of the likelihood that the transmitter will be in orbital motion around a star or planet. This shift will help us distinguish signals from cosmic noise and from terrestrial interference. The signal strength may vary with time, if coded information is being sent, and also possibly due to the changing orientation of the earth with respect to the transmitter. Finally, the transmission should come from the direction of a nearby star.

Summarizing this discussion, a receiver designed to detect interstellar radio signals at the surface of the earth should:

1. Operate at a frequency above 1,000 megacycles, with performance near 1,420 desirable.

2. Operate over a considerable range of band-widths, including 10 cycles per second and less.

3. Have receiver frequency vary considerably less than the band-width during the averaging time of the receiver, specifically not more than about one cycle per second in several minutes.

4. Have the most sensitive system now available.

5. Eliminate receiver noise effects as much as possible.

6. Discriminate against broad-band cosmic noise, if possible.

7. Preferably reject terrestrial interference that resembles the sought-after signal.

In an effort to detect interstellar radio transmissions, NRAO has established Project Ozma. It is named for the queen of the imaginary land of Oz—a place very far away, difficult to reach, and populated by exotic beings. A radiometer that fulfills the specifications above is now in the final stages of construction.... Almost all of the assembly and testing to date have been carried on by W. Waltman, Purdue University, and R. W. Meadows, of the United Kingdom's Department of Scientific and Industrial Research.

...The Ozma radiometer operates near 1,420 megacycles. It is essentially a highly stable narrow-band superheterodyne receiver, which utilizes the principles of both the Dicke radiometer and the d.c.-comparison type.

Two horns are placed together at the focus of the parabolic antenna, in order to eliminate terrestrial interference to some extent. These horns give the antenna two beams, one to point at the star under study, the other off into space near the star. As the electronic switch connects first one horn and then the other to the receiver, the telescope will look alternately at the star and at the sky beside it. Any radiation from the star will then enter in pulses whose duration is controlled by the switch. The synchronous detectors near the output end of the circuit will respond only to pulses synchronized with the switch, thus detecting only the desired signal. Receiver noise is eliminated, and also terrestrial disturbances.

Interference generally enters a radio telescope antenna through the horns directly, without a reflection from the paraboloid. In that case, both horns should receive the interfering signal with the same strength, and when the switch changes from one horn to the other there will be no change in level in the interference entering the receiver. As a result, the interference signal is not pulsed, and the synchronous detectors ignore it.

In the present receiver, immediately after the switch there comes a reactance amplifier, to be replaced later with a maser. The amplifier, which gives the radiometer high sensitivity, was built by Ewen-Knight Corp., while the electronic switch was made by H. Hvatum of NRAO.

The signal then undergoes four frequency conversions, this many being necessary because the final intermediate frequencies are very low, due to the narrow-band-width requirements. The frequency received by the radiometer is directly dependent on the frequencies of the four oscillators, whose output signals beat with the true signal to produce the intermediate frequencies. In our specifications all four oscillators must hold their frequencies constant to better than one cycle per second in 100 seconds, if the over-all received frequency is also to be that constant.

This is a most difficult requirement for the first oscil-

Box 24.1. Distance of Detection of Cosmic Radio Signals

$$R = 8 \times 10^{-6} \, (P_e A/T)^{1/2} \, (t/B)^{1/4} \text{ light-years}$$

R is the distance over which the signal can be observed;

P_e is the effective radiated power of the transmitter, in the direction of the earth, expressed in watts;

A is the effective area of the receiving antenna, in square meters;

T is the excess receiver noise temperature of the receiver used, in degrees Kelvin;

t is the receiver averaging time, in seconds;

B is the accepted band-width of the signal (that is, the band-width of the receiver used) in cycles per second.

lator, because its final frequency is about 1,390 megacycles, and it therefore must remain constant to one part in a billion. This accuracy is achieved by means of a special quartz crystal oscillator, the crystal being kept at a very constant temperature in an oven within an oven. The output of this oscillator is multiplied in frequency to give the desired final frequency.

A marker-frequency generator is used to provide weak signals from the output of the very stable oscillator at many fixed frequencies. These signals are inserted into the receiver for determining the exact frequency on which the receiver is operating, allowing the detection of Doppler effect.

After the fourth intermediate-frequency amplification, two filters pick out a broad band of noise, called the *comparison band,* and a narrow one designated the *signal band.* The gain of these filters is adjusted so that when very broad-band noise enters them their total outputs are equal. When these outputs are passed into the differencing circuit, its output is zero. However, a narrow-band signal fills only some of the frequencies of the filter for the comparison band, but all of those in the signal-band filter. The output of the narrow-band filter is then greater than that of the broad-band one, and there is a net output from the differencing circuit. This use of the d.c.-comparison circuit makes the radiometer respond only to narrow-band signals. The radiometer is set up for signals of about 40-cycles-per-second band-width. In the actual receiver, the electronic filters have variable band-widths that may be quickly adjusted to desired values.

The filters placed before the synchronous detectors pass only the frequencies to which the detectors will respond, and reject other frequencies that might cause them to operate improperly.

We see that an output from the final synchronous detector will occur only when receiving a narrow-band signal from a direction in which one of the antenna beams is pointing—the desired interstellar signal. The integrator only averages the signal strength over a chosen interval. The other two synchronous detectors and integrators connected directly to the comparison-band and signal-band channels monitor the performance of the radiometer.

Barring serious technical difficulties, our radiometer will go into operation on the 85-foot antenna early this year. The first objects to be looked at are Tau Ceti and Epsilon Eridani, two solar-type stars about 11 light-years away. Unidentified radio sources whose celestial positions are close to those of nearby stars will also be studied at an early opportunity.

It appears probable that this project or a similar one will someday succeed in detecting an artificial signal. Needless to say, the scientific and philosophical implications of such a discovery will be extremely great.

25

Frank D. Drake
Project Ozma

The question of the existence of intelligent life elsewhere in space has long fascinated people, but, until recently, has been properly left to the science-fiction writers. This is simply because our technology has not been capable of detecting any reasonable manifestation that might be expected from other civilized communities in space. Lately, however, astrophysical knowledge of the universe and our technology have advanced to a stage where these questions can no longer be ignored by scientists. I would like to describe briefly the astronomical picture connected with the development of life, and to follow this with some anthropomorphic arguments which help us in our search for other abodes of life. Finally, there will be given a brief description of some instruments that one might apply to this problem.

The astronomical results of recent years have shown that stars are being formed continuously in our galaxy, and apparently this has been going on since the very birth of the Milky Way. The original galaxy consisted almost entirely of pure hydrogen, but, as the stars have been born, lived out their lives, and died, the interstellar medium has been enriched with heavier elements. The stars that were born later in the life of the galaxy and those being born today come from material containing heavy elements. Solid bodies similar to the planets may be formed with these later-generation stars. The sun is one of these younger stars, being some five billion years old, as contrasted with the over-all age of the galaxy of some twenty-five billion years. Virtually all the stars in the vicinity of the sun are young stars containing enriched material from which solid bodies can be formed, and of the many stars in the vicinity of the sun, the sun is probably not one of the oldest members. This tells us that in our vicinity there are many stars, including some older than our own, which were born from material having in it the constituents from which planets are made.

Will planets be made? The answer to this is one of the big changes that has occurred in our thinking. During the postwar years the theories of von Weizsäcker and Kuiper regarding the origin of the solar system have replaced the old once-in-a-galaxy theories, such as those involving stellar collisions. These theories have been very successful in explaining observational facts. Although perhaps not accurate in detail, at least in their broad general form they are very probably correct. They solve the problem which has always plagued theories of the origin of the solar system—namely, the conservation of angular momentum in a condensing gas cloud. These theories do this by dumping the angular momentum into a second body or bodies, which orbit around the primary star formed in the stellar formation process. The angular momentum of a condensing star might be transmitted to the second body or bodies by means of magnetohydrodynamic or viscous effects.

There is, in fact, a great deal of observational evidence for this. One piece, of course, is the existence of our solar system where we find ninety-eight percent of the angular momentum in the planets themselves. Of course, this is not a convincing argument because there is a great deal of observational selection taking place here! However, when we look out into space, we find that forty percent of the stars occur as binary systems, suggesting that the dominant way in which angular momentum is dumped is by transferring it into the orbital motion of a second star. It is also impressive that the mean separation of binary stars is about twenty astronomical units, which is about the mean distance to the major planets of our solar system. It would appear that the difference between our solar system and the binary stars may be the result of a minor event which caused, in our case, several bodies to form instead of a single secondary body. Other evidence includes our many observations of companions of very small mass orbiting around nearby stars. The smallest such companion we know of, at present, is the companion of the star 61 Cygni, which has a mass of about 0.01 the mass of the sun. This is about the limit to which we can detect stars of such small mass. Jupiter, by the way, has a mass of about 0.001 the

Reprinted from *Physics Today,* 14 (1961), 40. Copyright © 1961 American Institute of Physics. Reprinted with permission.

mass of the sun, and so there is almost complete evidence here for a continuous gradation in the size of second bodies. It is of further interest to note that all stars similar to the sun rotate very slowly. They have apparently been very successful in ridding themselves of angular momentum. Struve proposed long ago that this slow rotation is evidence of planetary systems accompanying these stars. All of the above would suggest that the 60 percent of the stars that are alone in space may carry planetary systems with them.

Given a planetary system, the question is whether life will arise on it. Assume now that we have a planet that is at a temperature appropriate to life. It is believed that the early terrestrial atmosphere was a reducing atmosphere consisting in the main of such components as ammonia, methane, hydrogen, and water, a situation very similar to that of the major planets at present. Some years ago there was a very important experiment performed by Stanley Miller, at the University of Chicago. Miller took an imitation of the early earth's atmosphere, placed it in a container, and passed through it electrical discharges to simulate lightning. After many hours of this procedure he analyzed the contents of this container and found, to everyone's amazement, that large organic molecules had been built by this process. Molecules such as acetic acid were found, and, in particular, amino acids were formed in the jar, including the first amino acid, glycine. The amino acids are the basic constituents of protein material which is itself the basic building block of life. Miller had indeed shown that the basic building blocks of living things could be formed spontaneously in the early terrestrial atmosphere. Later this experiment was performed by Calvin in a slightly different way using nuclear radiation. The same results were obtained.

In recent months, Calvin has performed another remarkable experiment. He has analyzed chemically the constituents of the Calloway County meteorite which fell in Kentucky some years ago. In this meteorite, using the very sensitive methods developed by the organic chemists, there were found fragments of nucleoproteins (such as ribonucleic acid), which are the genetic material serving as the blueprint of living things. We now have evidence that at least fragments of genetic material are arriving from space in meteorites, and that the basic constituents of protein may arise spontaneously in reducing atmospheres. We end up with an over-all picture in which organic materials, the amino acids, and later the proteins were slowly built up in the oceans during the early stages of the earth's existence. Eventually the oceans consisted of a rich organic soup from which life emerged. The process of natural selection then led, after some five billion years of chemical and biological evolution, to our present civilization.

Two especially helpful points emerge from this discussion. First, it takes something like five billion years to produce an intelligent form of life. This means that for intelligent life to exist, the radiation from the star about which a planet orbits must stay nearly constant in intensity for some five billion years. Our new knowledge of the processes of stellar evolution eliminates many stars as possible abodes of intelligent life, because these stars are known to change intensity in times of less than five billion years. Secondly, biochemistry tells us that the temperature on a life-bearing planet must be in a fairly restricted range. This range can be defined roughly as the range within which water is a liquid. At much higher temperatures than the boiling point of water, all biochemical reactions become destructive, and, at temperatures much lower than the freezing point, the biochemical reactions become too slow. We conclude that there is only a certain thick shell around each star in which planetary temperatures will be appropriate to life. Stars of intrinsic luminosity not much less than that of the sun have such shells, or ecospheres, which are so small that the probability of there being a planet in the shell is remote. The overall conclusion is that single stars of near-solar luminosity are the most likely abodes of intelligent life.

In terms of the statistics of intelligent communities, the most significant fact is that the development of technological prowess on earth has occupied a very short time. Something like one hundred years is all that is required for a civilization to go from no knowledge of communication by means of electromagnetic radiation to complete mastery of such techniques. This is long on the time scale of a man's life, but is very short on the cosmic time scale—in fact, is about 10^{-8} the age of the galaxy.

On the cosmic time scale, which is what counts, a planet passes from no technical prowess to complete technical mastery in an almost instantaneous discrete jump. As this jump is made, a civilization rises above the level of scientific knowledge at which it can begin to communicate with similar civilizations over interstellar distances. The earth has just passed this point. In view of the continuous formation of stars, there should be a continued emergence of technically proficient civilizations. Because of the short time of transition from one state to another, were we able to examine all the civilizations in the galaxy, we would find very many less advanced than our own, very many more advanced, and only a very few in the transition stage. We should well expect most intelligent civilizations in space to be more advanced than our own. As Bracewell has stressed, the total number of civilizations using electromagnetic radiation depends not only on the number which come into being each year but also on the longevity of intelligent civilizations in a state where technology is practiced. It is possible that the mean longevity is quite limited. The use of technology on a planet may be terminated by such events as self-destruction by a civilization, a cosmic accident, or a change in philosophy which makes the use of technology passé. The total number of radio-using civilizations in a galaxy is obviously given simply by the product of the number of civilizations emerging each year and the mean longevity of civilizations in a state where electromagnetic communications are employed. If this longevity is small, the incidence of intelligent civilizations with which we may communicate will be very small. At present we must admit that the most likely early limitation on our own longevity is self-destruction. We conclude that the number of approachable civilizations in space depends not only on how many planets exist, but on another very important question, which is, "Is there intelligent life on earth?"

Let us be optimistic and assume there is, in which case there are many intelligent communities in space. We now ask how we may detect them. The manifestation of such communities for which we may best search now are electromagnetic radiations where information is being transmitted over long distances. This implies narrow-band transmissions. At present, our only narrow-band detectors are in the radio range, and we are therefore restricted to this portion of the electromagnetic spectrum. Knowing that, we must ask ourselves one other question: "Is our radio technology capable of detecting reasonable transmissions?" If we insert the technical parameters of our present-day radio receivers and antennas in the appropriate equations, we find we are capable of detecting reasonable transmissions—that is, those we ourselves radiate —over interstellar distances.

We have already seen that the best targets for our search are the nearest single solartype stars. It would further limit our search if we could logically deduce which frequencies might be best for our search. There are three hypotheses one might follow in this regard. The first, and simplest, says that we should search for radiation that other civilizations are using for their own purposes. Since, in most cases, technology will be virtually perfect, we may say that the frequencies used for long-distance transmissions will be those at which the natural impediments to transmission are at a minimum. The principal natural impediment to transmission is cosmic noise, which increases in intensity as we go to lower frequencies. A study of the intensity of cosmic noise suggests that frequencies above about one thousand megacycles per second will experience negligible interference from galactic noise. This doesn't help very much, of course, because it leaves a great deal of bandwidth. To improve this situation we may use an argument which has been presented by Morrison and Cocconi of Cornell. They proposed that other advanced civilizations might attempt to contact us in order to gather us into the group of intercommunicating civilizations that may exist. In such a case members of this group would attempt to signal us on a frequency at which an emerging civilization first uses high-sensitivity, narrow-bandwidth radio telescopes. With almost all civilizations,

this frequency will be that of the 21-cm line. Thus the second hypothesis suggests that frequencies near 1420.405 Mc/sec may be the best for our search. This, of course, agrees with the first hypothesis.

There is a third hypothesis, which is much the most economical one, and perhaps desirable because of this feature. This hypothesis suggests that other civilizations search the universe for radiation from abodes of life through passive observation with a very large radio telescope—in fact, a radio telescope so powerful that it may detect the very early transmissions of a civilization having just discovered radio. In this hypothesis only one large instrument is required, and discovery will be reliable and quick. Once a civilization is discovered it is expected that the large radio telescope would be used to send an introductory signal to the new-found civilization. The frequency of 1420 Mc/sec is still probably the most reasonable one at which to search. On this third hypothesis, we must be prepared to be disappointed in our search, for if the nearest such civilization is, say, 50 light years away, we must wait till the year 2030 before the replies to our early transmissions of the 1930's are returned.

From the above reasoning, 1420 Mc has been chosen as the frequency to be used in the initial stages of our attempt to detect extraterrestrial intelligent transmissions. . . . The receiver that has been built to meet the particular needs of this search uses two feed horns at the focus; the receiver switches between the horns, giving two beams, one of which is pointed on the target star, and the other in space along side the star. This allows us to distinguish true signals from the star from terrestrial interference which may creep in through side lobes. This switching also gives us the advantages of the Dicke radiometer. Since we use very narrow bandwidths, of the order of 100 cycles per second, we require very stable local oscillators in the superheterodyne receiver. In fact, our first local oscillator is phase locked to a highly stable quartz crystal oscillator. After the first conversion we use a standard high quality superheterodyne receiver, with four conversions, to bring the intermediate frequency to the very low frequency allowing us to filter out very narrow bandwidths. Two bands are used simultaneously, one wide, one narrow. Cosmic noise, which is broad band, will be present in both bands to the same extent, whereas a narrow-band signal will be in the narrow-band channel more strongly than in the wide-band channel. If we then difference the outputs from these two channels, cosmic noise will difference to zero whereas an intelligent signal will produce a net output. The device thus discriminates against cosmic noise. In order to achieve very high sensitivity, we employ a parametric amplifier at the front end of the receiver.

A search for intelligent transmissions has been conducted at Green Bank. We looked at two stars, Tau Ceti and Epsilon Eridani, which are the nearest solartype stars. After some 150 hours of observing, we obtained no evidence for strong signals from these stars.

I think you will see from the above discussion that both the scientific basis and the technological basis for this experiment are sufficiently sound. Nevertheless, our knowledge is not yet good enough to allow us to predict what the probabilities of success are. They are obviously quite low, and we cannot expect early success. We are faced with a sound, highly important experiment that requires the use of very expensive equipment for a very long period of time. Those who feel that the goal justifies the great amount of effort required will continue to carry on this research, sustained by the possibility that sometime in the future, perhaps a hundred years from now, or perhaps next week, the search will be successful.

26

Philip Morrison
Interstellar Communication

It is a great honor indeed to address the 1496th meeting of any organization, and particularly one founded by the man who was the second of the famous American physicists, after Benjamin Franklin, Joseph Henry.

I should like to put very clearly the thesis to which I wish to speak and concerning which I hope to present a very plausible case. I propose to assert that near some star rather like our sun there now exists a civilization with scientific interests and with technical possibilities much greater than those now available to us. Moreover, to the beings of such a society, our sun must appear as a likely site for a similar civilization. It is probable that for a long time they have awaited specific development of science near the sun and I believe that they look forward patiently to signals from our solar system which would make known to them a new society ready to enter the community of intelligence.

I would like to ask what sort of communication channel is open? What are the circumstances in those remote spaces, into which now, for the first time with some understanding, we begin to probe? Can we expect the extraordinary encounter with another civilization to which it seems to me we must inevitably come?

The idea is not a new one. It has had a lengthy history, and a kind of efflorescence in fiction in the recent past. There is also some scientific literature—about a half dozen papers in the last year or two. But the first references—and I hope to quote one of the most eloquent at the close of my remarks—go back at least to the thirteenth century, and indeed this view prevailed in those great times, at the turn of the seventeenth century, when that science which we today serve was in fact born.

Tonight, I would like first to speak to the question of whether or not we could recognize living things of a very different form from our own. I shall set up, if you like, a criterion of conservatism. Suppose we make a graph (Figure 26.1), plotting on the horizontal axis some measure of

Reprinted from the *Bulletin of the Philosophical Society of Washington*, 16 (1962), 58.

complexity. On the vertical axis we present the population of some region as a function of the complexity of systems in the population. In general, if we look at a sterile sample, grabbed at random from an imaginary world without life, or from a part of our own world without life, taking for instance a sample of sea, beach, or sky, we would find a curve similar to curve I. In a random distribution of atomic structures of all kinds we must expect that complex forms will very rarely exist. Whether indeed there is ever enough time to permit the chance existence of even one genuinely complex form such as we find in a fern or a tree or a man, I very much doubt. But, in principle, such a curve should contain an entry for any complex thing, *but* with a very small probability for extremely complex systems. This is the curve for a sterile world where no life has ever been seen.

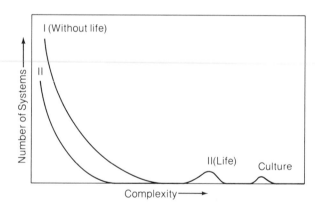

Figure 26.1.

The coming of life, I think, could be characterized quite objectively, without reference to its chemical nature or the place in which it is found, by modifying this curve in a very characteristic way. We will cut off the tail of the curve, replacing it by a rather rapid decline toward zero; but we will replace the missing tail by a pip (curve II). At first the pip is small and, as time proceeds, gradually increases, representing the accumulation of living forms, who derive their sustenance by operating on the environment. They prevent the establishment by chance processes of the intermediately complex products which, before the existence of life, would have filled in the part of the curve between the pip and the main section.

I suspect that such a two-piece curve would adequately characterize the existence of living forms. I would, however, go further. If we imagine now that living forms evolve, as they have evolved on our own planet, to a certain capability of manipulating the free energy of the environment beyond that which with their own tentacles, they can touch, then we will find the growth of still another pip. This part of the curve represent still more complex systems, and if our experience is any guide, it will grow still more, reflecting the cultural product of this second step. This is a third set of complexities. Here we include, for example, the books of the libraries of the world or, for that matter, the automobiles or the woven textiles of the architects' constructions. The presence of such breaks in the series of complexities, is, I think, a characteristic feature which we have to require for the existence of living forms, and then also for the existence of what we call cultural activities.

As you see, I have made no reference whatever to whether the systems are liquid, solid, or gaseous, whether they are magnetic or made of fluorine or carbon. We are dealing in matters in which we are at the moment simply too ignorant to be able to characterize them with even slight reliability. We are lucky if we can simply formulate a general program of such research.

I should like now to make a much more modest yet less general proposal. I should like simply to say that if to the process which unites familiar atoms into complexes, there is available an adequate stream of free energy, then we must expect to obtain curves similar to curve II of Figure 26.1.

There *are* more speculative thinkers who are not yet writing in serious journals but in, I assure you, quite serious works of fiction! Mr. Fred Hoyle, for instance, has described to us in his extremely suggestive and imaginative work, *The Black Cloud,* the emergence of a form of complexity built out of plasma, i.e., of magnetic fields, hot gas, and dust. I only mention this to show how far one can carry such sepculations.

No more imaginative discussion! Everything from now on is done with analogies of the most timid kind, avoiding any serious extrapolations. I shall look for precisely the material basis which we ourselves see on the earth, forming what I have called a scientifically capable society. I shall look first for the existence of the abundant light elements: hydrogen, carbon, nitrogen, oxygen, magnesium, phosphorus, sodium, potassium, chlorine, calcium, iron. Hardly anything else is needed. We know that in the process of nucleogenesis their incidence is high.

I would like to go so far as to assume that the chemical reactions which will lie at the base of the structures, unlike the plasma beings, are probably based on water in the liquid phase or, at the extreme of an extrapolation, to some other mode of utilization of the flexible and convenient hydrogen bond.

This places severe limitations upon the physical environment in which such forms could evolve; namely, the limitations through which we ourselves have evolved. If the temperatures are very much below zero degrees centigrade the process involving hydrogen-bond formation in solid water solutions are much too slow to give a very wide split oi the original curve in the times available. So, with too low temperatures, while something may go on, it may be very slow indeed. With too high temperatures, one cannot make any such structures very easily: the structures involving the rather weak but flexible and high manipulatable hydrogen bonds are next to impossible. Therefore we have a severe requirement on temperature.

Besides thermodynamic temperature, we have to worry about the presence of currents of free energy of such concentrated kind that they can destroy complex structures, even though they may not represent a large contribution to the over-all thermal content of the environment. Here, of course, I refer to fluxes of high-energy particles, ionizing radiation, and the like: quanta of energies large compared to energies of chemical bonds. Such bombardment must be severely limited compared to chemical formation rates or those structures cannot evolve. We know we must have something like the atmosphere of the earth for this protection. Open exposure to the indiscriminate currents and fluxes of space will prevent elaboration of molecularly based complexities of the sort I describe.

Let us consider the great nebula in Andromeda in the north sky. It is a circular mass of unresolved stars some 120,000 light-years across and appears to us as a tilted disk. We know that we live in a galaxy almost the twin of this one. Such a galaxy contains on the order of one hundred billions of stars, appearing to be merged together only because they are so far away that the photographic plate and the optical train through which the plate has been exposed cannot separate the images.

If someone looks at our galaxy from the outside, he sees such a relatively bland, slightly spiral, lightly marked disk of whirling stars, rotating once every three hundred million years. Not far from the center of this light patch there is the sun, an inconspicuous member of a population typical of the entire galaxy. Around the sun there whirl, among others, two planets which may be suitable for life; namely, the earth and Mars. On the earth is a clear and present instance of the development of the society of modest scientific capability in the very recent past; and on Mars, a strong indication that conditions favor the independent evolution of life, not indeed to any stage of technical capability but (very plausibly) to a stage of considerable chemical elaboration.

It is upon this sample of two that I think the argument we make must rest. I would like to stress the probable existence of some form of life on Mars—some form of growth of complexity at the expense of the free energy of the environment—because otherwise we are left to the uncomfortable sample size of one, the unique example of earth. I learned in statistics that it is very hard to make a conclusion from a sample so small as one. Anything can happen once, but about things that happen twice—one can at least say each is not unique. This is the main reason I would like to see growing in the Smithsonian Institution a little sample of that Martian vegetation which Dr. W. Sinton has made so plausible from his magnificent work with infrared spectroscopy.

Have we any certain knowledge of what goes on outside the solar system? I do not think so. But we have some conjectures and some plausible inferences which I should like to list briefly in order to establish the rest of my argument. We wish to look for those environments, resembling to a considerable degree the terrestrial environment and the Martian environment, which are hospitable enough to allow the elaboration of chemical complexity in the form of living beings. We are going to look therefore for terrestrial planets, with atmospheres, with free energy supplied by the sun, and with a temperature regime like our own.

In spite of considerable effort we have not yet found a really sound basis for the origin of the planetary system. We could not, for example, calculate clearly the distribution and mass and position *a priori* from the knowledge of the type of star we have. But we do have, besides some rather plausible inferences of this kind, some observations which are not difficult to describe.

Here again a graph (Figure 26.2), on which is marked the familiar letters which indicate to the astronomer, roughly speaking, a scale of temperature from about 50,000°K on the left to about 2500°K on the right, with our own sun being at the letter G. Next I would like to plot the measure of rotation of typical stars of these various classes. There is a great deal of rotation among the stars of classes of O, B, A, but we do not see much rotation in the F, G, K stars.

If we look at the distribution of the angular momentum in our present solar system we find that 99.5 per cent of the angular momentum lies, not in the sun, which has 99.9 per cent of all the mass, but in the planets which go about the sun.

Thus the angular momentum, which was presumably present in the whirling gases from which the sun condensed billions of years ago, now resides not in the massive sun but in the tiny planets. It is a plausible inference that the reason why the angular momentum of these young stars is still in the stars themselves, as we can see by the spectroscope, is that they have not made planets. Simi-

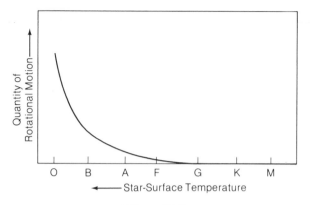

Figure 26.2.

larly the angular momentum which the older stars must certainly have had, if they were formed by the same processes that made the younger ones, has been whirled off perhaps in many forms; it is not unreasonable to say that some of these stars have bequeathed their spin to planets.

Here the argument is of course uncertain. We cannot be sure that the conditions for condensation of planets were right. We can be fairly sure that the angular momentum went off with much gas. But since the angular momentum is still resident in the planetary system of our sun, it is plausible that most of the stars of masses like the sun, might well be surrounded by a suitable cortege of planets, which bear a small part of the mass of the original cloud but an appreciable fraction of its original endowment of angular momentum. Near stars of F class and fainter we then expect to find an appropriate distribution of planets. Let us imagine some considerable fraction do have planets. (We will recall the uncertainty in a final factor eventually.)

There *must* be enough light; otherwise we would find too low a temperature for life. The planets which we impute to the stars must be in the right positions, receiving neither too much light, so that they are sun-baked like Mercury, or too little light, so that they are cold and sodden with mists of methane and hydrogen, like Jupiter. They must be somewhere analogous to the Mars-earth region.

In all this I follow the work of S.-S. Huang. He draws this inference: Since the light from the sun is received here diluted by the spread of the light flux in free space—if you like, by the inverse square factor—we can scale all stars and their planets to have the heat and light of earthly conditions, provided we scale the distance to the planet according the the luminosity of its star. The planets' distance being called R, R^2 will have to be proportional to the luminosity of the star.

Let us say we have a spread of possible distance in our system something like the spread between earth and Mars, or even between Mars and Venus. If we imagine that whirling planets have always formed disks, this allowable area is in turn proportional to R^2. (The area of an annulus is proportional to the square of its radius.) In such an area a planet near any star would find conditions tolerably close to our own.

Moreover, the calculations of William H. Guier and Robert W. Hart at the Johns Hopkins University Applied Physics Laboratory have demonstrated that the distribution of masses in any "solar" system ought to be rather flat in this region. It is reasonable not to make any further correction, but simply to say the area available for earthlike orbits is the only measure of how lucky a planet has to be in order to receive the right amount of light.

Now, we see that a very faint star may have planets feeling the same light intensity as we enjoy on earth, but only if they lie in a little disk hugging the star, to gain the benefit of the small warmth of their faint furnace. Our argument simply has shown that these habitable areas are proportional to the luminosity of the star. It might well be that there are such systems near faint stars, but they must be few, because the volume allowed in space for the statistical distribution of habitable planets is not large, since one must live so close to a faint star. Therefore, we can be pretty sure that the very numerous faint stars are not likely seats for planets endowed with the kind of warmth that we have here.

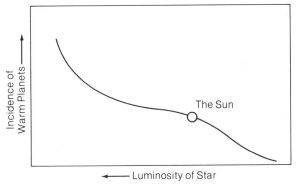

Figure 26.3.

One can calculate this nicely, using statistics on the distribution of stars in the galaxy, and we have done this, to find a curve rather like the one shown in Figure 26.3. The size of the zone of warmth, multiplied by the fraction of stars having that luminosity, gives the relative probability of finding a planet around a star, dependent on the luminosity of the central star. In the middle lies the sun, to the right are fainter and fainter stars, very numerous ones. To the left brighter stars, but conspicuous ones. On this consideration alone, each of these stars would be very likely to have planets because there is ample warmth; each has a big useful volume. But since they are few in number the total contribution cannot be large, and the brightest

ones of all still spin. They have no planets.

The faint stars are not very important then; but the stars at the center of the range are not so important either because the brighter stars, even though they are not very numerous, are so much brighter that they make up for their scarceness by their favorable chance to have comfortable planets. So, if we have no other criterion, we would say the most likely thing is that such planets will be found in that very wide zone of tolerance near the very bright furnace of the high-temperature stars.

But we have one more indispensable requirement—time. No doubt the spontaneous process of energy degradation and the transfers of free energy require a large amount of time—if you will, geologic amounts of time. Indeed, that is the story of paleontology. We have to allow billions of years for the elaboration of those many forms which necessarily precede the kind of complex beings we are looking for.

This time is not available in the case of the bright stars, because they burn themselves out, and move off the sequence, perhaps to go through all sorts of catastrophic changes. Only the conservative, smoothly flowing sources of radiation, the weak stars, will work. Suppose we want the star to remain without an appreciable change in temperature, during a time of three or four billions of years, like the time since the first signs of life appeared in our solar system. Then we must multiply this curve by the fraction of that time which the life of the star represents.

The very brightest stars do not live anything close to that time. Therefore, they contribute nothing. By the time we multiply this curve by an appropriate time factor we have produced the effect shown in Figure 26.4. This is the

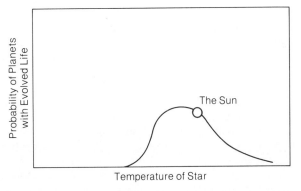

Figure 26.4.

probability, taking into account both light and time, among only those stars with planets. Recall that the brightest stars do not have planets because they too still rotate.

The sun lies tolerably close to the maximum of the final curve. About 90 per cent of all the stars that are plausible homes of life in our hypotheses are contained in the small range of surface temperature which astronomers would call late F to K classes. These stars vary from our sun's temperature by perhaps 10 per cent one way or the other.

We should probably exclude multiple stars (an argument also due to Huang), although multiple stars may well have planets. They illuminate their planets so differentially that instead of mere seasons, the planets undergo extremely complex rhythms of heating, unlike our rather smooth evolutionary history. A skeptic might well say this would be a kind of stimulation and challenge to early life. But consistent with our determination to extrapolate hardly at all, we shall exclude all multiple stars as being possible sites of life like our own. Maybe they are sites for other forms, more suitable to a climate which may change enormously in a few million years and then change back a few million years later, but we shall not discuss them further.

We can say that around the simple dwarfs of the main sequence, from what are called dG0 to dK2, we subsume perhaps 90 percent of all the possibilities for having planets with atmospheres and temperatures like the terrestrial planets of the solar system.

Now, the number even of these special stars is not small. The number of such eligible stars is a few hundred million in our galaxy alone. If we exclude the multiple stars, we cut this by a factor of three or four, to above a hundred million instead of several hundred million. If we allow that the galaxy is several times older than our sun and allow for the fact that some stars have played out within that time, we still come to many times 10^7, or say roughly one hundred million. Where are these stars located? They are stars of the disk population, found not far from the central galactic plane, anywhere from near the center quite far out to the rim.

One can compute that eligible stars are sitting 50 to 80 light-years apart throughout the whole bulk of the galaxy. A remote astronomer observing our galaxy sees a bright mass like the Andromeda nebula, with fifty to a hundred million star-spots at which he might plausibly argue that living forms occur. We now know he would be right about exactly one of those spots; namely, our own planet.

It seems to me an irresistible inference to say he may be right about many of those spots. There is no central feature; there is no great arrow in the heavens to mark where we live. We are but one mote in this enormous Keplerian ring that runs around the galaxy, democratically indistinguishable from our fifty million dG0 to dK2 counterparts.

Now we must ask: What is the history of life as we see it here? Can we not expect this to have some kind of counterpart in these other possible, still unknown seats?

I plot in Figure 26.5 time as abscissa and, as ordinate, the number, the population, the amount, or some other measure of quantity for a number of different interesting phenomena which we know to have gone on in time on the

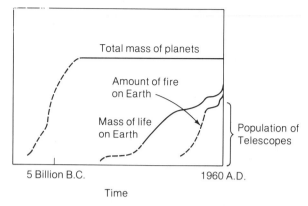

(The curves are on differing scales.)

Figure 26.5.

surface of a planet near our sun. Here we must be rather flexible. I will mention the significant features of the plots (which are not to be interpreted literally!).

At the right-hand end of Figure 26.5 is 1960. At the left is five billion B.C. I ask, for example, what would the expert observer, who knows everything, say about the plot of the planetary mass, the total mass congealed into good working planets around our sun? We know it has not changed very much for the last 4.5 to 4.6 billion years and that earlier there was a time when there were no planets at all, but only a kind of gas. So the planets' mass had to rise from zero up to its final value along a curve something like the top one in the figure.

Next, I plot on the same axes the total mass, not of planets, but of living forms in the solar system, i.e., the mass of life upon earth (Figure 26.5, middle curve). Life itself begins much later than the planets. Things rapidly grew, and built up smoothly until the time when the first land forms began after life had filled the seas for a thousand million years. Here occurs a little bump in the curve. Then came the flowering of land life, which represents an increase in total life, but not a great one. Even to this day most life probably exists not on the land but still in the sea, the first home of life. Again experts may disagree whether the lands hold 20 per cent or 80 per cent of life but, broadly speaking, that does not change the look of the curve.

As we come close to the present I would like to add one more small increase, not easy to calculate. In the last moments of geologic time, a little spike protrudes from the curve which represents the replacement of the forest by crop lands and the irrigation of new lands. It is the first effect of culture, the third rise in my original curve. The middle curve of Figure 26.5 then reports the total mass of living beings as a function of time.

I have also tried to estimate such a curve for flame. It is easy in principle for an observer to measure flame; he can distinguish the flame of a fire from most of the gases and glowing liquids of a volcano. When I thought this through, I was surprised to realize that long before there were men there was fire, burning in the grass lands and in the forests where lightning ignited it, over much of geologic time.

I have made a rough, but reasonable extrapolation, based on present experience in remote countries, which would lead me to believe that the plot of fire would look something like the bottom curve of Figure 26.5. It begins when the land forms begin. When there was no life on land there was not much fire. Fire grew nicely as the forests and grass became well developed; then it grew very much indeed when first men came on the scene. Thereafter it did not grow much until rather recently, when agriculture and then cities were invented. In the last couple of hundred years, when industry was developed, it went up again, though yet not very far. Of course this last time interval cannot be shown in the figure; there is actually a spike at the end.

Finally, I will show one more curve which is really the key to what I am driving at. It is a very easy curve to draw, the steepest possible. That is the curve for the population of telescopes in the solar system. Up to 350 years ago, there were absolutely none; and then, whatever number there are, now appear effectively all at once. You cannot fairly represent the time since Galileo by the thickness of a fine line; the curve is an absolute step function.

I take the population of telescopes to be a very good representation of the beginnings of a technically competent cultural inventory. Therefore, no matter what we think is the distribution of the histories of cultures in these other parts of the universe, since the rise time of science is so small compared to the spread in their starting times, to spread in the rates of evolution, and to every other cause of spread that we can imagine, the starting points of culture are distributed more or less uniformly over a time very large compared to the difference between Galileo or even the Chaldeans and 1960. That means if one simply assumes the cultures of these stars we talk about did not have *exactly* the same starting time and *exactly* the same evolutionary rate as our own—that is, unless the synchronization is exact to a wholly unreasonable degree—because of this very short rise, we can be sure, that if civilizations exist, then about half of them are far older culturally than our own. But what is the probability that these older cultures also have an appreciable longevity? They might, of course, give up science or even die out.

Here we come to points still harder to calculate than the very difficult problems of evolution and planetary formation with which I began. We are trembling on the edges of speculation which our science is inadequate to handle. Our experience, our history, is not yet rich enough to allow sound generalization.

I beg those who are historically minded and socially trained to consider whether any general remarks may be made to form some guide for use in this problem. How likely is it that populations of men, or manlike things, would evolve along that curious path which leads to the swift succession of those steeply rising functions which I think are characteristic of the artifacts of man? I do not know. We will say, then, that a certain number, say $\epsilon \times 10^7$ stars in the galaxy contain living forms superior in culture to us. Of those some may continue to exist, some may still be scientifically interested, some may have remarkable scientific ability.

The factor ϵ conceals the following probabilities which we know nothing about: (1) The probability that under the same conditions something like "men" will rise from other living forms; (2) the probability that those societies will remain interested; (3) that they maintain technical capabilities of an increasing sort; and (4) that those societies have a longevity great compared to the span of human history, if not comparable with the span of geologic time.

Let each person put in his own guess for ϵ. Those who are very pessimistic will say $\epsilon = 0$. I think if we approach the problem with the usual hopeful hypotheses of scientific investigation, we will say: "no reason to put it zero." I do not know what ϵ is, but we ought to try some schemes of measurement to find out. Therefore, we argue that near a number which may be somewhere up to 200 million stars, at most, certainly not much more, and perhaps as small as one, certainly not less, somewhere near this number of stars in our galaxy there are astronomers, telescopes, and the rest, and most of them understand much better than we do stellar evolution, planetary formation, radio propagation, etc. They do so not for any reason intrinsic to the mental forms with which they may describe these things, but for reasons intrinsic for the survival of these organisms in a bath of sunlight, protected by an atmosphere. These reasons force them, step by step, if they are to investigate their environment, to carry through the same sort of measurements and to obtain the same sort of information about the spaces between the stars and about the stars themselves as we have, only very likely much more.

That, then, is the situation in which we place ourselves when we look at the problem: Do these beings communicate, and how will they choose to send their communications?

First, what kind of communications would such advanced societies be likely to undertake? Would they go traveling? I submit that the motives for travel, even in a less advanced culture like our own, are becoming fewer and fewer from the point of view of the explorers of old time. Explorers seeking sources of raw material, like migrations of people seeking new crop lands, have relatively less importance each decade.

The major explorations of today, even the major travel of today, is for gathering information, even here on the surface of the earth. On a much larger scale, if one must dispatch a rocket ship to the Pleiades to bring back a carload of plutonium iodide, it simply is not worth it. If you are in a position to do something like that, you are in a much better position to make your plutonium, or to do without it. To dream of bringing back that cargo is to put the thinking of the merchant adventurers of the sixteenth century into the framework of a technical capability enormously greater than that of our own day.

There is only one real motive for travel (aside from ceremonial or symbolic travel, involved say in Mr. Khrushchev's visit to the United States) and that is to gain information. But to gain information, it is not necessary to travel; it is necessary only to *signal*. And the signal has one great advantage over every possible means of travel; information can be transmitted, as no travel can be carried out, at the speed of light itself.

The maximum rate of information gain will be obtained from a system which transmits, not things or people, but signals at the speed of light. Therefore, I think that, perhaps after a few temporary explorations in the near neighborhood, these $\epsilon = 10^7$ stars around whom these superior fellows are now living, have long been in intercommunication, over splendid channels of high complexity, using light-velocity signals, carried by fields of some sort, probably electromagnetic (although for all I know they may use neutrinos). The question is: Are they interested in doing anything besides that?

There is not much more sicence left to do at their level, if one studies, say, stellar evolution. What is still interesting is clearly the experience of our fellows, because we know that the most complex and the most unpredictable of these forms of complexity are the things that we plotted at the far right of Figure 26.1, in the cultural area. What are the novels? What are the art histories? What are the anthropological problems of those distant stars? That is the kind of material that these remote philosophers have been chewing over for a long time. Do they want to know about the earth? I would say if there are many of these stars, if ϵ is a largish number, comparable to 10^{-4} or 10^{-3}, then they do not specially want to know, because they have already seen many new societies emerge. But there may be, however, a little corner of interest still retained, and there are many societies who might be seeking.

If I may risk a somewhat frivolous statement, I will say that our earth is not the concern of the great enterprises of knowledge among those far societies, or even of their great enterprises of art; rather, it is the activity of a Department of Anthropology. They may well maintain a certain small subsidiary interest in looking around for new entrants into their great community.

Of course, if there are very few of them, if ϵ is a number comparable to 10^{-6} or 10^{-7}, then they will be strongly interested in finding us; but they will likely live very far away; the means of contact will be difficult, and even very

advanced civilizations will have a hard time making many round trips across the galaxy in search of this curious planet.

They would therefore try electromagnetic signaling as the simplest means to call the attention of even the most primitive fellows to what is going on. We need not look for sophisticated means of signaling. If we wish to land on the Queensland coast near Port Darwin and communicate with the Australians (I do not mean the Australian astronomers from Sydney, but the aboriginals), we would hardly set up a TV station and broadcast a program. We would rather use some simple audible means, like a steam whistle and a drum. Then these people who are sure to have that kind of communication will come to see what it is we have to sell, give, or trade, or what news we have to spread. So it is with the civilizations of the universe.

Their Department of Anthropology will maintain primitive signaling devices meant to catch those people who cannot do very much better. The anthropologists will feel that it would be nice to see how the primitives could enter their interesting society. Here the points of view diverge. I will mention the opinions of three different authors on this subject.

First, Professor R. Bracewell, who has what appears to me a rather tendentious scheme, not so good as one that will come later. He asks: How would I go about this? I would dispatch automatic probe ships to every plausible solar system in the neighborhood, to idle about each solar system like satellites. They would listen; when they heard radio signals or TV debates from their near neighbors, these satellites would mount up a big antenna and report home. He says, moreover, that such satellites would try to encourage communication directed at themselves by echoing what they heard; e.g., if they heard dot-dash they would echo back dot-dash; if they heard a commercial, they would echo back a commercial.

This, it seems to me, is a frightening degree of pessimism! I do not think that any drone in orbit would simply echo. Mr. Bracewell *does* point to the fact that echoes of mysterious origin are well known. They were unmistakably heard 25 or 30 years ago. Nobody knows their origin for sure. He thinks perhaps these were drone-orbiters echoing back to show they heard us. I would not, myself, build such unintelligent orbiters. If I heard a signal I would send back, not an echo, but something unmistakably meant to attract attention.

I agree his is a *possible* scheme. It does not depend on the abilities of the local people to do anything very good. If they can reach a neighboring drone by accident, that would be enough to tip off the news of their existence. However, it is very expensive to maintain the drones; to maintain them in space is perhaps relatively easy, but to maintain them in time, against the erosion of space, is very difficult, since they must sit in orbit for millions of years before they have the expectation of hitting the evolu-

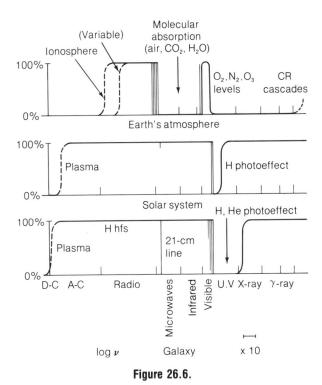

log ν Galaxy   x 10

Figure 26.6.

tion of science.

I think it is better to mount signal beams at home, beams of a simple and unmistakable kind, directed preferentially toward those points where we think listeners may sooner or later arise. The beams should be of a kind best suited to attract attention, and to carry the information over the distance of galactic space.

Figure 26.6 is a demonstration of the kind of transmissions which might succeed in space. Here you see the transmission in per cent plotted all the way from very slow frequencies, like turning on and off light switches, up to gamma rays. We notice the two famous windows through the earth's atmosphere explored by terrestrial astronomers. Here we have, at the bottom of Figure 26.6, the absorption of interstellar space. There are two very wide windows, but the ultraviolet and soft X rays are cut off by the atmospheres of planets, which will also cut off radiation in the millimeter and the decameter ranges.

If we look at these plots, then, it seems likely that we will want to use one of these windows which we ourselves find, either in the far gamma-ray region or in the radio region, or possibly in the visible. Certainly no one will use the UV and soft X-ray region. The designer of such equipment will choose some optimum. On which of these frequencies is the random noise of space most serious? Since he knows the conditions in space, he will choose rationally. If we too understand his rationale, we can predict his design.

There are two important kinds of noise in space. We

know the visual Milky Way and the radio Milky Way. We are looking at a sunlike star, because the planets of life are in orbit around such stars. It turns out that the noise from the star itself both in the visible and the gamma-ray regions is high, and the much more plausible channel appears to be in the radio region. Now in the radio region, at very low frequencies, the sky is very bright; at very high frequencies, the sky becomes dark, but the stars become bright. Therefore, there is an optimum for simple receivers, which are not capable of resolving star from sky, namely, the broad intermediate radiofrequency region, somewhere near a few thousand megacycles (Figure 26.7).

If we had to look at random for such signals, we would be searching indefinitely. But there is right here a unique frequency, as everyone knows, the one major spectral line in the radio band, the 21-cm emissions from neutral hydrogen atoms. This line, at 1420 Mc, or 21 cm, is a frequency for which there must be sensitive receivers in use on the part of anyone who would understand the nature of galactic space and matter in it. I suggest that for a signaling distance between ten and a few hundred light-years, this channel remains indispensable.

If you want to communicate to someone who does not know you are sending, you usually choose a frequency near the frequency he is already prepared to listen to. If I want people to listen to my illegal radio station, I always choose a frequency close to the frequency of the official broadcasting station, because I know listeners at that frequency can become listeners of a frequency close to it. They will hear me call from the Sierra Maestra. That is exactly what these remote people would do (though I would welcome detailed historical or social study of how one makes signals known to persons who do not expect to find them).

Therefore, set at 1420 Mc. Look as Dr. Frank Drake very courageously did last year, with his rather small

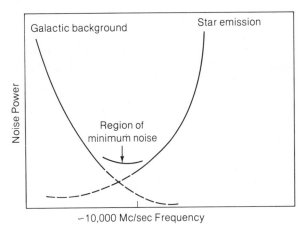

Figure 26.7.

mirror but good receivers, at a few nearby stars, to see if he was lucky enough to pick up the ethnological beams from τ Ceti or ε Eridani. He was not successful, but one cannot expect to be so lucky the first time. We cannot expect to have neighbors as close as 20 light-years. Maybe we will have to go to a thousand years if we are to find any at all. This is the Ozma project. I would very much like to lend my support to this investigation. I do not think it in the least foolish. I think it is worthwhile. I feel that there is no more philosophical or practical conclusion to be derived from astronomy than the conclusion that such signals would immediately bring.

I should like now to devote only a few paragraphs to suggest how the code would be sent. Writers often say it is indispensable for communication that the partners have something in common. Communication is not possible between completely isolated systems. This is indeed true but it is only a tautology. Communication is possible only when there is something in common but there is always one thing in common whenever there is communication; namely, the signal.

A signal is by definition some common physical properties of the transmitter and the receiver. Therefore, by denoting in the signal itself we can make communication; we can invent a language, so to speak, by pointing. What we point at is not some other object. We point at and with the signal itself.

How would I point? I am now going to present a little experiment. I am going to pretend that I am communicating to others without the use of language. I want to formulate it very simply, in a few moments, but realize that in fact this project could employ cryptographic computing machines, and many clever people; then, even a much more difficult problem could be solved in no time at all. The decoding probably would be relatively trivial. I can go a long way in three minutes. I will show that I can send signals which would elicit guaranteed response. But I cannot in this restricted space do anything very abstract. I must be allowed a few ground rules. Since the receiver would in reality be getting pulses at the rate of ten thousand or more a second, he would have much more information than I can illustrate. Therefore, I will use symbolic boxes, which I will call A, B, C, and so on (Figure 26.8). These lettered boxes stand for many pulses in a certain repeated pattern. Any pattern will do; if the receiver hears, say 200 pulses, then the same series comes again, I will call the whole series of 200 pulses pattern "A" (Figure 26.8a). I will use other distinct patterns "B", "C", and so on.

From these we will infer what the statements mean and what the meanings of the patterns are. Figure 26.8 represents the voltage on the output of the receiver. In Figure 26.8(1) is the first continuous sequence. Imagine the sequence occurs a few thousand times in a few seconds.

The next sequence (Figure 26.8(2)) too goes on a few

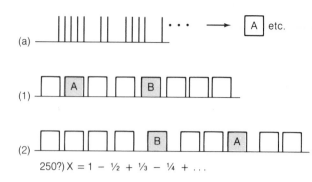

250?) X = 1 − ½ + ⅓ − ¼ + . . .

Figure 26.8.

hundred times or a few thousand times. Do I have any takers for what I would mean by A and B? It is, of course, clear that "A" is *plus,* "B" is identical with *equals.* I can skip the rest. I could have gone through all the algebraic symbols in the same way. When I have zero, minus, equals, I have no trouble signaling multiplication and division.

Now we receive another pattern block X and the pattern block for *equals,* followed by a series of pattern blocks and numerics. We evaluate this series and we find the series says 3.14159265358979323846. If X equals that series, what then is "X"? We have a name for it. X equals π. Other series are then transmitted to us and each of them defines the number π. These people have shouted at us for many seconds, "pi, pi, pi, pi," using infinite expansions.

Then we get the following signal. A narrow pulse, a long time with nothing, and another high narrow pulse. The next signal, the high narrow pulse, but the same time elapsing before another high narrow pulse. Again a high narrow pulse, and then a little pulse in the middle.

Then a high narrow pulse, and then two spaced pulses in the middle, like Figure 26.9, and then more and more of such spaced pulses.

Now, this is a curious thing. I will give a slight hint. In the first pictures, we saw pulses whose numbers changed, but their spacings showed no interesting features. Now, these pulses are distinct. There is a constant numerical pattern. Here is one pulse, then two, two, two, two, hundreds of these pairs of pulses, but spaced more and more.

Of course I get the cryptographers to work. They begin to do all kinds of operations on these spacings. Maybe if the group is clever, after a while, some physicist plots the way these spaces vary with the number of these high pulses. Soon they start coming together again as in Figure 26.9. Then the signal shouts "pi" at us again, and it is all algebraic forms, pi equals, pi equals, and the whole rigmarole starts again. What then is going on? As you have guessed, it is a circle. We now know the TV code.

Of course, they may not scan linearly. Maybe they scan in logarithmic spiral. It makes no difference to the

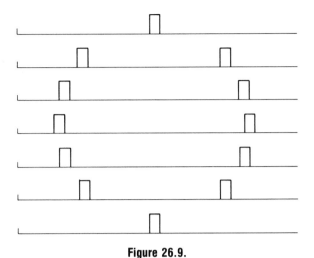

Figure 26.9.

method. As long as they supply us with a simple geometric patterns and some algebraic clue to it, we cannot take very long to make out the nature of their scanning raster. Once the television pictures enter, I retire from the field in favor of linguists, language teachers, and elementary-school teachers.

Can pictures alone convey adequate information? Teaching a language seems not at all hopeless. Guessing what will be in these pulses is not very profitable. We had better look for them, and not merely guess. (H. Freudenthal of Leiden has done the whole job even without pictures, by symbolic logic alone—in my view, too hard.) There is, at least, no intrinsic difficulty in communicating, to the level of being able to send decodable two-dimensional scans (even three-dimensional if we like) which a little bit of algebraic ingenuity and geometric intuition can lead anybody to decode.

We will have, say, two weeks of pulses to work on; it is not a hard problem. Then we would see displayed before us animated films, so to speak, of whatever it was they wanted to teach us. I do not think it would be very long before we would have a rudimentary language. This is the first communication, which would last a few years. We would be doing at least pretty good secondary-school work before we could hope to have acknowledged the first signal.

Someone said to me, "This would tend to divide scientific questions in two kinds: Those that can be answered on earth within twice the transit time, and those that should be put on this channel—very much in demand—and sent off to Them to get the answers."

I have spoken too long, but I think that one can demonstrate the probable existence of these beings, in what number I do not know. Their communications are most plausible; and it is even likely they will try to communicate with us. We should listen only, and not yet try to do anything more ambitious. It is worthwhile at least mentioning the problem of communicating to the Andromeda galaxy, that neighbor collection of a hundred billion stars. It is very hard for me to believe that nowhere in its great disk, containing a hundred billion stars, 250 million of them likely to have planets like terrestrial ones, is there any scientifically competent civilization. I can think of only one or two ways to signal them, which sound far beyond the capacity of men. Maybe one of the stars can be modulated by interposing an opaque screen. It would have to weigh about 10^{20} grams (the mass of a comet), distributed in micron-size particles over a five-degree zone of a sphere surrounding the star, and moving in an orbit like the orbit of a planet.

If this could be modulated every six months or so, taken away and put back again, or changed to affect the interstellar intensity, we could make it beam a series of algebraic equations at us. Perhaps in that remote galaxy, some patient signalers have for fifty million years tried to modulate a star. These ideas are real ones, and not meant wholly lightly. I should like to conclude with that impression. I think I am not producing science fiction, but legitimate speculation of demonstrable plausibility.

I should like to close with an early reference to these ways of thought. Teng Mu, a scholar of the Sung Dynasty in China, wrote this. I cannot close more fittingly than by reproducing the words now seven hundred years old, of a man who thought as we think, but who lacked the technical capability to verify in real life what his imagination was capable of foreseeing:

Empty space is like a kingdom, and earth and sky are no more than a single individual person in that kingdom.

Upon one tree are many fruits, and in one kingdom there are many people.

How unreasonable it would be to suppose that, besides the earth and the sky which we can see, there are no other skies and no other earths.

27

Alastair G. W. Cameron
Communicating with Intelligent Life on Other Worlds

Are there extraterrestrial civilizations scattered here and there throughout our galaxy? And will man eventually be able to converse with them? For centuries the possibility that intelligent beings live on other worlds has been a fascinating but elusive topic for speculation. In recent years, astronomical findings have put this unsolved problem into clearer perspective. Lately, engineering and electronics advances have made it possible to consider soberly the prospects of communication over interstellar distances.

In this article we shall first attempt to estimate the number of extraterrestrial civilizations in our galaxy. Then, we shall examine potential ways of communicating with them. The gaps in our knowledge are enormous; nevertheless, the shape of the problem can be explored. For definiteness, let us put the reasoning in quantitative form.

Symbolically, the number N of civilizations in our Milky Way system that are technically competent to communicate with us may be expressed by the product of seven numbers:

$$N = S \times P \times E \times B \times I \times F \times L.$$

The first factor, S, is the total number of stars in our galaxy that are of kinds suitable to be centers of planetary systems. It is multiplied by P, the fraction of these stars that have planets, and by E, the average number of planets, per planetary system, that are sufficiently earthlike to support life. The next factor, B, is the probability that any one such planet will develop biological species. Further, I is the number of these species attaining enough intelligence to be capable of interstellar communication. This must be multiplied by a psychological factor, F, which is the fraction of intelligent species that will be psychologically motivated to communicate. Lastly comes the time factor L—the lifetime of the communicating species, expressed as a fraction of the total

Reprinted by permission from *Sky and Telescope,* 26 (1963), 258.

time during which the planet has sustained life.

This is a long chain of probabilities. There is scientific evidence available that bears on some of them, but for others only guesses are possible now. Even so, it is worthwhile to trace the chain link by link.

First of all, how many stars in our galaxy have planets? This question is closely related to the still unsolved problem of the origin of the solar system. Up until about 1940, many astronomers believed that our system originated in a rare accident in which the sun passed very near another star; in this near collision a streamer of hot gas was pulled out of the sun, later condensing into planets. This kind of theory was dropped after several fatal objections were realized: for example, the streamer would expand explosively instead of condensing.

Since 1950, the more plausible theories of solar system origin invoke a disk of gas and dust that once surrounded the sun, and explain the planets as condensations in its material. Presumably this disk was formed at the same time as the sun itself. Current estimates give our galaxy an age of something like 15 billion years, and about 4.5 billion for the solar system. Conditions in the galaxy when our planetary system was formed may, therefore, have been not too different from those today. Hence astronomical information about star formation at the present time has a direct bearing on the origin of planetary systems.

Star formation is a continuing process in the Milky Way. Most of the stars in it are older than the sun, the oldest perhaps 15 billion years of age. The hot, blue supergiant stars in Orion are very young, with ages in the millions and tens of millions of years. As an interstellar cloud contracts to form a star, the conservation of angular momentum requires an increasingly rapid rotation. The main differences among current theories of star formation are their various ways of explaining how the accelerating rotation at least partially disrupts the condensing mass before it can shine as a star.

In F. Hoyle's picture, the condensing protosun became rotationally unstable when it was about as large as the present orbit of Mercury. Contracting further, it released a

Figure 27.1. The 300-foot (91-meter) telescope at Green Bank, West Virginia, is one of the largest steerable radio telescopes in the world. Similar telescopes might be used in a long-term project to search for radio waves emitted by extraterrestrial civilizations.

gaseous disk which, expanding, remained magnetically coupled to the still rapidly rotating sun. From this disk the planets condensed to retain most of the angular momentum of the solar system.

Another theory maintains that a condensation would not exchange angular momentum with its surroundings by magnetic coupling. In this view, the protosun began to shed a very massive nebula when it was larger than the present orbit of Pluto. But the condensation process is assumed to have been very inefficient, only one percent of the matter in the solar nebula going to make up the planets, and some escaping from the solar system and carrying away angular momentum.

Both these theories, which represent limiting cases, suggest strongly that the occurrence of a planetary system may be a normal stage in star formation—not a rare accident. This statement is true primarily for isolated stars; if twin protostars condense from the interstellar medium to form a binary system, stable planetary orbits are unlikely.

Limits may be imposed on the kinds of stars that can have planets harboring civilization. It is known that the massive O- and B-type supergiants are radiating energy at such enormous rates that they can have existed as stars for only a relatively short time. In fact, a star whose mass is 10 or 20 suns must burn all its hydrogen to helium in a scant three million years or so, if it continues to shine as a supergiant. Life on a planet belonging to such a star could have existed no longer than that. On our own earth, the emergence of human intelligence has come only after about two billion years of evolution. Consequently, it seems very unlikely that any O or B star's planets would be the abode of intelligent life. The argument can be extended to rule out all stars with masses of more than about 1¼ suns, if it is postulated that a planetary system must exist as long as our own before intelligent life appears.

On the other hand, dwarf stars having 0.1 solar mass or less radiate so little heat that they are unsuitable as homes of life. Thus S is roughly equal to the number of single stars in our galaxy with masses between 1¼ and 0.1—perhaps 4×10^{10}. The fraction P of these stars that have planets is equal to one (all of them), if the idea is correct that development of a planetary system is a normal stage in star formation.

Our next step is to estimate the average number of planets per system that are capable of supporting life. Here the range for conjecture is wide. One line of argument is to consider as suitable only planets whose blackbody temperatures permit water to exist in liquid form. On the assumption that the radial spacing of the planets in other systems resembles that in ours, a rough calculation gives the number of such planets per system as 1.4. But there is a further restriction, imposed by mass. Any planet as massive as Jupiter would presumably retain a hydrogen-rich atmosphere inhospitable to life, whereas a planet less massive than Mars might lose too much of its atmosphere. My personal guess is that considerations of mass will reduce to 0.3 the average number E of planets per system that are capable of supporting life.

What is the likelihood of biological forms appearing on any particular one of the suitable planets? In recent years some laboratory experiments have given fresh insight into the origin of life. If a mixture of methane, ammonia, and water vapor in a sealed tube is bombarded by X-rays or ultraviolet light, or subjected to an electric discharge, quite complex organic molecules are formed. Among these are amino acids. Many biologists accept the possibility that such processes taking place on a planet with oceans and a reducing atmosphere could result in life. We may go a step further to suppose that on any planet where conditions are suitable life will sooner or later appear. This amounts to putting B equal to one.

If life has started on a planet, how many species will eventually become technically competent to communicate with other worlds? On Earth there is only one such species, man, at the present time. We do not know how many other species with the necessary technological ability may evolve during the next few billion years. As an order-of-magnitude estimate, let I equal one.

The most elusive of all our probabilities is that describing the likelihood that any technologically competent civilization would actually undertake space communications. Even within the narrow sample of known human societies, there is a bewildering range of psychological motivations. Out of ignorance, we tentatively write 0.5 for the factor F, since it lies somewhere in the range zero to one.

Lastly, the longevity of the communicating civiliza-

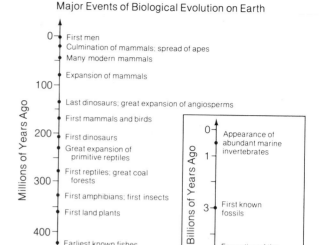

Figure 27.2. This diagram by Su-Shu Huang chronicles some of the major stages in the development of life on the earth. The insert is a compressed scale that includes some very early events in our planet's history. Is this time scale typical for planets in our galaxy?

tions has to be considered. While life has existed on Earth for two billion years, radio engineering is only a few years old! Any estimate of the lifetime of a technologically advanced civilization should allow for the possibility of self-destruction. However, if a species learns to control its own biological and social evolution, its technology may last for a geologically long time. My arbitrary choice for the average duration is a million years. Now, on astrophysical grounds it is likely that most stars with inhabitable planets are somewhat older than the sun. Perhaps, then, a typical inhabited planet has lasted three billion years since intelligence has had a chance to develop there. Dividing this number into one million years gives L as 0.0003.

At this point in our discussion, we have adopted some numerical value for each of the seven factors on the right side of the equation. Substituting these values gives approximately 2,000,000 for the number of civilizations within our galaxy that are now capable of communicating with us.

Of course this is a highly uncertain estimate, depending on a long series of assumptions. The numerical result itself is less significant than the listing of the steps from which it came. Nevertheless, let us use this figure of 2,000,000 civilizations at face value.

How far apart in space are these civilizations? Near the sun there is about 0.03 star per cubic parsec. Multiply this by the ratio of the number of stars that are potential abodes of life to the total number of stars in the galaxy. It turns out that the average distance between neighboring sites of

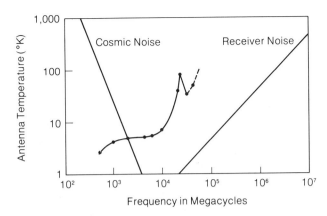

Figure 27.3. Data for selecting the best frequency for interstellar communication. Connected dots show the interference of our atmosphere.

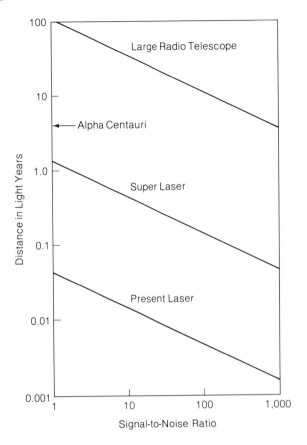

Figure 27.4. The distances to which light and radio signals can penetrate space with a given signal-to-noise ratio. Only large radio telescopes can hope to reach beyond the nearest stars. Diagrams after "Interstellar Communication."

intelligent life is roughly 90 parsecs or 300 light-years. How can signals be transmitted over such large distances and have a chance at the other end of being recognized as artificial?

The choice of operating frequency depends to a large extent on signal-to-noise ratio. If a visible signal were sent from Earth, the principal source of noise would be the background light from the sun. At radio frequencies, the background noise will be mainly the general radio emission from the galaxy. This emission lessens with increasing frequency until, at higher frequencies the so-called quantum noise inherent in any detector builds up rapidly. In between, there is a range of frequencies where noise is a minimum. In presenting this argument, G. Cocconi and P. Morrison point out that the minimum region contains the 21-cm. line emitted by neutral hydrogen in interstellar space.

A technologically advanced civilization elsewhere in the galaxy, if interested in radio astronomy, would also build receivers to observe this line, because of the enormous amounts of neutral hydrogen between the stars.

It has been objected that the 21-cm. line is not where we should listen for intelligent signals, because the interstellar hydrogen produces a large noise peak at just that place in the radio spectrum. The objectors urge that twice or half the frequency would be more logical for surveillance. Against this argument are two facts: the size of this hydrogen peak depends upon direction in space, and its position is shifted by the Doppler effect.

Then, too, there is the problem of where to look. Where should the antenna be aimed among the hosts of stars in order to receive the signals that may be coming from only one of them at any particular time? Altogether, the listener for interstellar signals faces a needle-in-a-haystack situation.

It has been suggested that laser systems could be used to transmit optical signals at the wavelength of some Fraunhofer line in the spectrum of the parent star. But it would be a more difficult feat to send a recognizable laser beam at optical frequency to nearby Alpha Centauri than to send recognizable radio pulses with a giant antenna to distances of 100 light-years. The former task is well beyond present capabilities; the latter is on the verge of feasibility.

* * *

In the recent book *Interstellar Communication*, B. M. Oliver describes a procedure for sending pictures or diagrams. Suppose the transmission has 250 pulses, mingled with blanks that extend the message to 1,271 items or bits. The listener on a planet of, say, Epsilon Eridani is expected to recognize that 1,271 is the product of two prime numbers, 41 and 31. He would thereupon rearrange the message as a 41-by-31 rectangular array of marks and spaces, to reconstruct the picture sent from Earth. The first exchanges of information between two widely separated civilizations might in this way be made by a kind of interstellar television.

28

N. S. Kardashev
Transmission of Information by Extraterrestrial Civilizations

1. The principal factors which exert a determining effect on the range of space radio communications are the transparency of the interstellar medium to radio signals, the level of the equipment noise and space noise, and the power of the transmitters. The greatest possible range for establishing space communications could be set most likely in the range from 10^9 to 10^{11} cps (Oliver, 1962). The absorption coefficient of the interstellar medium is negligibly small at those frequencies. The equivalent noise temperature may be represented in the form $T_N = T_n + T_t = T_q$, where T_n and T_t are respectively the temperature due to synchrotron radiation and due to background thermal cosmic radio emission, and $T_q = hf/k$ is the equivalent noise temperature due to quantum fluctuations in the minimum detectable signal (h and k are the Planck constant and the Boltzmann constant). The expression for T_N gives an estimate of the limiting sensitivity which might be achieved in the case of an ideal noise-free receiver and observations outside the earth's atmosphere. In Figure 28.1, we find plots of T_N as a function of the frequency in accord with up-to-date radio astronomy data (Turtle et al., 1962; Wilson, 1963; Altenhoff et al., 1960). The paramount role in the establishing of long-range communications within the confines of our galaxy will evidently be played by thermal and nonthermal radio-frequency emission from the galactic disk (over a range of $\pm 50°$ in longitude on either side of the center of the galaxy). In that case, we have

$$T_N = 2 \cdot 10^{27} \cdot f^{-2.9} + 10^{19} f^{-2} + 4.8 \cdot 10^{-11} f. \quad (1)$$

In dealing with the problem of possible success in setting up communications between the galaxies, we must take into consideration the brightness temperature of the background at high galactic latitudes, which is due to synchrotron radiation from the halo and from the metagalaxy. In this case, we have

$$T_N = 10^{26} f^{-2.9} + 4.8 \cdot 10^{-11} f. \quad (2)$$

In both cases, the noise temperature will display a deep-sloping minimum in the decimeter and centimeter wavelength ranges, which renders this range more suitable for space communications over exceptionally vast distances.

2. Let us evaluate the information content of communications channels for application to this problem.

The upper bound of the rate of information transmission at a specified average transmitter power and specified noise distribution is determined by the corresponding Shannon theorem (Goldman, 1953):

$$R = \int_{f_1}^{f_2} \log_2 \left[\frac{S(f) + n(f)}{n(f)} \right] df, \quad (3)$$

where $S(f)$, $n(f)$ are the functions of the spectral power density of the useful signal and of the noise, respectively. By solving the appropriate variational problem, we may show the maximum rate of information transmission to be achieved under the condition

$$S(f) + n(f) = n(f_1) = n(f_2). \quad (4)$$

Here, f_1 and f_2 are the bounds of the transmitter transmission band. It is accordingly quite clear that the spectrum of the artificial source must display the shape of the curve in Fig. 28.1, but with the sign reversed,

$$S(f) = n(f_1) - n(f),$$

i.e., the spectrum of the artificial radio emission must feature a maximum, and in a region of frequencies lower than this frequency will fall as $a - bf^{-2.9}$, whereas, in the region of higher frequencies, it will fall as $a - cf$ (a, c, b here are constants dependent upon the power and bandwidth of the transmitters and on the noise distribution).

Reprinted from *Soviet Astronomy-AJ*, 8 (1964), 217. English translation © 1965 American Institute of Physics. Reprinted with permission.

To make a rough estimate of the rate of information transmission, there is of course no need to take the logarithmic term in Shannon's theorem into account. The rate of information transmission $R = f_2 - f_1 = \Delta f$ then, i.e., it will be equal to the system bandwidth. Let P be the transmitter power, and we shall assume the radiation isotropic. There appears to be no point, in our view, in discussing the possibility of establishing outer space communications when high-directivity antennas (Cocconi and Morrison, Reading 20) are being used for transmission, since the probability of success in establishing such communications is virtually nil. The high directivity of the radiation can be conveniently used, in all probability, only after a two-way communication exchange has been set up, and if we bear in mind the fact that the distances between the two civilizations in question may be comparable to or even more enormous than the dimensions of our galaxy, then this second mode of communications may be brought to fruition much later, and in the meantime the need for isotropic radiation geared to establish contacts with new potential listeners will not have abated. Let the transmission be carried out over a bandwidth Δf, and let the maximum expected distance of the transmission be r, with A the effective area presented by the receiving antenna, T_N the noise temperature referred to the input of the receiving antenna. We shall assume the transmission to be reliable when the condition $100\,kT_N = PA/4\pi r^2 \Delta f$ is fulfilled, i.e., when there is a 100-fold excess of signal over noise. Then

$$\Delta f = PA \,/\, 400\pi r^2 kT_N. \qquad (5)$$

Consequently, the channel capacity will not be less than Δf bits per second. The radio emission flux per unit frequency interval will not be less than

$$F_f = kT_N/A. \qquad (6)$$

3. The most important parameter to deal with in this problem is the power P, about which several hypotheses are in order. Calculations show that the total quantity of energy expended by all of mankind per second at the present time is about 4×10^{19} erg, and the annual increase in this energy expenditure is placed at 3–4% over the next 60 years, on the basis of statistical findings (Putnam, 1948). Now let this increment represent an annual increase in energy consumption by a factor of $1 + x$, so that in t years the increase will be $(1 + x)^t \approx e^{tx}$ times $(x \ll 1)$. Assuming $x = 1\%$, we find that the energy consumption per second will be equal to the output of the sun per second, 3200 years from now, i.e., 4×10^{33} erg/sec, and that in 5800 years the energy consumed will equal the output of 10^{11} stars like the sun. The figures arrived at seem to be inordinately high when compared to the present level of development, but we see no reasons why the

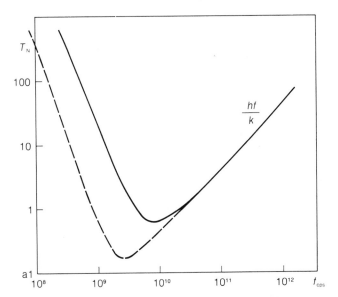

Figure 28.1. Noise spectrum outside the confines of the earth's atmosphere: ——) in the direction pointing toward the center of the Galaxy; ————) in the direction toward the galactic pole.

tempo of increase in energy consumption should fall substantially than predicted. Moreover, the availability of a large amount of information forthcoming from other and more highly developed civilizations might contribute to a staggering increase in energy consumption.

In line with the estimates arrived at, it will prove convenient to classify technologically developed civilizations in three types:

I—technological level close to the level presently attained on the earth, with energy consumption at $\approx 4 \times 10^{19}$ erg/sec.

II—a civilization capable of harnessing the energy radiated by its own star (for example, the stage of successful construction of a "Dyson sphere" (Dyson, Reading 22); energy consumption at $\approx 4 \times 10^{33}$ erg/sec.

III—a civilization in possession of energy on the scale of its own galaxy, with energy consumption at $\approx 4 \times 10^{44}$ erg/sec.

4. Estimates of the possibility of detecting a type I civilization (Drake, Reading 24) and related experiments in the "Ozma" project in the USA have revealed the extremely low probability of any such event. Consider the possibility of detection and reception of information sent by type II and type III civilizations. First of all, we assume here that one of the principal tasks of such communication efforts would be the transmission of information from a more highly developed civilization to a less highly developed one. Starting from the present level of development of radio physics as point of departure, we see that in principle it is possible to build antennas, within the next two decades, with an effective area of 10^5 m^2 and with

Table 28.1

NUMBER OF BITS PER SECOND $\sim \Delta f$

Type of civiliza-tion	Trans-mitter power	r = 100,000 light years	r = 10 million light years	r = 10 billion light years
II	4×10^{33} erg/sec	3×10^9	3×10^5	Transmission of a large quantity of information impossible $3 \cdot 10^{10}$
III	4×10^{44} erg/sec	2.4×10^{15}	2.4×10^{13}	

receiving apparatus featuring a noise temperature $T_N \approx 1°K$. If a transmitter is designed for this system to receive and record information, Equation (3) will show that the radio emission flux at the receiving point will be not less than 1.4×10^{-26} W/m$^2 \cdot$cps, an amount which is well within the recording capabilities of presently existing radio telescopes. [Of course, far simpler equipment will be required to detect the signals since, in contrast to the reception of information, here we may utilize the averaging techniques common in radio astronomy, and the radiometric gain $\sqrt{\Delta f \tau}$ (where τ is the build-up time) may enhance the sensitivity by 3 to 4 orders of magnitude over a fairly wide signal band.]

We now cite some estimates of the quantity of information obtainable, in line with formula (2). For a type II civilization, we have to take into account the frequency variation of T_N as in accord with formula (1). $T_N = 1°K$ is assumed for a type II civilization. Table 28.1 lists estimates for three distances corresponding to the transmission of information within the confines of the galaxy, within the confines of a local system of galaxies, and within the confines of the portion of the metagalaxy accessible to observation. The estimates arrived at show that should there exist even one type II civilization within the confines of the local system of galaxies, there will be a realistic possibility of securing an enormous quantity of information. The same holds for the existence of even one single type III civilization in the portion of the universe accessible to observation. For purposes of comparison, let us estimate the time required to transmit 10^8 printed and manuscript-form publications now available on the earth through a channel of 10^9 cps bandwidth. Assuming that each written work contains an average of about 10^6 bits of information, we find that the total quantity of 10^{14} bits of information may be transmitted in 10^5 sec, i.e., in a single day. However, it is quite evident that there is no need to transmit all of the 100 million publications in order to broadcast the principal data on the status of science, technology, and culture on the earth, for such information would contain a colossal amount of "redundancy." Apparently, all of the basic information could be compressed in 10^5 books of 10^6 bits each, which would come to 10^{11}

bits and would take only 100 sec to be transmitted via the same communications channel. Finally, it is entirely reasonable to assume that type II and type III civilizations would be in possession of information many orders of magnitude in excess of what we have available at the present time. For that reason, they would have to be broadcasting practically continually, and this would also be the case for increasing the possibility of reception by type I civilizations. Moreover, in order to improve the reliability of the information received and in order to afford some opportunity to make connections with new subscribers, there would have to be periodic repetitions of the programs broadcast.

Note again that, in all likelihood, a type I civilization would be capable of sending a return signal only after its energy consumption had increased measurably. Consequently, the communications would be a one-way affair at the start, and the problem of how long it takes the signals to propagate would be a secondary one.

As is evident from the above estimates of transmitter power for type II and type III civilizations, the figures are very close to the power of synchrotron radiation from nebulas formed in supernova explosions, or from radio galaxies. Calculations of the optimum transmitter bandwidth show that the transmission spectrum may also closely resemble the spectra of discrete radio sources. Several criteria could be singled out which would be useful in discriminating artificial radio sources from the vast number of radio stars accessible to observation.

The artificial sources would evidently (1) have to have very small angular dimensions (at least in the case of type II civilizations); the angular dimensions would have to be of the order of the angular dimensions of stars, i.e., less than $0''.001$. The now known natural radio sources must have appreciably larger angular dimensions, larger than $0''.01$ in fact, according to theory (Slysh, 1963); (2) they would have to possess circular polarization, so that the effect of the Faraday rotation of the plane of polarization in the interstellar medium would not distort the information received; (3) they would have to exhibit variability in time without leading to statistical fluctuations; this is obviously a criterion of outstanding importance, but it is possible that a regularity in time variations of the signal might be revealed only in the course of observations using special equipment of wide bandwidth (10^9 cps) and of sufficient sensitivity to operate with very short time constants; (4) finally, it is to be anticipated that certain details would be present in the spectrum of the source suspected of artificiality which would have been designed for the express purpose of emphasizing its artificial origin, in particular, we might anticipate such a feature in the environs of the 21-cm wavelength line. Information transmission at that frequency within the confines of this galaxy would be inadvisable, since the signal will be strongly absorbed by neutral hydrogen. It would be reasonable, for that reason, to eliminate a band of 1-2 Mc width, say of rectangular shape, in the continuous spectrum of the artificial source, to lay special emphasis on the unusual nature of the radiation. Considerations involving the anticipated shape of the complete spectrum of the artificial source also deserve close attention. These arguments were cited above when the estimate was made of the quantity of information which could be transmitted. The most characteristic feature of the spectrum is the linear dependence of the flux on frequency in the high-frequency region of the spectrum (cf. Figure 28.2).

It is consequently of the utmost importance to carry out a program of studying and searching discrete radio sources, in the immediate future, with due attention paid to the spectral features mentioned here. Note that even at the present time we have knowledge of about twenty to thirty or so radio sources, with the upper limit of the angular dimensions in the range of 1 to 10 sec or arc (Allen et al., 1962). Some of these sources have been identified with peculiar optical objects the nature of which is still obscure (Schmidt, 1963a; Oke, 1963; Greenstein and Matthews, 1963). Most of them have not yet been successfully identified with optical objects.

For example, two sources of radio-frequency emission from outer space, CTA-21 and CTA-102, were recently discovered at the California Technological Institute (Harris and Roberts, 1960), and display angular dimensions not less than 20″, and have not been identified with a single one of the optical objects in the Palomar sky charts and, even more intriguing, these sources exhibit a spectrum highly similar to the anticipated artificial spectrum. Figure 28.2 presents the observational data reported in (Conway, Kellerman, and Long, 1963) on these two sources, alongside the theoretically predicted spectrum. For purposes of comparison, the spectrum of a typical natural radio source, Virgo A, is added (the scale is compressed by a factor of 10 on the ordinate axis in the case of Virgo A).

The most promising region for a search for artificial radio signals is apparently that in the direction toward the center of the Galaxy, since the density of the stellar population is greatest there along the line of sight. It would also be appropriate to investigate the closest galaxies, and in the first instance the large nebula in the Andromeda constellation and the Magellanic Clouds, as well as the closest radio galaxies NGC 4486 and NGC 5128.

In conclusion, we should like to note that the estimates arrived at here are unquestionably of no more than a tentative nature. But all of them bear witness to the fact that, if terrestrial civilization is not a unique phenomenon in the entire universe, then the possibility of establishing contacts with other civilizations by means of present-day radio physics capabilities is entirely realistic. At the same time, it is very difficult to accept the notion that, of all of

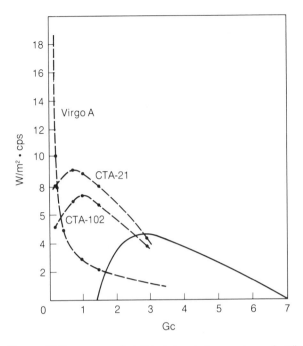

Figure 28.2. ————) Anticipated emission spectrum of radio transmitters of extraterrestrial civilizations; — — — —) spectrum of radio sources CTA-21 and CTA-102, suspected of being artificial radio sources, and spectrum of a typical natural radio source Virgo A.

the 10^{11} stars present in our Galaxy, only near the sun has a civilization developed. It is still more difficult to extend this inference to the 10^{10} galaxies existing in the portion of the universe accessible to observation. In any case, the deciding word on this question is left to experimental verification. In particular, we may anticipate that space rockets will clear up the question of whether or not life exists on other planets in the solar system in the years to come. The discovery of even the very simplest organisms, on Mars for instance, would greatly increase the probability that many type II civilizations exist in the Galaxy. Radio astronomical searches could of course play a decisive part in resolving this problem.

29

Carl Sagan
On the Detectivity of Advanced Galactic Civilizations

Mankind now possesses the technological capability of communicating at radio frequencies, over distances of many thousands of light years, with technical civilizations no more advanced than we. But before a program is initiated to search systematically for such signals it is important to demonstrate at least a modest probability that one technical civilization exists within such a range. The possibility that much more advanced civilizations exist—societies which can be detected over much larger distances—will be discussed presently. The pitfalls in placing numerical values on the component probabilities of N, the number of extant technical civilizations in the Galaxy, are numerous and treacherous; nevertheless, there does seem to be a limiting factor whose significance has not always been appreciated.

While much more sophisticated formulations are now available (see, e.g., Kreifeldt, 1973) the first algebraic expression for N, due in its original formulation to F. D. Drake, will serve our purpose:

$$N = RL \qquad (1)$$

Here R is the rate of emergence of communicative technical civilizations in the Galaxy, and is a function of the rate of star formation, the fraction of stars which have planets, the number of planets per star which are ecologically suitable for the origin of life, the fraction of such planets on which the origin of life actually occurs, and the fraction of such planets on which intelligence and eventually technological civilizations actually emerge (see, e.g., Shklovskii and Sagan, 1966). L is the mean lifetime of such civilizations, and is strongly biased towards the small fraction of technical civilizations which achieve very long lifetimes—lifetimes measured on the geological or stellar evolutionary time scales (Shklovskii and Sagan, 1966). But such civilizations will be inconceivably in advance of our own. We have only to consider the

Reprinted from *Icarus*, 19 (1973), 350. Copyright © 1973 by Academic Press, Inc.

changes in mankind in the last 10^4 years and the potential difficulties which our Pleistocene ancestors would have in accommodating to our present society to realize what an unfathomable cultural gap 10^8–10^{10} years represents, even with a tiny rate of intellectual advance. Such societies will have discovered laws of nature and invented technologies whose applications will appear to us indistinguishable from magic. There is a serious question about whether such societies are concerned with communicating with us, any more than we are concerned with communicating with our protozoan or bacterial forebears. We may study microorganisms, but we do not usually communicate with them. I therefore raise the possibility that a horizon in communications interest exists in the evolution of technological societies, and that a civilization very much more advanced than we will be engaged in a busy communications traffic with its peers; but not with us, and not via technologies accessible to us. We may be like the inhabitants of the valleys of New Guinea who may communicate by runner or drum, but who are ignorant of the vast international radio and cable traffic passing over, around and through them.

A convenient subdivision of galactic technological societies has been provided by Kardashev (Reading 28). He distinguishes Type I, Type II and Type III civilizations. The first is able to engage something like the present power output of the planet Earth for interstellar discourse; the second the power output of a sun; and the third the power output of a galaxy. By definition, Type I civilizations are capable of restructuring planets, Type II civilizations of restructuring solar systems, and Type III civilizations of restructuring galaxies. I believe that a civilization of approximately Type II has, with an exception to be described later, reached our communications horizon. For computational convenience, I also assume that a civilization which has emerged to Type II technologies has also successfully passed through the critical period of probable technological self-destruction—the period in which terrestrial civilization is now immersed.

These ideas can now be restated as follows: let f_g be the fraction of technical civilizations which survive for geological or stellar evolutionary time scales, L_g, and let L_d be the mean time to self-destruction of those Type I civilizations which do not achieve Type II technologies. Then,

$$L \sim (1 - f_g) L_d + f_g L_g \qquad (2)$$

Accordingly, the total number of extant civilizations in the Galaxy,

$$N \sim R[(1 - f_g) L_d + f_g L_g], \qquad (3)$$

is different from the number of civilizations within our communications horizon,

$$N_c \sim N_1 \sim R[(1 - f_g) L_d]. \qquad (4)$$

The ratio of these lifetimes

$$N_c/N \sim [1 + f_g(L_g / L_d)]^{-1}, \text{ for } f_g \ll 1 \qquad (5)$$

$$\sim (L_d / f_g L_g), \qquad \text{for } f_g L_g \gg L_d. \qquad (6)$$

Equations (5) and (6) are independent of R. Of the civilizations within our communications horizon only

$$N'_c \sim R f_g (1 - f_g) L_d \qquad (7)$$

are destined to have lifetimes $\gg L_d$.

We now specialize to some illustrative numerical cases. I emphasize that values differing by several orders of magnitude from the ones I choose are certainly conceivable and may even be probable. We adopt (Shklovskii and Sagan, 1966) $L_g \sim 10^9$ yr, $f_g \sim 10^{-2}$, and $R \sim 10^{-1}$yr^{-1}. I further assume $L_d \sim 10^3$ years. From events of the past few decades a case can be made for L_d 1–2 orders of magnitude smaller; the resulting conclusions will be correspondingly more pessimistic. Independent of the choice of L_d, as long as $L_d \ll L_g$, we find $L \sim 10^7$ years, and $N \sim 10^6$ galactic civilizations (Shklovskii and Sagan, 1966). Assuming such civilizations are randomly distributed, the mean distance to the nearest is a few hundred light years, and searches for such civilizations, using existing technology, would seem to be in order. However, if we count only those civilizations within our communications horizon, we find, with the same choice of numbers,

$$N_c /N \sim 10^{-4} \quad \text{and} \quad N_c \sim 100.$$

In this case the distance to the nearest communicative civilization is $\sim 10^4$ light years—well beyond easy detectability, assuming that our communicant is at approxi-mately the same technological level and we have no prior knowledge of where to look. And of these 100 societies only $N'_c \sim 1$ is likely to avoid self-destruction.

Almost all of these 100 civilizations of Type I or younger must have technologies significantly in advance of our own, and it may very well be possible to make contact with them. But the prospects are very much dimmer than in the case of 10^6 communicative galactic civilizations. The situation can be improved somewhat by taking $L_d <$ the interval to the communications horizon, rather than equal to it as we have assumed here; but we have been optimistic in our choice of L_d and I find it difficult to imagine that many civilizations $> 10^3$ years in our technological future would be anxious to communicate with us.

The situation seems to be that Type II civilizations may be, in terms of contemporary terrestrial communications technology, at small distances from us—but, in the same terms, noncommunicative; whereas Type I civilizations may be communicating—but tend to be too far away for us to detect easily. The operational consequence is that the detection of civilizations of Type I or younger is more difficult than has generally been assumed, and that such an enterprise will require much more elaborate radio systems—for example, very large phased arrays—than currently exist, and very long observing times to search through the $\sim 10^9$ stars which must be winnowed to find one such civilization.

On the other hand somewhat more serious attention must be given to the question of Type II and Type III civilizations—the level where, according to the previous argument, most of the technical societies in the universe are. A Type II civilization can communicate with the Earth from our nearest galactic neighbors; a Type III civilization can communicate across the known universe—and this employing only laws of nature which we now understand. If only a tiny fraction of such civilizations are interested in antique communications modes they will dominate the interstellar communications traffic now accessible on Earth. The best policy might therefore be to search with existing technology for Type II or Type III civilizations among the nearer galaxies, rather than Type I or younger civilizations among the nearer stars.

30

Gerrit L. Verschuur

A Search for Narrow Band 21-cm Wavelength Signals from Ten Nearby Stars

The search for artificially generated radio signals originating on planets orbiting nearby stars has often been discussed but no very sensitive experiments aimed at detecting such signals have ever been reported in the scientific literature. Cocconi and Morrison (Reading 20) originally suggested that another civilization desirous of communicating with neighbors in space might use the 21-cm band because the rest frequency of the neutral hydrogen ground-state transition was a naturally occurring standard of which any advanced civilization would be aware. Drake (Reading 25) has reported an experiment to search for such signals, but the details of that experiment were never published in terms of the sensitivity attained.

Troitskii *et al.* (1971) have more recently reported on observations at 927 MHz in the direction of 21 stars but their detection limit was about 2×10^{-21} W/m² in their bandwidth of 13Hz. This paper reports on the negative results of a search for narrow band radio signals around 1420 MHz in the direction of ten nearby stars, primarily chosen on the basis of their distance and similarity to the Sun and, in two cases, chosen because they may have objects of planetary mass orbiting them.

In justifying an extensive program to search for the radio signals from other civilizations one naturally has to make several assumptions about the likelihood that any given nearby star has a planetary system, that life will have evolved, that it is a technological society and that they would bother to transmit signals toward other planets. Even making favorable assumptions concerning the first three of these categories, it is this author's belief that any detection of signals from another civilization will most likely be an accidental one in the sense that we will pick up signals not meant for us. For this reason it is unlikely, on the basis of our present knowledge of the way we operate on Earth, that a wavelength around 21-cm is the wavelength at which to search because it should be borne in mind that any advanced civilization is also likely

to have instituted a series of protected bands for radio astronomy research purposes and might therefore not be transmitting signals in the λ21-cm band at all! However, the present observations were nevertheless made and the most sensitive equipment at this wavelength presently available was used. The results indicate the type of integrations required in order to substantially improve the limits now set to the level of any possible extra-terrestrial transmissions around λ21-cm.

In this paper the λ21-cm hydrogen line profiles toward the ten stars studied are shown. These may be used as zero-epoch profiles for subsequent searches of this nature. Furthermore, examples of the 20 MHz coverage obtained in the direction of two of the stars are also given. The experiment was performed at various times with both the 300-ft telescope (where observations were fitted in as part of a more extensive program to study galactic neutral hydrogen clouds) and the 140-ft telescope of the National Radio Astronomy Observatory in Green Bank, West Virginia.

I estimate that in the present series of experiments a 5-min integration is equivalent to about 4 days of integration for the equipment used as part of Project Ozma by Drake (Reading 25).

The 300-ft Telescope Data

The 300-ft radio telescope was used to observe three stars with a feed tracking a star for 4 min each day for about a month. A comparison feed tracked a position 20 min of arc away from the star at the same time. The signals were sent through two parametric amplifiers whose system temperatures were 110°K and fed into each of two 192 channel digital spectrometers. The overall bandwidth of the 192 channel receivers were variously set to cover 130, 65, 32 and 16 km/sec centered about the velocity of the star with respect to the local standard of rest. (1 km/sec ≡ 4.74kHz at this wavelength.) For the narrowest

Reprinted from Icarus, 19 (1973), 329. Copyright © 1973 by Academic Press, Inc.

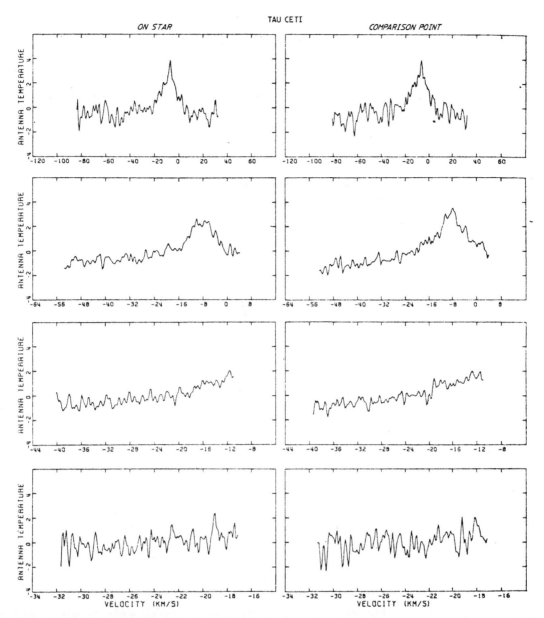

Figure 30.1. Data obtained with the 300-ft telescope in the direction of Tau Ceti and a comparison point 20 min of arc away (observed simultaneously) with a range of bandwidths. Velocities are with respect to the local standard of rest, with regard to which the stellar radial velocity is −26 Km/sec.

bandwidth observations the presence of signals at the velocity of the star would have been revealed. However, if the alien transmitter were broadcasting at a frequency corresponding to the laboratory frequency on their planet, and if their planet has an orbit similar to that of the Earth around the Sun then their signals would have been Doppler shifted out of the narrowest bandpasses used in this experiment. The narrowest bandwidth I used gave a channel width of 0.49kHz.

The observations of these three stars were fitted into an extensive observing program at 21-cm during which galactic neutral hydrogen clouds were being studied. Other stars were not included because they would have interfered with the routine observations of the main observing program and even for these three stars it was not possible to track each of them for the maximum time allowed by the tracking feed system of the 300-ft telescope.

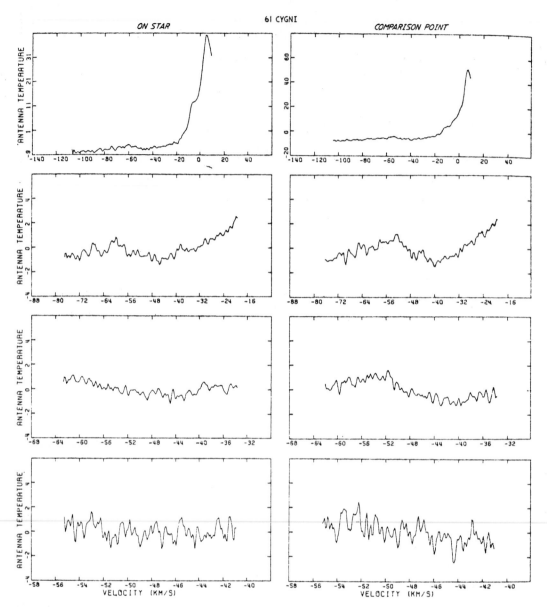

Figure 30.2. Data obtained with the 300-ft telescope in the direction of 61 Cygni and its comparison point. The radial velocity of the star is −48 Km/sec.

In Table 30.1 the observations made with the 300-ft telescope are summarized. A total of 2 hr of data are available on 61 Cygni, 66 min on Tau Ceti and 71 min on Epsilon Eridani. While tracking the star and its comparison point, profiles were recorded every 10 sec of time and examined separately to see if any signals greater than the peak-to-peak noise in this interval (4K) were present. None were found. The observations in the direction of any particular star with a particular bandwidth were subsequently added together in order to improve the chances of detecting a weak continuous signal which might have a very narrow frequency structure. The results of some of these integrations are shown in Figure 30.1 for Tau Ceti, in Figure 30.2 for 61 Cygni and in Figure 30.3 for Epsilon Eridani. The data for both signal and comparison feeds are shown and have been smoothed a little, using a Haning weighting function to reduce the noise level. The plots are given in terms of antenna temperature and the conversion to brightness temperature may be made using an aperture efficiency of 0.6 for the stars around −9 to −16° declination and 0.75 for the star at declination 38°.

No signals of the order of 1 K or greater were found in

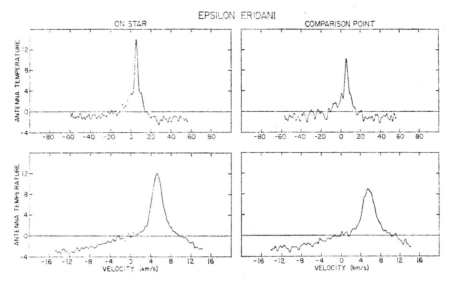

Figure 30.3. Data obtained with the 300-ft telescope in the direction of Epislon Eridani. The radial velocity of the star is +0.7 Km/sec.

any of the observations on these three stars. This converts to an upper limit of 1 kW/Hz for the power of a transmitter on a 300-ft diameter telescope beamed at the Earth from a distance of 10 light years. Therefore, in the narrowest band data we would have detected a 500-kW transmission (see below). Very roughly one can also state that the limit of detection was about 10^{-26} W/m²/Hz, or 5×10^{-24} W/m² for the narrowest band observations for Tau Ceti and 61 Cygni, or nearly three orders of magnitude below the levels set by Troitskii *et al.* (1971) at 927 MHz.

The 140-ft Telescope Data

Narrow band observations in the direction of HD 165341 (70 Oph) were performed on June 25 and 26, 1972. In Figure 30.4a the results of two 50-min integrations in the direction of this star are shown while in Figure 30.4b one of these profiles is plotted so as to show the full height of the hydrogen line. The receiver used was a cooled parametric amplifier (system noise temperature 48K) combining with the Mk II, 384 channel digital spectrometer. The bandwidth used was 6.9 kHz per channel and the peak-to-peak noise in a 50-min integration was 0.1°K. We can say that no narrow band signal of the order of 0.07°K was seen in a 350 km/sec range about the local hydrogen profile in a 100-min integration. The velocity of the star is 10.7 km/sec with respect to the local standard of rest.

A limit of 0.07K converts to 0.25 flux units on the 140-ft telescope (i.e., 1.7×10^{-23} W/m² in our bandwidth) which, in turn, allows us to set a limit of 3 Megawatts for an alien transmitter radiating anywhere in the 2.3 MHz

bandwidth seen in Figure 30.4, from a 100-m diameter telescope beamed at the Sun (see below) from a distance of 5 pc.

On August 23, 1972 observations were made in the direction of the other eight stars listed in Table 30.2. A 20-MHz frequency range centered about the hydrogen line rest frequency was observed in the direction of each star, although small gaps occurred in this coverage, due to edge effects in the digital correlator profiles. The receiver used was the cooled parametric amplifier and the 20 MHz was covered with a channel bandwidth of 7.2 kHz. The observations were made by setting the spectrometer so that the 384 channels covered a total bandwidth of 2.5 MHz and the receiver was frequency switched to two reference frequencies, centered 2.5MHz above and below the signal band. After a 5-min integration the central frequency and reference frequencies were shifted by 2.4MHz. This meant that a positive going signal in the signal band on one local oscillator setting would be expected to appear negative going on an adjacent setting because it was then likely to be in the reference band. However, no such signals were detected. Sometimes a further 5-min integration was done around the velocity of the star and, in the case of Barnard's star, a total of 20-min was obtained at the central velocity setting.

In Figure 30.5 the complete data on Barnard's star and Tau Ceti are shown, again with the neutral hydrogen line clipped. Only a straight baseline was taken out of each integration covering 2.5MHz, and the curvature noticeable between −860 and −1360 km sec⁻¹ is due to the shape of the receiver bandpass. The data on the other six stars listed in Table 30.2 are essentially the same in appearance to that shown for the two stars in Figure 30.5.

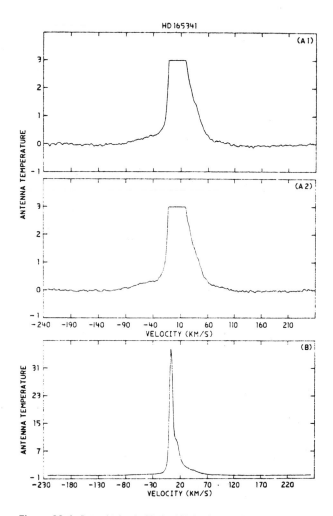

Figure 30.4. Data obtained with the 140-ft telescope in the direction of HD 165341 (70 Oph). A1 and A2 are two dependent 50-min integrations and B shows the same data as A2 with the scale altered to include the full hydrogen emission line.

We conclude that no narrowband signals of the order of 0.30K or greater were seen in the direction of any of these stars.

Limits to the Power of the Alien Transmitters

Let the receiving antenna have a dish diameter of D_r meters. Then the area of this dish is

$$A = 7.85 \times 10^3 \, D_r^2 \text{ cm}^2 \tag{1}$$

Let the limit of detection in antenna temperature be $L(K) \equiv 4.7L$ in flux units on the 140-ft telescope. Therefore the

limit of detection in watts (P_L) is given (for the 140-ft) by

$$P_L = 3.717 \, L \, D_r^2 \times 10^{-26} \text{ W/Hz} \tag{2}$$

If an alien transmitter on a planet R parsecs away is used in conjunction with a 100-m telescope having a bandwidth of 9 min of arc, then the area (B) subtended by this beam at the Earth is

$$B \simeq 4.9 \times 10^{31} \, R^2 \text{ cm}^2 \tag{3}$$

Therefore the fraction of the power received by the 140-ft telescope is given by A/B and the upper limit to the power radiated at the alien planet is then $P_r = P_L \times B/A$, i.e.,

$$P_r \sim 2.3 \times 10^2 \, R^2 \, L \text{ W/Hz} \tag{4}$$

In the last three columns of Table 30.2, I list the detection limit (in T_A) for possible signals in the velocity ranges indicated in the second last column. These limits we determined by inspection of the data. In the last column the upper limit to the power being transmitted in the 7-kHz bandwidth of a 100-m dish at the distance of the respective stars is listed. Equation 4 above was used to derive these limits. In order to derive the limit to the flux being received one needs to convert the antenna temperature to brightness temperature by dividing the former by 0.76 for the 140-ft telescope and to get flux units one multiplies this number by 3.64. For example, for the velocity interval around zero velocity on Barnard's star our power limit received is 4×10^{-23} W/m² in the 7.2-kHz bandwidth. Alternatively, we can set a limit of 665kW to a transmitter in a 100-m dish on one of the planet's orbiting Barnard's star, which is known to have at least two planets of Jupiter-like mass (van de Kamp, 1969a). The other star thought to have planets is L-726 (Fredrick and Shelus, 1969).

Our power limits for transmitters around the two stars thought most likely to have Earth-like planets, Tau Ceti and Epsilon Eridani, are ~6 Megawatts in our bandwidth.

Comments on the Hydrogen Line Profiles

In Figure 30.6 the 21-cm profiles obtained in the direction of each of the stars are shown drawn to the same scale. Both Figures 30.4 and 30.6 reveal some interesting, but expected, aspects of the neutral hydrogen profiles observed in the direction of the stars.

Ross 154: We observe weak emission with $T_A \sim 0.15$K at -90 km sec⁻¹ in the direction of this star ($l = 11°$, $b = -10°$). This is in good agreement with the weak signals around this velocity found by Dieter (1972). This emission is an aspect of the high velocity cloud phenomenon in this part of the sky. (See Verschuur, 1973a).

Table 30.1.

DETAILS OF THE OBSERVATION ON THE 300-FT TELESCOPE

Star	Position (1950)		Comparison point (1950)		Dates (1971)	Frequency covered (kHz)	Total integration time (min)	Heliocentric velocity of star	Velocity w.r.t. l.s.r.
	Right ascension	Declination	Right ascension	Declination					
Tau Ceti	$1^h42^m30^s$	$-16°11'00''$	$1^h44^m20^s$	$-16°30'30''$	Oct. 7–21	156	33	−16 km/sec	−26 km/sec
					Oct. 22–25	78	12		
					Oct. 26–31	312	18		
					Nov. 1	625	3		
61 Cygni	$21^h01^m50^s$	$38°30'22''$	$21^h06^m50^s$	$38°10'47''$	Oct. 4–20	156	69	−64 km/sec	−48 km/sec
					Oct. 21–24	78	18		
					Oct. 25–30	312	28		
					Oct. 31	625	4.6		
Epsilon Eridani	$3^h30^m34^s$	$-9°37'15''$	$3^h32^m20^s$	$-9°56'45''$	Oct. 7–20	156	36	+15 km/sec	0.7 km/sec
					Oct. 22–30	156	32		
					Oct. 31	625	3.6		

61 Cygni: A weak peak in the emission profile at −108 km sec^{-1} is visible. Comparison with the data of Verschuur (1973a) suggest that this would be due to emission from spiral arm γ at a distance of about 13 kpc from the Sun at this longitude (82°) and therefore expected. The matter giving rise to this emission would have a z distance of some 1.3 kpc.

Ross 248: This profile appears to show emission out to +50, maybe +70 km sec^{-1} which for its location (*l* = 110°, *b* = −17°) is a velocity forbidden on normal models of galactic rotation. However, in view of the wings visible in the profiles of the next two stars we must conclude that this type of signal is probably due to emission in the far out sidelobes of the 140-ft telescope.

Tau Ceti and L-726: These two stars are at latitudes +73° and −76°, respectively, and at about the same longitudes (174°, 175°). Both profiles show weak emission extending to about ±80 km sec^{-1} which, if one roughly fits a Gaussian, suggests a broad feature with $T_A \sim 0.1$°K at about zero velocity. It is probable that the emission from other parts of the galactic plane in sidelobes 30 dB down from the main beam would produce just such a signal.

Wolf 359: The profile in Figure 30.6 suggests the presence of considerable emission out to −100 km sec^{-1} which is surprising. It may be due to the side lobe problem mentioned above. The emission out to −50 km sec^{-1} is associated with the so-called intermediate velocity matter

Table 30.2

Name	HD	Type	Distance (parsecs)	α 1950			δ			*l*	*b*	Radial velocity of star			Velocity interval (l.s.r.) (km/sec)	Power limit (Megawatts)[b]
												Heliocentric (km/sec)	l.s.r. (km/sec)	Detection limit (T_a)[a]		
				h	m	s	°	′	″							
Barnard's Star	—	M5	1.83	17	55	24	4	24	34	30.8	13.9	−108	−90	0.12	−340 to 135	0.665
														0.20	−2380 to 2160	1.109
Wolf 359	—	M8	2.38	10	54	12	7	20	37	244.1	56.3	+13	+ 9.3	0.15	−250 to 50 10 to 250	1.407
														0.20	−2270 to 2270	1.876
L-726-8	—	M6	2.42	1	36	24	−18	13	00	175.1	−75.7	+29	+19.3	0.30	−2270 to 2270	2.909
LAL 21185	95735	M2	2.52	11	00	31	36	21	46	184.7	65.5	−86.5	−84.3	0.15	−250 to −100 30 to 240	1.577
														0.30	−2270 to 2260	3.155
Ross 154	—	M6	2.85	18	46	42	−23	52	57	11.4	−10.3	−4	7.5	0.30	−2270 to 2270	4.035
Ross 248	—	M6	3.16	23	39	24	43	55	08	110.0	−16.9	−81	−73	0.12	−250 to −50 20 to 240	1.984
														0.30	−2270 to 2270	4.960
EPS ERI	22049	K2	3.31	3	30	34	−9	37	15	195.8	−48.1	+15	0.7	0.30	−2270 to 2270	5.426
61 Cyg (A-B)	201092	K3-K5	3.40	21	04	50	38	30	22	82.3	−5.9	−63.5	−48.2	0.15	−270 to −30 20 to 190	2.871
														0.30	−2320 to 2230	5.743
TAU CETI	10700	G8	3.75	01	42	30	−16	11	00	173.5	73.3	−16.2	−26.1	0.30	−2290 to 2250	6.986
70 Oph (A-B)	165341	K1-K5	5.04	18	02	54	02	30	30	29.8	11.8	−7	+10.7	0.07	−232 to 255	2.944

a. By inspections of data (≈ peak-to-peak noise).
b. In the 7-kHz bandwidth radiated from a 100-m telescope at the star.

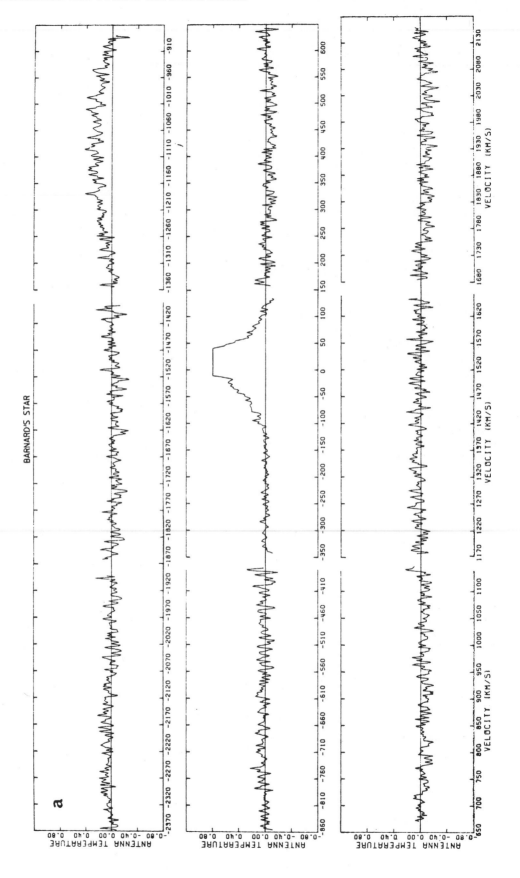

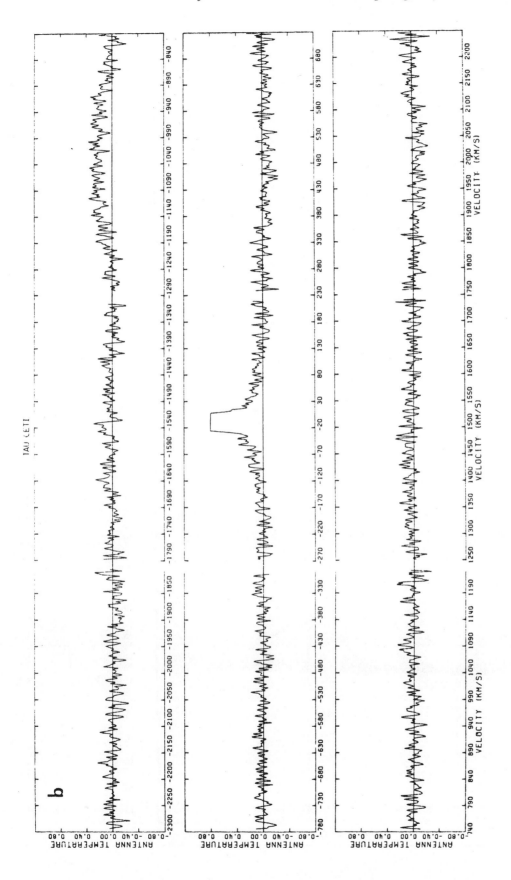

Figure 30.5. Data obtained with the 140-ft telescope in the direction of Barnard's star and Tau Ceti. The integration time was 5 min except in the central portion of the coverage on Barnard's star. Gaps in the data reflect regions of the spectrum where edge effects in the spectrometer were larger than the noise level.

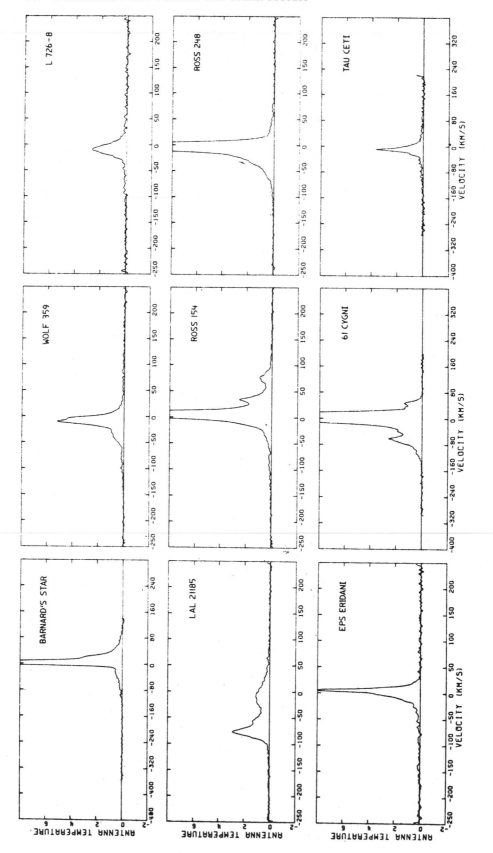

Figure 30.6. The 21-cm emission profiles in the direction of each of the stars studied with the 140-ft telescope shown on the same scale.

known to exist near this part of the sky ($l = 244°$, $b = +56°$) although this such emission has not actually been reported for such large longitudes.

HD 165341 (=70 Oph): the weak emission out to -90 km sec^{-1} is expected for its position in the sky originating in very distant spiral arms at high z (Verschuur 1973a) although this too may be due to side lobe radiation in view of its low level.

Eps Eridani, Barnard's Star, and LAL 21185: These profiles do not show anything unexpected for their location in the sky.

Conclusions

These data obtained in the direction of ten nearby stars have not revealed the presence of any signals which might be attributed to an artificial source, either on Earth or in the direction of the stars. Substantial further improvement in sensitivity, frequency coverage, and number of stars observed is possible with existing equipment.

31

Robert E. Machol
An Ear to the Universe

Right now, electromagnetic waves from a civilization outside the solar system may be passing through this page. Although such waves may be at levels too low to be detected with present equipment, there are good reasons for believing they exist. Thirty years ago, most astronomers thought that man was alone in the galaxy. Today, it is believed by most (but not all) knowledgeable scientists that in our galaxy there are billions of stars with planets on which life could originate. It is probable, therefore, that there exist a great many civilizations capable of communicating—or at least making themselves known to us.

Today, in fact, astronomers are more likely to discuss ways of discovering extraterrestrial civilizations than to debate their existence. One reflection of the seriousness U.S. scientists accord to the search is a project at the National Aeronautics and Space Agency's Ames Research Center to determine the feasibility of looking for signals. Known as the Search for Extraterrestrial Intelligence (SETI), the project is small (funded at a rate about 0.0001 that of the Apollo Project) but work has proceeded for nearly five years, beginning in 1971 with a summer study called Project Cyclops under the direction of Bernard M. Oliver of Hewlett-Packard.

The SETI team is presently studying advanced technologies to determine whether they can provide improvements and cost reductions. In addition, the team is performing preliminary system design, elaborating search strategies, and supporting some related development efforts (especially in data processing). Finally, the team is studying schemes for detecting signals—a prerequisite to communication. The plan right now is only to listen—transmitting a response would involve a whole added set of problems—but deciding how best to listen is, in itself, a project.

At various times, a number of signal types have been mentioned as possible communication media. Current thought is that only electromagnetic radiation offers a reasonable hope of providing a detectable signal. Within the electromagnetic spectrum, microwave frequencies are the most promising region for extraterrestrial signals. Efforts have been focused on this region because it contains the hydrogen line at 1.420 GHz—the most prominent radio line in the sky—and the strong hydroxyl line at 1.667 GHz. It is thought that these frequencies and the range between them must appear especially interesting to any life on the chemistry of water (hydrogen plus hydroxyl). Consequently, where are we most likely to "meet"? At the water hole!

X rays or higher-frequency radiation have been rejected as possibilities because of their high energy per quantum. Ultraviolet radiation at about 10^{-7} meters would be only barely possible: such a signal beamed at the earth from a nearby star for the purpose of making contact could be detected because the blackbody radiation from the star—with which the signal must compete—is very low at this frequency, and because high antenna gain is possible. A 1-kW laser operating at 100 nanometers with a 10-meter transmitting dish could produce 25 photons per second in a 10-meter receiving dish at a distance of ten light-years—perhaps the maximum range for such a system. The dishes would have to be precise and the beam would be so narrow that, if pointed at the sun, it would miss the earth by a wide margin.

Ultraviolet signals are occasionally sought using spaceborne telescopes, but the Ames group is concentrating on systems with greater range, in the belief that the probability of success is low if only the nearest stars are examined. After all, there are about a million times as many stars within 1000 light-years as there are within ten light-years.

Visible and near-infrared radiation appear impossible because of the overwhelming background radiation in this range from the nearby star. Far-infrared radiation is considered possible by some Soviet scientists, for reasons the Ames scientists do not find compelling.

The microwave region permits signal detection (or

Reprinted from *IEEE Spectrum*, 13 (March 1976), 42. © 1976 IEEE. Reprinted with permission.

Figure 31.1. Antennas designed to receive intelligent interstellar signals might appear, from the ground, like a forest of huge round trees. Control and processing equipment could be housed in a central building.

communication) at minimum energy. At frequencies below the microwave region, the background galactic noise becomes prohibitive; at higher frequencies, the quantum noise (high energy cost per photon) becomes prohibitive.

The minimum energy microwave range is broad, extending from about 1 to 60 GHz. But the earth's atmosphere attenuates the upper range, narrowing the window for ground-based systems to about 1 to 10 GHz. If antenna cost and sensitivity to Doppler frequency shifts are taken into account, the window narrows even further to 1 to 2 GHz. Fortunately, this includes the 1.420- to 1.667-GHz water hole. Although assumptions about the kind of signal a strange civilization might send us are fraught with uncertainty, and there is danger of being too anthropocentric, the microwave region seems to be a reasonable place either for a meeting or for eavesdropping.

As originally proposed in the Cyclops report, a phased array of paraboloidal dishes, each 100 meters in diameter, could be used to detect 1- to 2-GHz signals. There might ultimately be a thousand or more such dishes, but the search could start as soon as a few had been constructed. The dishes would be spaced with their centers about 300 meters apart so that they do not cast shadows on each other at low elevation angles. An antenna of 1000 dishes needs many square kilometers—it could perhaps be located in the desert in the southwestern United States.

The complete SETI system might ultimately cost many billions of dollars, depending on how many dishes were built. The cost of the antennas dominates the cost of the system; therefore, the cost will be lower if a signal is detected early during construction.

Drifting in a sea of frequencies

A complicating factor in the detection scheme is frequency drift. Drift may come from instability in the transmitter or receiver, or from revolution and rotation of the earth and of the planet on which the transmitter is located. (Uniform relative motion between a star and our sun will cause a fixed frequency shift but no continually

Table 31.1.

STAR SPECTRUM CHARACTER BY CLASS

Spectral Class[a]	Spectral Characteristics
O	Very hot stars with He II absorption
B	He I absorption; H developing later
A	Very strong H, decreasing later; Ca II increasing
F	Ca II stronger; H weaker; metals developing
G	Ca II strong; Fe and other metals shown; H weak
K	Strong metal lines; CH and CN bands developing
M	Very red, TiO bands developing

a. Sequence of letters remembered by generations of beginning astronomy students by the mnemonic: "Oh, Be A Fine Girl, Kiss Me!"

changing frequencies.)

It is easy to compensate for Doppler drift rates produced by the earth's rotation and revolution—since it is known where the receiver is looking, the necessary shift in the local oscillator frequency is also known (although slightly different degrees of shift are required at opposite ends of a wide band such as the water hole).

It is conceivable that the extraterrestrial transmitter may be purposely shifted in frequency. If two people are searching for one another in a forest, it is better for one to stand still than for both to be moving about. But the worst possible situation is for both to stand still. With this analogy in mind, the transmitting civilization might decide to sweep slowly through the 242 MHz of the water hole. At a sweep rate of 1 Hz, the process would take almost eight years.

For eavesdropping, Doppler drift compensation cannot be expected and some drift in the received signal must be expected.

Pointing and processing

The difficulty of the problem may now be grasped. If we wish adequate S/N(signal-to-noise power ratio) in our receiver, we must have low noise power, which requires a narrow band—probably less than 1Hz. But this means about 10^9 channels even within the water hole, and we should probably search all of them. Furthermore, we need a strong signal, hence a large antenna, and therefore a narrow receiving beam; which may mean 10^9 different pointing directions and we may have to search most of them. Furthermore, we do not know the drift rate, the polarization, or the modulation, adding further to the volume of the multidimensional parameter space that must be searched. Finally, we do not know *when* to look, and may wish to reexamine the same signal "point" in this space.

As indicated below, each "point" must be examined

for several minutes, so it is hopeless to search them all sequentially. Fortunately, it is possible to examine many such "points" simultaneously by using currently feasible data processing techniques.

Essentially, this examination requires a high-resolution spectral analysis, obtained by taking the Fourier transform of the raw received radiation. A spike in the frequency-domain representation (the power spectrum) indicates a sine wave in the input and therefore a possible signal from space.

Analog Fourier transform processors are most advanced at the moment—especially optical processors that use lenses and photographic film, but processors based on surface-wave acoustics are also being developed. Digital processors employing the fast Fourier transform (FFT) are not as yet so well developed as the analog, but it seems likely that they will soon be more powerful and less expensive.

The effectiveness of a processor can be measured by its time–bandwidth product (TBP), which is the ratio of the total bandwidth searched (here called the frequency band) to the width of the "channel" (here called the frequency bin). TBP is therefore equal to the number of channels that the system is capable of analyzing simultaneously. (The name, "time–bandwidth product," derives from the fact that the bandwidth, instead of being divided by the width of the frequency bin, can be multiplied by its reciprocal, the coherent integration time.)

Since the noise power is directly proportional to the width of the frequency bin, it is desirable to reduce that width as much as possible. Bin widths below 0.001 Hz are probably not of interest even if we could build them— partly because of modulation limitations, but mostly because the receiver requires at least $1/B$ seconds to settle down to a bin width of B hertz, which means that the search time would be excessive at such small bin widths. Bin widths of 0.1 Hz are technically feasible, but staying within the frequency bin during the 10-second coherent integration time requires drift rates less than 0.01 Hz.

Another lower limit on the width of the bin is the loss that occurs if it is narrower than the transmitted signal. One can only guess at the bin width needed to eavesdrop on communication signals, but there is good reason to believe that widths far below those of usual communication channels might be useful. For example, typical broadcast signals on earth, with bandwidths measured in kHz or MHz, have a large part of their power in the carrier, which often is stable to a fraction of a hertz.

A data processor with a TBP of, say, 10^6 could examine simultaneously one million 1-Hz bins over a 1-MHz band, or one million 10-Hz bins over a 10-MHz band, or 500 000 10-Hz bins in each of two orthogonal polarizations over a 5-MHz band (which would make it possible to examine all polarizations simultaneously).

Selectively searching the most likely stars would take a lot less effort than searching the entire sky. On the other hand, there are several reasons why a signal might be missed by a selected-target search. Other kinds of stars might have planets suitable for life—notably, M stars, which are far more common. Or a signal might come from a star farther away than 1000 light-years—even from another galaxy—with sufficient strength to be received. It would be worthwhile to receive such a signal, even though it couldn't be answered in any reasonable length of time. Or a signal might conceivably originate from some object not visible as a star from earth—perhaps a Jupiter-sized isolated mass.

If it were decided to search the entire sky (or entire promising areas of the sky) in a reasonable length of time, it would be necessary to look in several directions at once, and this means adding data processing capacity. For example, the proposed Cyclops dish array can accept signals from another main beam next to its original main beam if a duplicate phasing system and data processing system are added. In fact, another beam can be added for each phasing system and data processor added, up to the number of dishes in the array.

Equivalent to adding processors is recording the received radiation of each antenna element after it has been put through its IF amplifier, and processing it later. The beam can then be "pointed" in a sequence of directions by the processor. Non-real-time data processing would offer other important advantages if the state of the art in recorders could be sufficiently advanced. The hard copy that it produced would be invaluable if the receiver eavesdropped on an intermittent or highly directional signal. Within the data processor, the receiving antenna polarization could be adjusted; the beam could be repointed to put the target exactly at its center; and a filter could be matched to the signal. The net effect would be to increase the sensitivity (or range) of the system.

If weak signals are to be received, a large, high-gain antenna is needed, which implies a narrow beam. But the narrower the beam, the longer it takes to search the sky. The 300-meter dish at Arecibo, Puerto Rico, for instance, if operated at the water hole, would have to be pointed in 100 million different directions. For an even larger antenna, such as the Cyclops array, the number of pointing directions increases in proportion to the effective antenna area.

The number of pointing directions can be reduced by broadening the beam, by selecting a limited number of promising directions, or by looking in many directions simultaneously. A broad beam might be useful for a quick search for strong signals, but it would not aid in the ultimate search.

Box 31.1. Drake's Equation

Astronomer Frank Drake of Cornell University, in his *Intelligent Life in Space* (Macmillan), wrote, "At this very minute, with almost absolute certainty, radio waves sent forth by other intelligent civilizations are falling on the earth. A telescope can be built that, pointed to the right place and tuned to the right frequency, could discover these waves." Drake's "almost absolute certainty" is based on an equation he developed for N, the number of technologically advanced extraterrestrial civilizations that are communicating:

$$N = Rf_sf_pn_ef_lf_if_cL$$

In Drake's equation, R, the average rate of star formation in the galaxy, is equal to about 20 stars per year. While f_s, the fraction of stars that are suitable "suns" for planetary systems, is about 0.1; most stars belong to class M (see Table 31.1) and are probably too small, whereas a few, such as O and B stars, are almost surely too short-lived. The fraction of "good" stars with planetary systems, f_p, is now thought to be around ½—a value that has helped change scientists' minds about the possibility of extraterrestrial signals. Thirty years ago, f_p was estimated at a pessimistic 10^{-9}.

The mean number of planets suitable for life, within such planetary systems, n_e, is probably greater than 1. The value depends on what is considered necessary for life, but even if liquid water is assumed necessary, there are probably several planets per solar system that are neither too hot (like Mercury) nor too cold (like Neptune). Even secondary satellites, such as the large moons of Jupiter and Saturn, may be among the suitable habitats.

Estimates of f_l, the fraction of such planets on which life actually originates, have changed recently, and have given more reason for optimism. Thirty years ago, it was hard to conceive how the organic matter of life could originate. Since then, it has been demonstrated that a mixture of methane, ammonia, water, and hydrogen gases (which are constituents of Jupiter's atmosphere) is transformed by ultraviolet light or electric sparks into a wide variety of amino acids, and precursors of nucleotide bases, sugars, and other components of living matter. Some key remaining steps are obscure—it is still not known how replicating substances such as DNA–RNA are synthesized, or how a cell or sex develops. But in an ocean full of organic molecules and with lots of time, these steps do not appear improbable.

The fraction of such planets on which, after the origin of life, some form of intelligence arises—represented by f_i in Drake's equation—is also thought to be high. Evolution doubtless takes place wherever life arises, and intelligence has survival value.

It is intriguing to speculate on the value of f_c, the fraction of such intelligent species that develop the ability and desire to communicate with other civilizations. The intelligent beings might be like porpoises and therefore be unlikely to build radiotelescopes, or they may not have developed writing and therefore cannot accumulate knowledge, or they may have rejected technology and developed a pastoral society, or they may have developed a rigid 1984-type civilization in which such a thing as interstellar communication is discouraged. Nevertheless, few people are so pessimistic as to put the value of f_c below 0.1.

The mean lifetime, L, in years, of a communicative civilization might be zero if its members destroy each other before they can communicate, or it may be millions or billions of years if they solve their sociological problems. Because the value of L depends on the nature of a civilization that is surely very different from ours, and because even our own society isn't clearly understood, this factor is the most uncertain on the righthand side of Drake's equation.

Since R is about 20, and since the product of the next six factors is probably not much below 0.05, N may be numerically almost equal to L—which means that there may indeed be a large number trying to establish contact.   *R.E.M.*

Box 31.2. Why Search?

Why—when millions are starving; when man is dangerously polluting his environment and exhausting the earth's resources; when the human race may extinguish itself in atomic war—why should money be spent seeking signals from intelligent life on another world? There is no reason, if one believes that our civilization is at the end of the road after 4½ billion years of evolution. In fact, there is then little reason to carry out any large-scale socially supported projects such as the space program or research into biochemistry, medicine, cosmology, particle physics, or the mysteries of the brain.

But there is every reason for the search if one believes that the human race has not yet attained the peak of its evolutionary development, and that it may survive long enough to inherit a future as far from modern man's comprehension as the present world would have been to Cro-Magnon man. The search is prerequisite to communication, and communication will yield a vast array of benefits.

It would be an extraordinary coincidence if some other race were exactly as advanced as we were—more probably it is millions of centuries older or younger. But if it is as much as one century behind us, it is not yet using radio. It follows that any civilization that we contact is likely to be enormously more advanced. Thus, we may be able to use "their" knowledge and take a giant leap in our own science and technology. Far more important, we may discover the social structures most apt to lead to self-preservation and genetic evolution. We may discover new art forms that lead to a richer life. At the very least, communication will end the cultural isolation of the human race, enabling us to participate in a community of intelligent species. Then, perhaps, a spirit of adult pride in man will supplant childish rivalry among men.

It is likely that interstellar communication has been going on in our galaxy ever since the first civilizations evolved some four or five billion years ago. The participants in this heavenly discourse will have accumulated an enormous body of knowledge handed down from race to race—a sort of cosmic archeological record of our galaxy. If such a heritage exists, it will illuminate the human race's future. It would certainly be worth the cost of the search many times over.

By responding to a beacon, we reveal our existence and advertise earth as a habitable planet. Will then hordes of superior beings invade earth and annihilate or enslave mankind? If, as seems probable, interstellar travel is enormously expensive even for an advanced culture, then only the most extreme crisis would justify a mass interstellar migration—and even then it seems likely that the migrating population would seek uninhabited worlds to avoid adding the problems of conflict to those of the journey itself. In any event, our not responding would not really protect us, since a race bent on invading us could find us through our present radio emissions.

The possibility that mere communication with a clearly superior race could be so damaging to our psyches as to produce retrogression instead of advancement, even with the best intentions on the part of the alien culture, has been suggested. But there will be time for mankind to prepare itself. The round trip for communication may take a century or more, so the information exchange will not be a rapid-fire dialog. We may receive a vast amount of information, but the rate of reception, the gaps in the picture, and the effort needed to construct a model of the other race from the data should prevent any violent cultural shock.

No one can assert that interstellar contact is totally devoid of risk. But the potential benefits outweigh the risk. To obtain the benefits we must take that first step: We must search.

Bernard M. Oliver
Hewlett-Packard

Where or when?

A signal meant to attract our attention might be omnidirectional and continuous. Or "they" may choose to use high gain and point sometimes in one direction and sometimes in another. They may illuminate a given direction for only a few seconds, returning periodically (maybe once a year). Or they might illuminate a given direction for a long time and come back only after millenia have passed. Or they may have selected a few stars, of which we might hope to be one, for special and perhaps even continuous attention.

It has been suggested, for example, that a beacon from a binary-star system might be a fan-shaped beam whose plane is perpendicular to the orbital plane of the stars and includes the line joining them; in that case we should look at such a system only when that line is pointing toward us.

For eavesdropping, a signal from a planet rotating at the same speed as earth would intersect the SETI receiver for 1000 seconds if its beam were at least 4 degrees wide along the chord of intersection. Such an occurrence might seem like an extraordinary coincidence. On the other hand, if there are many such beams and if they are looked for long enough, at least one of them is bound to show up.

In any event, the question of *when* to look seems almost unanswerable. It may be necessary to prepare for all possibilities, including reexamining promising directions and/or frequencies repeatedly.

Trading off

In a working SETI system, many tradeoffs will have to be made to ensure an acceptable probability of success at a reasonable cost. For example, attempting to improve signal-to-noise ratio by the brute-force method of adding dishes would cost billions of dollars per decibel in an ultimate system. It would be far better to expend a more modest amount on reducing the noise temperature of the system. Fractions of a degree are significant, especially near the theoretical lower limit of 3 K (the sky temperature) for a space-based antenna, or somewhat higher for an earth-based antenna.

Data-processing equipment can be traded for time by doing things sequentially instead of simultaneously, and time can be traded for antenna area since signal-to-noise ratio goes up linearly with antenna area and with the square root of integration time. The exact compromise depends on the objectives of the system (for example, how much more is it worth to get the first positive result in, say, 10 years instead of 20 or 1000 years), on whether an area search is chosen in preference to a selected-target search, and on the maximum frequency-drift rate decided on. It might even be reasonable to suppose that the total expenditure for data-processing equipment should be of the same order of magnitude as the antenna expenditure.

Consider one of the possibilities: a Cyclops-like system consisting of 1000 100-meter dishes with a bin width of 0.4 Hz, a system noise temperature of 10 K, and coherent integration time of 2.5 seconds. Total integration time of 250 seconds in each pointing direction will be required to obtain adequate probability of detection and false-alarm rate. Such a system could detect a transmitter putting out 1 GW of equivalent isotropic radiated power at a distance of several hundred light-years. The system could, during the course of about 25 years, examine the million or so most likely stars (operating at a moderate duty cycle of 1/3), provided the frequency drift rate did not exceed 0.15 Hz. And it could do all this with only two processors, one for each polarization.

The system could search the entire sky in a slightly longer time—but would need 2000 data processors instead of two.

Or the system could look at the million selected targets and allow for frequency drift rates up to 1 Hz. But it would need 26 data processors, 13 for each polarization. For the same maximum drift rate, the system could look at the whole sky if it had 26,000 data processors.

In these examples, if each of the data processors had a TBP of 10^9, the system could cover the entire water hole. To search a wider band of microwave frequencies requires more data processors, or more powerful data processors, or more time, or some combination of these. To search promising directions more than once, the time of the search or the power of the data processors must be raised accordingly. These equipment numbers could be cut by a factor of 3 with non-real-time data processing if the data processors operated continuously while the antenna worked only one third of the time.

The first search for extraterrestrial signals of intelligent origin was made in 1960. Project Ozma, directed by Frank Drake, presently at Cornell University, searched 100-Hz bins over a 400-kHz band near the hydrogen line with a 26-meter dish, looking in two directions (at two promising stars about ten light-years away) in a single linear polarization. The system had a noise temperature of 350 K.

Today, with state-of-the-art receivers and existing radiotelescope antennas, the system sensitivity could be increased by at least 60 dB compared with Ozma: 15 dB by reducing noise temperature, 25 db by reducing bin width, and 20 db by using the 300-meter antenna at Arecibo. Furthermore, many more directions and frequencies could be examined. But to do this, more powerful data processors are needed. Thus, the NASA—Ames SETI group has turned its attention to the development of data processors. The group is developing a digital FFT processor and correlator with a TBP of 2×10^6, which, it is believed, can ultimately be expanded to a TBP of $2 \times$

Box 31.3. Why Not Travel?

Why not visit instead of sending messages? Because interstellar travel—while not theoretically impossible—is extremely difficult.

Theoretically, if a spaceship could accelerate at 1 *g* for one year, it would reach approximately the speed of light. If it accelerated at 1 *g* for 4.6 years (according to its own clock) and then decelerated at 1 *g* for 4.6 years, it would arrive at a point 100 light-years away. Within 100 light-years there are more than 1000 stars sufficiently like our sun (types *F, G,* and *K*) to be interesting. The ship could travel to one of these, explore it, and then return in a total ship time of only 20 years (of course more than 200 years would have passed on earth before the ship's return—the "twin paradox" of relativity).

But from the viewpoint of energetics, it is not so easy. Consider the energy required for a much more modest journey—round trip to the nearest star, 4.3 light-years away, in ten years. For the spaceship propulsion system, assume the theoretical ultimate: a 100-percent-efficient matter-annihilation engine whose exhaust velocity is equal to the speed of light. The trip would consume 34 pounds of fuel (half matter, half antimatter) for each pound of payload. For a 1000-ton spaceship (including crew, living quarters, engines, life-support systems, and whatever else is needed), 34,000 tons of fuel would be required. The total energy expended in the trip would be about 10^{18} kWh—far in excess of all the fossil and nuclear energy expended in human history.

For more moderate amounts of energy, the trip time and the size go up rapidly. A fusion rocket would take 80 years for the round trip and require a minimum theoretical mass ratio of 80; a fission rocket would require many hundreds and a conventional rocket many thousands of years. This is only to visit the nearest star, and we would expect to have to visit thousands—perhaps hundreds of thousands—of stars before detecting intelligent life.

Would even a highly advanced civilization be able to surmount these obstacles? It doesn't seem likely. Certainly, we on the earth lack by many orders of magnitude the ability to explore other solar systems.

In comparison, the energy requirements for radio communication are surprisingly modest. It would take of the order of 1 kW to power a transmitter that could be detected from as far away as 1000 light-years, using transmitting and receiving antennas no larger than the 300-meter dish at Arecibo, and a bandwidth of 1 Hz at S-band.   *R.E.M.*

10^9. The expanded processor is expected to be available in a few years at a price between $10 million and $100 million. As soon as even a more modest processor is ready, the Ames group hopes to begin a serious search with it.

If such a search is unsuccessful, the next step would be to construct larger antennas. But perhaps a signal will be detected!

32

David C. Black and Graham C. J. Suffolk
Concerning the Planetary System of Barnard's Star

It is now generally believed that the phenomenon represented by the Solar System, a star attended by a retinue of planets, is a relatively common one in the galaxy. Other planetary systems can be useful in understanding our own Solar System. Although it is unlikely that these systems will ever be studied in the same detail as our own, they provide information such as the spectral types of stars which seem to have planetary systems and in certain cases, information as to the gross structure of the planetary systems themselves. Unfortunately, it is difficult to observe distant planetary systems. Planets are intrinsically faint objects which emit little or no radiation of their own. Consequently, one must rely on indirect methods to detect their existence. The most successful such method to date is to look for perturbations or "wobbles" in the motion of the parent star which may be attributed to a dark companion(s) in orbit about the star.

The star which affords the most detailed inference of the dynamics of an extrasolar planetary system is Barnard's Star, a red dwarf of a spectral type dM5. This object has the highest recorded proper motion, 10.3 sec of arc per year, rendering the task of delineating perturbations in the star's motion somewhat easier. Van de Kamp and co-workers at Sproul Observatory perceived the importance of such observations and have been carefully observing Barnard's star during the period from 1916 to 1919 and again from 1938 until the present. These observations led van de Kamp (1963) to suggest a dynamical interpretation of the star's proper motion wobble as resulting from a Jupiter-like object orbiting the star with a period of 24 yr in a highly eccentric orbit. In 1969(a) van de Kamp refined his earlier work and suggested that the companion to Barnard's Star was a planet of 1.7 Jupiter mass, orbiting once every 25 yr in a somewhat more eccentric orbit than the 1963 solution. Later in 1969, van de Kamp published an alternate dynamical analysis of Barnard's Star. In this instance he considered two planets

in corevolving, nearly coplanar, orbits about the star. He found that two objects of mass 1.1 and 0.8 Jupiter masses, revolving with periods of 26 and 12 yr, respectively, fit the observed motion as well as the previous one-planet models.

We wish to reconsider here the published observational data on Barnard's star. The motivation for this reconsideration is that the available data are of sufficient quality to provide an indication of the number of significantly massive planets in the Barnard's Star system, contrary to the findings of van de Kamp (1969b).

Data Analysis

The data used for this analysis were gathered by the Sproul Observatory and tabulated by van de Kamp (1969b), and consist of yearly mean remainders in right ascension (R.A.) and declination (Dec.) during the interval from 1938.49 to 1968.40. A more detailed discussion of the observations and their reduction may be found in van de Kamp (1969a). That paper erroneously lists a value of +7 for the remainder in declination for the year 1953.51.

The analysis of the data undertaken here is predicated on the assumption that planets massive enough to affect detectable perturbations in the motion of Barnard's Star are in circular orbits. This choice of orbit geometries is not without justification. Our own planetary system is characterized by nearly circular orbits, this circularization presumably arising mainly from the presence of gas in the primitive solar nebula. It is likely that a nebula existed around Barnard's Star (as the planets themselves may indicate and that a circularization process acted there also.

The data were analyzed in two ways. First, remainders in both R.A. and Dec. were Fourier analyzed to determine what, if any, common orbital periods were present in the data. This has the drawback that periods less than 4 yr are not detectable in light of the yearly character of the data; and since the observations only span 31 yr, the temporal

Reprinted from *Icarus*, 19 (1973), 353. Copyright © 1973 by Academic Press, Inc.

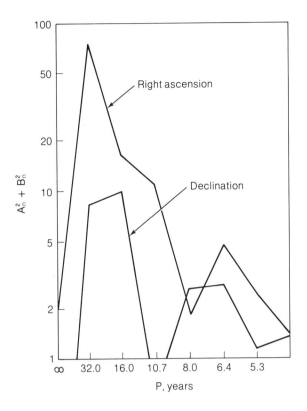

Figure 32.1. Power ($A_n{}^2 + B_n{}^2$) spectrum from a Fourier analysis of the remainders in right ascension and declination as a function of orbital period.

$$X_k = \sum_{i=1}^{N} a_i \sin \omega_i(t_k + \phi_i) \qquad (1)$$

$$Y_k = \sum_{i=1}^{N} b_i \sin \omega_i(t_k + \psi_i), \qquad (2)$$

where the index i runs over the number of planets, the index k refers to the 31 discrete positions in time at which the yearly means are prescribed, a_i and b_i are amplitude coefficients, $\omega_i = 2\pi/\tau_i$ is the orbital frequency of the i-th planet while ϕ_i and ψ_i are the zero phases of the i-th body in R.A. and Dec., respectively. The system represented by (1) and (2) is coupled through the ω_i's and contains $5N$ parameters.

The function minimized was

$$\Phi = \sum_k \left\{ |\bar{X}_k - X_k| + |\bar{Y}_k - Y_k| \right\},$$

where $\bar{X}_k$ and $\bar{Y}_k$ are the mean yearly remainders in R.A. and Dec., respectively. The procedure then was to determine those values of the parameter set $\left\{ a_i, b_i, \omega_i, \phi_i, \psi_i \right\}$ which minimize Φ. The functional minimization was accomplished by means of an IBM System/360 scientific subroutine using the method of conjugate gradients.

Results

Minimizations were carried out for 25 different sets of starting values for the parameters $\left\{ a_i, b_i, \omega_i, \phi_i, \psi_i \right\}$ with $N=2$. Shown in Table 32.1 are the orientations of the initial and converged orbits from 9 of the 25 trials. These nine were selected because they represent the extreme cases as well as contain the essential behavior found in all 25 minimizations. A wide-range of values for the inclination (i) and node (Ω) of the initial orbits was employed in the analysis. Specifically, $62° \leqslant i_s \leqslant 78°$, $66° \leqslant \Omega_s \leqslant 111°$ and $52° \leqslant i_s \leqslant 105°$, $40° \leqslant \Omega_s \leqslant 134°$ for companions B1 and B2, respectively. Such a range of starting configurations is necessary to adequately test for uniqueness of the converged configurations. Although a unique set of converged orbits was not obtained for both companions, several strong trends are evident in Table 32.1.

Perhaps the most consistent result is the narrow range of orbital planes defined by companion B1. Taking into account all 25 of the analyses, we find $\langle i \rangle = 76° \pm 2°$ and $\langle \Omega \rangle = 107° \pm 2°$, the uncertainties being $\pm 1\sigma$. A measure of the "reality" of this orbit for B1 is provided by Trial 5. In this case, the companions were initially placed in nearly coplanar orbits characterized by $i \sim 69°$ and $\Omega \sim 55°$, more than 50° from the average value of Ω for B1. However, when the residuals were minimized, the orbit of B1 converged to that defined by the average values of i and Ω. It should be noted that in contrast to the large readjustment

resolution of the power spectrum is marginal in the range from 10 to 30 yr. However, as can be seen in Figure 32.1, certain features do appear in the power spectra of the R.A. and Dec. data which are of use in the more detailed analysis to follow. Most of the amplitude in the R.A. and Dec. motion is supplied at periods in the range 10.7–32.0 yr. The time resolution is not sufficient to indicate the exact periods, but an inflection in the R.A. data for periods between 10.7 and 16.0 yr strongly suggests that at least two distinct periods are present, one with $16.0 \leqslant \tau \leqslant 32.0$ and a second with $10.7 \leqslant \tau \leqslant 16.0$ yr. Van de Kamp's two-planet solution yielded orbital periods of 26 and 12 yr, consistent with these intervals. There is also a peak at about 7 yr period which is common to both power spectra, suggesting the interesting possibility of a third body orbiting the star. We wish, however, to confine our attention in what follows to the family of two planet solutions.

The above analysis only serves to indicate the number and approximate values of orbital periods associated with the objects perturbing Barnard's Star. In order to obtain a more precise delineation of the dynamics of the system, an additional analysis was performed. Letting X denote the motion in R.A. and Y that in Dec., we have

Table 32.1.

STARTING AND CONVERGED COMPANION ORBITS

Trial	Companion	i_s	Ω_s	i_c	Ω_c	Mean Residuals (.1 μ) R.A.	Dec.	P(yr)
1	B1	72°	111°	77°	106°	2.7	2.8	26.2
	B2	55	46	60	56			12.0
2	B1	69	104	77	106	2.7	2.8	26.2
	B2	69	40	60	54			12.0
3	B1	69	104	75	108	2.7	2.8	25.9
	B2	52	133	57	53			11.8
4	B1	70	104	76	106	2.7	2.8	26.3
	B2	61	90	63	56			12.0
5	B1	75	52	76	107	2.7	2.8	26.4
	B2	64	57	60	57			12.0
6	B1	62	90	74	111	2.8	2.9	25.7
	B2	64	57	68	49			12.2
7	B1	78	67	81	110	2.8	2.9	26.4
	B2	64	57	52	51			12.3
8	B1	66	103	76	106	3.0	2.8	26.2
	B2	60	63	57	70			12.2
9	B1	72	111	76	107	2.7	3.2	26.1
	B2	105	134	90	84			11.9

of the orbit of B1, the converged orbit of B2 remained similar to its assumed initial orbit.

The set of converged orbital configurations for companion B2 are not, in general, as consistent as that defined by companion B1. However, if one considers only those configurations which yield the lowest average residuals in R.A. and Dec., one finds $\langle i \rangle = 60° \pm 2°$ and $\langle \Omega \rangle = 55° \pm 2°$. Trials 2, 3, and 4 represent convergence to this average orbit from very dissimilar initial orbits, the change in Ω from initial to converged value being 80° in Trial 3. It is particularly instructive to consider Trial 4 *vis-à-vis* Trial 5. In both of these minimizations the companions were placed in starting orbits which were nearly coplanar. The assumed starting orbits for the Trial 4 are essentially those of the solution given by van de Kamp (1969b), while the initial orbits in Trial 5 are similar to the average orbit defined by companion B2. In each case, the coplanar configuration was found to be unstable with respect to the data and a major readjustment of the orbital configuration resulted from the minimization. The converged configurations in Trials 4 and 5 were nearly identical and consistent with the average configuration found for each companion separately.

The average values for the parameters $\{\alpha_i, b_i, \omega_i, \phi_i, \psi_i\}$ which gave the best fit to the data are given in Table 32.2. The uncertainties listed for each quantity are statistical and do not include the uncertainty in the data. All 25 trials are involved in the averaging of the parameters for B1, while the averages for B2 involve only those minimizations which gave the lowest average residuals (i.e., 2.7 in R.A. and 2.8 in Dec.). The average orbit for B1 found

here agrees very well with that given by van de Kamp (1969b), however, there are clear differences in the elements of B2 as given here and by van de Kamp.

Discussion

The results obtained here indicate that two important conclusions may be drawn concerning the planetary system of Barnard's Star. The first is that *at least* two massive planets ($M \sim m_J$) orbit the star. Van de Kamp (1969b) concluded that the average residuals from his two-planet solution were no better than those provided by his one-planet solution. The average residuals obtained for the orbits given in Table 32.2 are lower than those found by van de Kamp. However, the most compelling evidence for at least two planets is provided by the well-defined orbit of B1 and the clear presence of a high frequency ($P \sim 12$ yr) oscillation in the remainders in Dec. (cf. van de Kamp, 1969b). The orbit of B1 indicates that most of its effect on the wobble of Barnard's Star will be in R.A. and that this effect will be characterized by a period of ~ 26 yr. It should be noted that if one considered a single planet in an elliptical orbit, one would expect harmonics of the fundamental period to appear in the observed motion. However, as van de Kamp (1963, 1969a) has found, the amplitude of these harmonics is not sufficient to account for the observed high frequency perturbation in Dec. There appears to be no way that B1 could give rise to a significant wobble in Dec., which the data require, with a period ~ 12 yr.

Related to the statement that at least two massive planets are required to account for the motion of Barnard's Star is the conclusion that the planets are not in coplanar orbits. As minimizations 4 and 5 indicate, the data appear to be unstable against coplanar orbits. The orbital configurations which provide the best fit to the data (Table 32.2) indicate that the relative inclination between the orbit of B1 and that of B2 is $\geq 50°$. It must be emphasized that the orbit given for B2 is the one which produced the lowest average residuals found in this work. That does not guarantee that it is in fact the orbit of B2. The technique

Table 32.2.

ORBITAL ANALYSIS OF BARNARD'S STAR

	B1	*B2*
Mass (M_J)	1.15 ± 0.15	0.7 ± 0.1
a (.1 μ)	9.7 ± 0.2	3.4 ± 0.4
b (.1 μ)	3.7 ± 0.3	2.7 ± 0.2
ϕ	1947.3 ± 0.2	1948.1 ± 0.1
ψ	1937.3 ± 0.3	1946.3 ± 0.1
P (yr)	26.2 ± 0.2	12.0 ± 0.1
$\langle i \rangle$	76° ± 2°	60° ± 2°
$\langle \Omega \rangle$	107° ± 2°	55° ± 2°

employed in this work is to locate minima of the function Φ in the 10-dimensional space defined by the parameters $\{\alpha_i, b_i, \omega_i, \phi_i, \psi_i\}$. Uncertainty in the data and limitations in the numerical technique are such that the minimun of Φ which is obtained may only be a local minimum, not an absolute minimum. The converged solutions 6-9 in Table 32.1 represent cases where other local minima were located. Thus, although the orbit given for B2 is the optimum one based on this analysis, one cannot in fact give a definite specification for the orbit of B2. This does not invalidate the requirement that at least two planets be present and that their orbits are relatively highly inclined. Possible cosmogenic implications of this planetary system are discussed elsewhere (Black and Suffolk, 1973b).

The discussion to this point has assumed that the data are free of any major errors. One type of uncertainty which is frequently discussed in this context is systematic error. However, systematic errors in general will not alter the conclusions reached here. The type of error which would possibly invalidate the results obtained here is a periodic error in Dec. with a period of order 12 yr and amplitude comparable to the remainders found by van de Kamp. The existence of such an error in the observations cannot be ruled out *a priori,* but no corroborative evidence for its existence has been given.

Acknowledgments

We thank A. Summers and J. Kennedy for their assistance in the numerical analysis. Support for this work was provided by NAS Postdoctoral Resident Associateships and is gratefully acknowledged. One of us (GS) also acknowledges the support of an ESRO Fellowship during the completion of this work.

33

Ronald N. Bracewell and Robert H. MacPhie
Searching for Nonsolar Planets

Explaining the planets has long been a goal of human curiosity, but how they originated is still controversial today. The erratic planetary motions, which are what drew attention to the planets in the first place, are by contrast now well understood, and the resolution of that puzzle played a pivotal part in the development of modern science. Elucidation of the origin of the planets may also prove to have deep meaning, although some doubt this, thinking that planets represent a more or less natural by-product of star formation. However, we have to admit that there is no generally accepted evidence for nonsolar planets. What is a nonsolar planet? Here the term is used to describe a hypothetical body, bearing a reasonable resemblance to our nine planets but associated with a star other than the Sun. Just how close the resemblance would need to be, whether we would accept highly eccentric orbits, low masses, unconsolidated aggregates, lack of a common orbital plane, or other extreme departures from our planetary system, is a matter that must await observation, for as yet the catalog of nonsolar planets contains no entries.

Discovery of such bodies would be of value for at least two reasons. First, the theory of stellar evolution would benefit from knowledge about other planetary systems and in particular the origin of our solar system would be illuminated, and second, the discussion of extraterrestrial life would be influenced. We are in a period of ferment over the possibility of making contact with advanced civilizations but we are suffering from lack of sufficiently detailed astronomical knowledge about our galactic environment. If we could say, "Here is a list of nearby stars that have planets" it would greatly help planning for future courses of action that might be taken to establish contact. The Soviet astrophysicist Shklovskii (1978) doubts whether there are any advanced civilizations within 1000–3000 parsec of us, and there is an interesting argument favoring uniqueness (Bracewell, 1974, 1978b).

Reprinted from *Icarus*, 38 (1979), 136. Copyright © 1979 by Academic Press, Inc.

The discovery of nonsolar planets would be a step toward settling a matter that is of philosophical importance to mankind (Oliver and Billingham, 1972; Morrison *et al.*, 1977).

This paper is concerned with a method that might permit the detection of nonsolar planets. As background we consider three alternatives that also might succeed. It turns out that all four methods present challenges to existing technology and therefore need to be subjected to comparative study as a preliminary to decisions for further action that might be taken subsequently.

Possible Search Methods

There are two existing observational techniques which, if sufficiently refined, might detect nonsolar planets. These are long-focus astrometry and radial velocity measurement. A third technique, which is not unknown, but certainly is not in everyday use in astronomy, is apodization.

Astrometry

Refracting telescopes with very long focal lengths (of the order of 15 m) produce plate scales in the range of 15 arcsec/mm that permit extremely small angular displacements of a stellar image relative to background stars to be detected. If the observations available over a year are combined a precision of 0.003 parsec is attainable. A purpose of such observations is to study binary stars by detecting their orbital motion about a joint centroid (van de Kamp, 1975; Gatewood, 1976).

If our solar system were to be observed from a distance of 10 parsec, the light from the planets would be difficult to detect but the Sun would appear to execute a complicated motion due to the gravitational attraction of the planets from their changing directions. Under the influence of Jupiter, which would dominate the other planets, the Sun would move along a projection of an epicycloid with a period of 12 yr and a lateral amplitude up to 0.0005 arcsec.

Thus the technique of astrometry is a candidate for the detection of nonsolar planets and needs to be studied to see how improvements of an order of magnitude or more can be introduced. Needless to say, emphasis on atmospheric refraction will be important because the twinkling of starlight involves random fluctuations of 0.1 arcsec in the direction of arrival.

Radial Velocity Measurement

Transverse motion is not the only concomitant of orbital motion; there is also radial motion, which results in Doppler shift. Thus, if the Sun and Jupiter were observed from a distance there would be a sinusoidal variation in radial velocity with an amplitude of 12 m sec^{-1}. For six years the radial velocity would exceed the mean and for six it would be below the mean. The amplitude of the Doppler shift, for example, of the prominent Fraunhofer line at 656.28 nm, would be 0.00003 nm. The mean radial velocity, which would be of the order of 10 km sec^{-1}, would shift the spectrum considerably more. Observations extending over the best part of the 12-yr period, or even more, depending on the precision achievable, would be required to demonstrate convincingly the existence of a periodicity in the radial velocity.

Current practice achieves precisions of several hundred meters per second, so here again improvements of little more than an order of magnitude (Serkowski, 1976, 1978) would offer the possibility of success with a nonsolar planet. It is not clear whether technology or astrophysics will prove to be limiting. It is conceivable that bulk motions of the stellar envelope with velocities of tens of meters per second, could foil attempts to exhibit a 12-yr periodicity. But the method might succeed on some stars while at the same time revealing new astrophysical phenomena on others.

Apodization

We have seen how two current techniques of astronomy could potentially advance to the level of detecting nonsolar planets. They are indirect techniques in the sense that the planet itself is not observed but only its influence on a second body. One would therefore not be sure of the exact nature of the body whose presence was inferred. Apodization is a method that faces up to the problem of direct detection of a nonsolar planet (Ken Knight, 1977). The basic problem is that diffraction of the starlight at the edge of the telescope aperture obscures the planetary light, a phenomenon that would not be avoided even if the Earth's atmosphere were absent. In the visible range Jupiter at 10 parsecs delivers only one photon per second to a telescope aperture of 1 m, whereas the Sun delivers a billion, many

of which reach the photographic plate at the point where the planetary image would be situated. To overcome diffraction it is merely necessary to shade the aperture so that the transmission of light falls off gradually from center to edge making the edge imperceptible. Theoretical studies indicate that the problem is purely a technological one, of undoubted difficulty, but inviting attention (for detailed discussion see B. M. Oliver in Black and Piziali, 1978). The apodized telescope would need to be in space orbit. It would not require observations covering years but would give prompt answers. All three of the methods so far discussed depend on observations made in visible light.

Infrared Radiometry

By going to infrared wavelengths one realizes an increase in planetary radiation relative to the stellar radiation. Although the thermal infrared radiation from the planet is not strong it nevertheless occurs near the peak of the black-body spectral distribution, whereas the infrared band is well down from the peak of the stellar spectrum. A drawback is that the observations must be made from earth orbit in order to avoid the infrared radiation from the Earth's atmosphere. However, many of the technical problems of infrared equipment in orbit have already been considered in connection with the infrared astronomy satellite (IRAS) (Hedden, 1976). Two concepts are discussed here. One involving long-baseline interferometry was proposed to the NASA-Ames Workshop on Extrasolar Planetary Detection, May 20–21, 1976, by Dr. Mark A. Stull and Dr. Thomas A. Clark. The other involves a much shorter, spinning interferometer, suitable for shuttle launch (Bracewell, 1978a) into earth orbit.

Infrared Considerations

The Infrared Advantage

It can be calculated that, at the distance of 10 pc, the light from a planet like Jupiter is approximately 2×10^{-9} of that from a star like the Sun. If Ω is the solid angle that the planet subtends at the star and A is the albedo, the ratio of the light energy L_p received from the planet to the light energy L_s received from the star is given by

$$L_p/L_s = A\Omega/4\pi.$$

At maximum elongation, when the planet presents one bright and dark half in our direction, it is estimated that A might be around 0.2. Then we obtain

$$L_p/L_s = 0.2 \times 10^{-7}/4\pi = 1.7 \times 10^{-9}.$$

The corresponding ratio at infrared wavelengths is quite

different because the planet is shining in its own right at a level appropriate to its temperature. In the long-wavelength Rayleigh-Jeans regime, brightness B is given by

$$B = 2kT\lambda^{-2} \text{ W m}^{-2} \text{ Hz}^{-1} \text{ sr}^{-1},$$

where $k = 1.38 \times 10^{-23}$ JK^{-1} is Boltzmann's constant, T is temperature, and λ is wavelength. Let P_s and P_p be the infrared powers received on earth in a narrow band from the star and planet, respectively. Then

$$P_p/P_s = \alpha(T_p/T_s)(D_p/D_s)^2,$$

where T is temperature and D is diameter. The factor α involves the surface emissivities, limb darkening and other refinements. Taking α to be unity and using the standard data of Table 33.1, we find

$$P_p/P_s = 2 \times 10^{-4}.$$

There will be a component of stellar infrared radiation reflected from the planet, but it will be negligible.

These elementary calculations show that there is an intensity factor of 10^5 favoring infrared over visible radiation as a medium for planetary detection (Bracewell, 1978a).

Choice of Wavelength

In the preliminary calculations establishing the existence of the infrared advantage it is sufficient to use the Rayleigh-Jeans approximation to the Planck formula for the brightness of a black body. We now need the full expression

$$B = 2hc\lambda^{-3}[\exp(hc/k\lambda T) - 1]^{-1} \qquad \text{W m}^{-2} \text{ Hz}^{-1} \text{ sr}^{-1},$$

where $h = 6.6 \times 10^{-34}$ Js is Planck's constant, and $c = 3 \times 10^8$ m s^{-1} is the velocity of light. At a

Table 33.1.

REFERENCE DATA FOR STANDARD STAR–PLANET SYSTEM

	Star	Planet
Temperature	6000°K	128°K
Diameter	1.39×10^9 m	1.4×10^8 m
Angular diameter	0.00096″	0.000097″
	(4.7×10^{-9} rad)	(4.7×10^{-10} rad)
Distance	3×10^{17} m	
	10 pc	
	33 ly	
Angular separation	0.53″	
	(2.6×10^{-6} rad)	
Linear separation	7.78×10^{11} m	

wavelength of 40 μm the brightness falls below the Rayleigh-Jeans value by a factor of 5.5 so that the overall infrared advantage would drop to about 1.8×10^4. At a wavelength just under 10 μm the advantage of 10^5 applying at the long-wavelength end is wiped out. Choice of wavelength must therefore be to the long-wavelength side of about 20μm if the infrared advantage is to be realized.

As with apodization, infrared interferometry does not depend on observations spread over the orbital period of the planet.

Suppression of Starlight by Interference

Long Baseline Interferometry

The techniques of very-long-baseline interferometry in radio astronomy suggest an application to the detection of a planet using infrared radiation. Table 33.1 gives the parameters of a star-planet system similar to the Sun plus Jupiter observed from a distance of 10 parsec (32.6 light years) in a direction perpendicular to the planet's orbital plane. In subsequent computations these parameters will be adopted. The basic phenomenon noted is that as the elements of an interferometer are moved apart fewer and fewer radiating objects remain detectable and they are the ones of least angular diameter. Since the star is much larger in diameter than the planet there should be a range of interferometer element spacings that tend to suppress the star. As a numerical example consider that the star is a uniformly bright disk of radius b. As far as a two-element interferometer is concerned the star might as well be a line source whose intensity I versus angle θ is given by

$$I = I_0(1 - \theta^2/b^2)^{1/2}.$$

The fringe visibility V expressed as a function of interferometer baseline length s measured in wavelengths will be given by

$$V = (\pi bs)^{-1}J_1(2\pi bs).$$

With a baseline spacing of 40 km, which implies two orbiting collectors connected together by an electromagnetic link, and a wavelength of 40μm, the baseline is 10^9 wavelengths long. The angular spacing of the interferometer fringes is then 10^{-9} rad. The star diameter is 4.6 times greater, or we could say that just over four fringes fall across the star (Figure 33.1). Consequently, as the star changes its direction relative to the fringe pattern little rise and fall in received power is noticed. In fact we can calculate the degree of modulation from the visibility formula above and find $V = 0.028$. (In this calculation the envelope of the Bessel function is appropriate.)

By contrast, the planet has an angular diameter

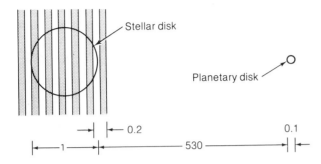

Figure 33.1. Geometry for very-long-baseline approach. Dimensions in milliarcseconds.

4.6×10^{-10} rad, which is less than the fringe spacing. As the fringes sweep over the planet the modulation of received power is substantial and is described by a visibility that may be calculated as before and comes to $V = 0.76$. Thus an advantage of $0.76/0.028 = 27$ has been gained. This is offset by the Rayleigh-Jeans correction factor of 5.5 with the result that the factor 10^5 previously favoring the star is reduced by a factor $27/5.5$ to 2×10^4.

Thus even though the notion of placing many fringes on the star seems qualitatively attractive it does not give dramatic quantitative results. The sample calculation given above may be refined, however, and in particular will be improved when limb darkening of the star is taken into account.

Use of Interferometric Null

A means of enhancing the planetary radiation relative to that from the star would be to place a null of an interference pattern on the star. Of course there will be practical questions related to pointing the null and making sure that the interferometer has the necessary degree of symmetry to ensure that an effective null is achieved. However, first we investigate the potential gain from this principle to see whether it is encouraging. Let the intensity of the interference pattern having fringe spacing Φ be described as a function of direction by $\sin^2(\pi\theta/\Phi)$.

If the star is situated at $\theta = 0$ then the response to the planet will be maximized if the planet lies at $\theta = \Phi/2$, as in Figure 33.2. Since the angular separation of star and planet is 2.6×10^{-6} rad this gives $\Phi = 5.2 \times 10^{-6}$ rad or 1.06 arcsec. The interferometer baseline necessary to produce this fringe spacing is $10^6/5.2$ wavelengths, which at a wavelength of 40 μm, comes to 7.7 m.

We now calculate the star-to-planet power ratio P_s/P_p at 40 μm when a null is centered on the star and the planet is on an interference maximum. We get

$$P_p/P_s = \int\int_{\text{planet}} \sin^2(\pi\theta/\Phi)d\Omega / 5.5(T_s/T_p)$$

$$\times \int\int_{\text{star}} \sin^2(\pi\theta/\Phi)d\Omega.$$

To find the ratio of the two integrals we first note that the integral over the planet is just the solid angle subtended at the Earth by the planet, namely $\pi(0.000049/206264)^2 = 1.74 \times 10^{-19}$ sr. The integral over the star reduces, for a uniform disk of radius b, to

$$2\int_0^b 2b(1 - \theta^2/b^2)^{1/2}\sin^2(\pi\theta/\Phi)d\theta$$

$$\approx 4\int_0^b (b^2 - \theta^2)^{1/2}(\pi\theta/\Phi)^2 d\theta$$

$$= 4(\pi/\Phi)^2(\pi b^4/16) = 8.75 \times 10^{-24}.$$

We now find $P_p/P_s = 80$. This is indeed a favorable result because it shows the power received from the planet exceeding that from the star by a factor of 80.

Spinning Interferometer

In situations where small signals have to be detected in the presence of unwanted signals that might cause confusion it is a well-established practice to modulate the desired signal. Consider the interferometer previously described to be spinning with angular frequency ω about an

Figure 33.2. Intensity of interference fringes and line-integrated brightness (W m^{-2} Hz^{-1} rad^{-1}) of star and planet (not to scale).

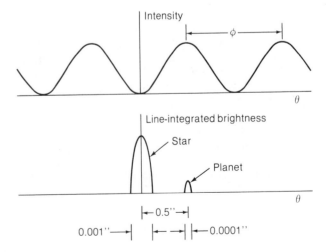

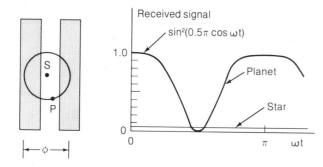

Figure 33.3. As planet P traces locus through fringes of spacing Φ flat-topped signal rich in harmonic at 4ω is received. Precisely centered star S produces small constant signal.

axis passing through the star. Then the signal from the star will not vary in strength but the signal from the planet would rise and fall with a fundamental frequency 2ω as in Figure 33.3. The waveform would not be strictly sinusoidal but of a characteristic flat-topped form $\sin^2[(\pi/2)\cos\omega t]$ containing a noticeable amount of the 4ω harmonic. The ratio a_2/a_1 of the second harmonic at 4ω to the fundamental at 2ω is 0.31.

Very faint signals can be recovered, if they are modulated at a known frequency 2ω, by synchronous detection that filters out that frequency. Alternatively, where the modulation is substantially nonsinusoidal, the received signal can be broken into segments of duration $1/2\omega$ and averaged.

Pointing Error

In the presence of pointing error the power received from the star will not be steady but will rise and fall in a characteristic manner depending on the amount of error and will not in general possess the same form as the planetary signal. The planetary modulation can be made markedly different if the interference fringe pattern is compressed so that the planet is κ fringe spacings away from the star instead of half a spacing. Such compression can be obtained by lengthening the interferometer baseline or by shortening the wavelength.

The signal from the planet, which was previously of the form $\sin^2[(0.5\pi)\cos\omega t]$ will now become $\sin^2[(\kappa\pi)\cos\omega t]$. The second harmonic content of this waveform is distinctly enhanced; in fact it is possible to choose κ so as to suppress the fundamental component at 2ω completely. On the other hand, the signal due to the star, which is closer to the rotation axis, will be more nearly sinusoidal with a fundamental frequency 2ω. Hence, by singling out the 4ω component, a frequency that is very precisely known, it should be possible to gain an advantage. One way of using this advantage would be to relax the pointing

accuracy requirement on the spinning interferometer. Further aid in disentangling star from planet could come from measurements at two wavelengths chosen to capitalize on the differences in spectra of star and planet. spectra of star and planet.

The basic pointing accuracy required is 1 msec of arc, or about the angular diameter of the star. It would not be surprising if stability of this order could be achieved with a massive rotating body in space. The most obvious approach is continuous attitude control based on an optical star-tracking telescope as in the proposed satellite gyro test of general relativity (Everitt, 1977). The telescope axis would be fixed on the interferometer null and share as much as possible of the infrared optics. As the principal angular perturbations are calculable a variant is to correct the attitude at discrete intervals and allow the interferometer to perform approximate force-free (Poinsot) motion in between. The various possibilities require study in relation to the orbits in which the interferometer might be placed. Internally damped Poinsot motion offers the possibility (given adequate signal-to-noise ratio) of allowing pointing error to be determined and corrected for by computation as an enhancement to pointing control.

Signal-to-Noise Ratio

The observing time required to detect a planet depends on the signal-to-noise ratio. The time would be shorter if the collecting area A were greater or if the bandwidth $\Delta\lambda$ were wider. For numerical calculations it will be assumed that $A = 1$ m^2 and $\Delta\lambda/\lambda = 0.1$. Three important factors dominate the discussion, the first being the failure of current infrared detectors to achieve ideal performance. However, any calculations based on today's detectors will be outdated by the time a spinning infrared interferometer could be launched because intensive effort is going into the development of better infrared detectors for a variety of important applications. Therefore it makes sense to assume that infrared technique will not be limiting several years from now. Even so, the infrared optics will need to be cooled to liquid helium temperatures and carefully shielded from terrestrial radiation. Such techniques will be introduced on the infrared telescopes now being planned (Witteborn and Young, 1976; Hedden, 1976).

Photon Noise

The number of photons collected per second on a receiving area A is given by

$$F = 2\eta c\lambda^{-3}[\exp(hc/k\lambda T) - 1]^{-1}(\Delta\lambda/\lambda)\Omega_p A,$$

where η = effective overall emissivity of planetary surface, $\Delta\lambda/\lambda$ = fractional bandwidth, Ω_p = solid angle of planet seen from Earth, and T = planetary temperature.

Adopt the following values appropriate to a planet like Jupiter at a distance of 33 light years: $\eta = 0.2$, $T = 128°K$, $\lambda = 40\mu m$, $\Delta\lambda/\lambda = 0.1$, $\Omega_p = 1.7 \times 10^{-19}$ sr, $A = 1\ m^2$.

Then $F = 2$ photon per second on the average. But the random rate of arrival means that for every n photons received there is photon noise measured by an r.m.s. uncertainty $n^{1/2}$. To make a signal level reading to 10% accuracy will take 50 sec and to 3% accuracy will take 500 sec, neglecting thermal noise. To establish a graph showing the rise and fall of the planetary signal, say about 70 readings to 3% accuracy, therefore requires 35,000 sec or 10 hr of observing time.

Zodiacal Light

The unavoidable photon noise would determine the detection time were it not for infrared emission from the particles that contribute to the zodiacal light. Only about 1 μm in size and spaced about a kilometer apart they nevertheless produce a strong infrared background against which a planet has to be detected. Rocket observations reported (Soifer *et al.*, 1971; Pipher *et al.*, 1971) in the range 5 to 23 μm are fitted by a grey-body emission spectrum corresponding to a color temperature of 280°K and an optical thickness of 3×10^{-8}. This means a photon flux density of 6×10^8 photons $sec^{-1}\ cm^{-1}\ \mu m^{-1}\ sr^{-1}$. With the numerical data given, the photon noise due to zodiacal light particles will be 150 photons sec^{-1} compared to a mean planetary rate of 2 photons sec^{-1}. An observation extending over a month will now be required to detect the planetary signal and measure it to a few percent accuracy.

Confusion Due to Field Stars

A 14th-magnitude field star situated in a direction only 0.5 arc-sec away from the star under investigation could simulate a planet, but if such a situation were encountered the field star could immediately be identified as nonplanetary by its infrared spectrum. Thus at 10 μm the field star would be much stronger than at 40 μm, whereas a planet would be undetected. However, the situation can arise only infrequently, about once in 20,000 in the galactic plane.

Features of a Spinning Infrared Interferometer

Optical System

The infrared radiation collected by the two apertures needs to be brought to a common location where cancellation by interference can take place. In principle this could be done by a pair of lenses with common focal points and

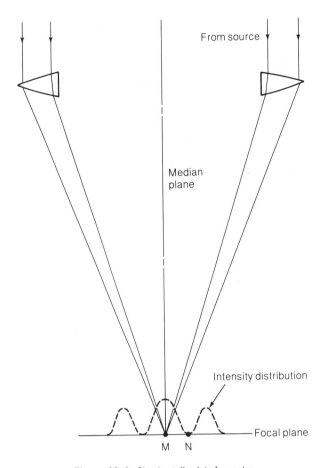

Figure 33.4. Simple stellar interferometer.

focal planes as shown in Figure 33.4. Because of transmission losses in the lens material it is desirable to avoid large lenses and to utilize reflection optics. However, the simple arrangement allows certain factors to be explained clearly. For example, we see from symmetry that, if there is a point source in the median plane, an interference maximum occurs where the median plane cuts the focal plane at M, whereas what we need is a null. There is a null line nearby at N but we cannot use it for the following reason. As the wavelength of radiation changes the maximum at M remains in the median plane while the minimum shifts. Therefore, if a band of radiation is employed, as is necessary for sensitivity, the null tends to be washed out and the suppression of the starlight is impaired. What we want is a central wideband null whose location is *independent* of wavelength. This is just what results from the combination of two off-axis paraboloids P_1 and P_2 with foci respectively at F_1 and F_2 and with a beam-splitter H (Figure 33.5). Focus F_1 receives one-half of the power from P_1 and also, after reflection at H, one-half of the power from P_2. Because of the phase reversal at reflection, the contributions at F_1 from a point source in the median plane cancel. A similar cancellation occurs at F_2, where a duplicate infrared detector may be employed if desired.

The reflection system differs in another way from the lens arrangement. There is no maximum adjacent to the null; the null spreads throughout the focal patch, whose size is set by diffraction at about 0.2 mm. This is several times greater than the distance MN and eases the requirement on the infrared detector which, in the lens arrangement, would have to be small compared with MN in order to be on the null.

In the illustration only two curved mirrors are shown for simplicity but more sophisticated optical systems such as the Gregorian arrangement with several additional reflecting surfaces as in the planned shuttle infrared telescope facility (SIRTF) (McCarthy, 1976) can be contemplated.

Tracking Control

A standard star tracker is a small telescope that forms a star image on a sensor array that can detect when the image is off center and actuate a servomechanism to bring the telescope axis back onto the star. The actuators may move the telescope on its platform or, by means of jets or inertia wheels, may move a whole satellite so as to maintain orientation. Such a star tracker could be mounted rigidly to a rigid interferometer frame. Any flexure introduces the possibility that the star might move off the interferometer null while remaining correctly positioned on the axis of the star tracker. To avoid this situation it is desirable that the tracking system utilize as much as possible of the interferometer optics. If, for example, the visible starlight falls on the parabolic mirrors, interference takes place at F_1 and a servomechanism could lock onto a null. Naturally it costs more to figure mirrors to the higher quality required for visible light but perhaps only some central portion of each mirror would need upgrading. The advantage of being able to work with visible light has been recognized in the 92-in. infrared telescope recently in-

stalled by the University of Wyoming (Gehrz and Hackwell, 1978). To have detectors for both infrared and visible light simultaneously present it would be necessary to separate the two by a 45-degree plate S that transmits visible light and reflects the infrared to a new focus F_3 where the infrared detector is placed. After the interferometer is in orbit a collimation adjustment will be required to bring the tracking axis into the median plane of the interferometer with the aid of a variable delay in the visible beam. This is shown schematically by a tiltable transparent plate C. A visible-light detector at F_1 feeds a servomechanism that controls the interferometer in yaw (rotation about an axis perpendicular to the paper). Control in pitch (rotation about axis F_1F_2) depends on the fact that the interferometer spins about the axis to the star, no additional detector being required. Detection of the null would be facilitated by modulating the path difference in the interfering beams, for example with a transversely vibrating wedge W. The same effect can be achieved by vibrating plate H axially along the line F_1F_2.

Thermal Considerations

Infrared sensitivity requires that thermal radiation to the detector be minimized by cooling the interior walls of the interferometer tube, the mirrors and other components. Liquid or superfluid helium temperatures are indicated for the critical components but less critical parts may run at higher temperatures such as the triple point of neon at 30°K provided the temperatures can be stabilized so as not to cause unacceptable thermal distortion. It will be appreciated that the interferometer pathlengths of 125,000 wavelengths need to be kept equal to a fraction of a wavelength, or a part in several million. It is true that by careful attention to symmetry the designer can relieve the situation because *uniform* temperature change does not

Figure 33.6. Infrared interferometer combined with interferometric star tracker using visible light.

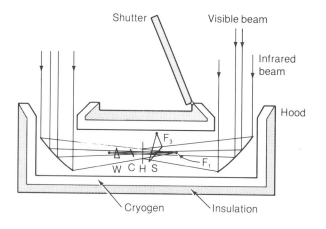

Figure 33.5. Interferometer using reflection optics exhibits wideband null.

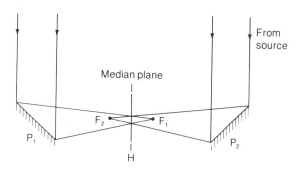

disturb the interferometric balance, nevertheless temperature stability is vital. The baseline path difference, as measured in visible wavelengths, seems to make even higher demands, but in fact the fundamental requirement is that the absolute baseline length be held to a small fraction of the *infrared* wavelength. One way of understanding this is to note that if the star-tracking interferometer were to slip a fringe and lock onto the next null the operation of the infrared interferometer would hardly be affected.

In Figure 33.6 two cavities are shown. One contains cryogenic fluid and an outer cavity contains insulation. The insulation might consist of many windings of very thin conductor such as aluminized mylar, slightly crinkled to allow vacuum spaces. Convection is eliminated, temperature uniformity is encouraged by the conduction along the sheets, radiation transfer, which varies as $\epsilon T^3 \Delta T$, is inhibited by the low temperature T, the very small temperature difference ΔT between layers, and the low emissivity ϵ, and conduction can occur only through point contacts of the mylar film.

Windows to the mirrors are a potential route for heat to enter and must be provided with carefully designed hoods. In addition shutters are required to keep out solar and terrestrial radiation in certain orientations. The representation shown can be regarded only as schematic.

Acknowledgments

Funds for the support of this study have been allocated by the NASA-Ames Research Center, Moffett Field, California, under Interchange No. NCA2-OR745-716. The authors benefited particularly from discussions with Dr. David C. Black and Dr. Russell Walker.

34

Thomas B. H. Kuiper and Mark Morris
Searching for Extraterrestrial Civilizations

A concerted national, and perhaps international, effort to find radio signals from extraterrestrial civilizations (ETC's) will probably be started within the next few years. It will be the outcome of nearly 20 years of smaller searches at various radio observatories (Sagan and Drake, 1975) and about a decade of preliminary planning for a dedicated facility (NASA, 1977). One major study conducted by Stanford University and NASA Ames Research Center (Oliver and Billingham, 1972) resulted in a design for an array of up to 1000 antennas of 100-meter aperture. The search for extraterrestrial intelligence (SETI) has been identified in "Outlook for Space" (NASA, 1976) as one of the possible future tasks for the National Aeronautics and Space Administration (objective 125) and the large receiving array as one of the prerequisites (system 4010).

In planning for SETI, there are two possible approaches. One is to develop scenarios for the presence and behavior of ETC's based on our present understanding of the universe (and perhaps an anthropomorphic view of the nature of "civilizations"), and then to devise a strategy that appears to maximize the likelihood of contact. An advantage of this approach is that it makes it possible to demonstrate the feasibility of interstellar contact at our level of technological development. This is the approach that was used in Project Cyclops (Oliver and Billingham, 1972) and further pursued in the Science Workshop on Interstellar Communication. It has led to the concept of a very large antenna array for SETI. An alternate approach to SETI, which is being developed at the Jet Propulsion Laboratory (Gulkis et al., 1976), is to make no assumptions about extraterrestrial signals other than that they may exist, and to survey exhaustively in frequency over the whole sky for anomalous radio emission. This approach has also received the endorsement of the Science Workshop on Interstellar Communication (B.C. Murray, private communication). With this philosophy, one does not

ask whether a particular receiving system is adequate, but rather whether the proposed program will extend our observational base to new frequencies, sky directions, or sensitivities. From this point of view, almost any SETI activity can make a valuable contribution.

The apparent premises used in scenario approaches to SETI planning can be reduced to the assumptions that (i) we are not unique as a technological civilization in our galaxy, (ii) interstellar travel for the purposes of "manned" exploration or colonization never reaches the level of feasibility or practicability, and (iii) interstellar beacons meant for contact between civilizations are likely enough to warrant intensive searches. A direct conclusion from these assumptions is that the likelihood of intercepting a beacon increases with increasing sensitivity of the detecting apparatus, and thus with increasing aperture size of the receiving antenna. Our purpose in this article is to examine critically these three interrelated assumptions in an attempt to outline a productive strategy for SETI. The scenario approach is pursued further to investigate the question of what sensitivity is required in the detecting apparatus in order to draw significant conclusions within the framework of our rather general scenario.

We follow the principle that, with complete lack of factual information on the existence or behavior of ETC's, the most meaningful way to proceed is to make a minimum number of assumptions and only assumptions that are consistent with (i) our understanding of natural physical processes, and (ii) the expectation that the behavior of extraterrestrial beings or civilizations may be extrapolated from the gross behavioral patterns of terrestrial animals and of human beings in particular. We consider this to be a proper application of Occam's razor, in that it puts the least number of unknown or unknowable parameters into a model for interstellar intercourse. Of course, if observations prove to be inconsistent with such a model, then it is proper to introduce additional or alternative hypotheses. By proceeding in this way, we hope to outline a maximum likelihood situation regarding the presence of other civilizations in the galaxy and their

Reprinted from *Science*, 196 (1977), 616. Copyright 1977 by the American Association for the Advancement of Science.

ability to contact ours. By referring to historical trends in human civilizations, we make the implicit but quite plausible assumption that all civilizations have, in principle, similar origins in the natural selection process and that the behavior of organisms is thus determined in large part by natural forces which are similar everywhere. By proposing a framework that is most consistent with available knowledge, we can suggest guidelines for a search for extraterrestrial intelligence that, whether successful or not, has meaningful consequences *at our level of understanding of the universe*.

Interstellar Travel Reexamined

It is often assumed that interstellar travel is, if not strictly impossible, highly impractical and therefore very improbable (Marx, 1963). The contention of this article is not that the practice of interstellar travel is an inevitability for all technologically advanced civilizations, but that the probability is high enough that, given a modest number of advanced civilizations, at least one of them will engage in interstellar travel and thus colonize the galaxy. Indeed, if various optimistic estimates for the rate of emergence of technological civilizations (Oliver and Billingham, 1972; Ponnamperuma and Cameron, 1974; Shklovskii and Sagan, 1966) are correct, then this probability would only have to exceed 10^{-8}.

The plausibility of interstellar travel has been argued previously (Spencer and Jaffe, 1963b; Powell, 1972; Bracewell, 1974). The Project Cyclops report (Oliver and Billingham, 1972, p. 34), on the other hand, follows the analysis of Spencer and Jaffe (1963b), but comes to the entirely different conclusion that "interstellar flight is out of the question not only for the present, but for an indefinitely long time in the future." Spencer and Jaffe had shown that a multistage deuterium fusion rocket would require a mass ratio (fuel to payload) of 9 for a one-way trip (with deceleration) at one-tenth the speed of light (0.1c). The Project Cyclops report makes the unnecessary assumption that a return trip need also be made, which not only increases the mass ratio to 81, but also increases the travel time by a factor of 2. Without this assumption, it becomes possible to travel to about ten stars at 0.1c within one human lifetime (although we do not need to make the further assumption that the trip be limited to a single human lifetime). Even if a return voyage were intended, the spacecraft could be self-sufficient enough to replenish its fuel supply at the destination.

There are two potential objections to these arguments for the feasibility of interstellar travel. The first lies in the anticipation of nuclear fusion engineering capabilities. However, in keeping with the principle of minimum assumptions, no fundamentally new physics is being advocated, and since no known physical principle states that a deuterium fusion rocket is impossible or unfeasible (it appears to be primarily an engineering task), there is every reason to expect its realization within the next century or so. Indeed the feasibility of interstellar propulsion has been seriously investigated in Project Orion and Project Daedalus (Gatland, 1974; Bond and Martin, 1975; Spencer and Jaffe, 1963a). Note also that, for the sake of a plausibility argument, we have confined ourselves to fusion energy propulsion, whereas other possible forms of propulsion can be envisioned.

The second possible objection is that the distances between suitable habitats might be prohibitively large; that is, the travel time is long compared to the lifetime of an individual. There are several answers to this. First, there is no reason to suppose that there are no intelligent beings having significantly longer lifetimes than ours. Second, a suitable habitat need not be narrowly defined: given presently foreseeable technological capabilities and the knowledge through preliminary unmanned exploration of what to expect, we suppose that a reasonably large fraction (>0.01) of planets could be rendered inhabitable. Third, given a somewhat more advanced technology, it may be possible to construct spaceships of sufficient sophistication to permit several generations to live in circumstances not significantly less comfortable than those encountered in crowded cities. Fourth, we cannot rule out the possibility that long voyages might be undertaken in a state of suspended animation. Finally, it is not clear that colonists need confine themselves to planets. With technology given a free reign, one can extrapolate O'Neill's (Reading 57) discussion of space colonies to consider orbiting habitable vessels about new stars (Jones, 1976).

This objection applies to any discussion of SETI. If habitats suitable for colonization are so rare as to preclude interstellar travel any *any* civilization, then one can argue that planets capable of engendering intelligent life are probably also rare, to the point that radio beacons are, at best, few and far between. The most urgent and the most technologically feasible development that will have a bearing on the presence of ETC's (aside from actually detecting their signals) is the measurement of the distributions of planets around a sizable number of stars.

Colonization of the Galaxy

Therefore, given advanced technological civilizations, which are capable of transmitting powerful radio beacons which we could detect across interstellar distances, it is likely that a significant fraction of these civilizations will physically expand by colonizing their neighboring galactic environments (Viewing, 1975; Hart, Reading 42;

Jones, 1976).[1] The argument may be advanced that, for economic or other reasons, ETC's may simply choose not to engage in colonization. This, however, would be counter to the trends evidenced by life-forms on our own planet, where every species extends itself as far as is physically possible. Given man's historically proved urge to explore, expand, and colonize, we make the minimal assumption that this trend will not be halted or reversed at our present stage of development. That is, the technological trend will continue, and manned visits to neighboring stars are likely to occur. This tendency is extrapolated to be the same for all technological civilizations.

Having argued that interstellar travel is a likelihood for technological societies that do not annihilate themselves, we now proceed to calculate how quickly a technological civilization might populate the galaxy. The minimal assumption here is that interstellar travel does not stop after the first jump from the home planet to the nearest stars, but rather continues to occur in steps as each colony becomes a new center of expansion after a suitable regeneration period. Hart (Reading 42) gives an estimate for the effective expansion velocity, but we estimate more conservatively that the regeneration period after each 10–light year step is 500 years.[2] The effective expansion velocity becomes $0.016c$ for an actual travel velocity of $0.1c$, and is rather insensitive to the actual travel velocity assumed. At this effective expansion velocity, a technological civilization will populate the galaxy in a time $t_c \sim 5 \times 10^6$ years. Jones (1976) has considered the expansion rate in some detail, and arrives at similar values for t_c. Viewing (1975) arrives at much larger values of t_c by making extremely conservative assumptions about the regeneration period and the travel velocity, but as will be evident from the discussion below, our estimate can be revised upward by two orders of magnitude without affecting the conclusions.

The primary uncertainty in any discussion of SETI is the space density of technological civilizations in the galaxy. The relevant quantity, N, is the number of technological civilizations existing at any given time which are or have been able to maintain their technology long enough to begin the expansion process by establishing several nearby colonies. Once this occurs, self-annihilation of the home society would probably not affect the expansion process significantly, since the colonies are necessarily quite independent of the home planet (Von

Hoerner, 1972, 1975a).[3] Estimates of N vary greatly [for example, see (Oliver and Billingham, 1972, p. 24) and references therein], so we will not make any assumptions about this number, but rather address ourselves concurrently to the full range of possibilities. If N is very small, then (i) the galaxy is probably not significantly populated or explored, and (ii) radio beacons from intelligent civilizations probably do not exist in sufficient quantity to justify a massive search. However, if N is not small, then the probability is near unity that the galaxy is completely colonized because $t_c \ll \Delta t$, where Δt is the dispersion in formation times of stars of any given type (population I). Indeed, $\Delta t \approx$ the age of the galaxy.

In summary, the above arguments yield two quite different possibilities: (i) technological civilizations that last long enough to begin the colonization process are rare, in which case the galaxy is essentially unpopulated and the number of radio contact beacons is negligible, and (ii) there are several such civilizations, in which case the galaxy is fully explored or colonized.

Case (ii) has several immediate implications, including the following.

(1) A "galactic community" would exist, in which one or several different civilizations communicate with each other throughout the galaxy, and we are located within the "sphere of influence" of one or more of these civilizations.

(2) The solar system would probably have been visited.

(3) An advanced civilization would probably have representatives somewhere in the solar neighborhood. This encompasses Bracewell's (Reading 21) suggestion that an advanced civilization may already have sent an unmanned probe into our system to contact us when we reach some developmental threshold.

Implications for a SETI Strategy

The discussion above has definite implications for a radio search for beacons from extraterrestrial intelligent beings. In case (i), we see that the value of a radio search is problematic. Case (ii) is the one of interest. If the galaxy is thoroughly populated, as case (ii) implies, there is little reason to expect extraterrestrial beings to be transmitting radio beacons across galactic distances (either beamed or isotropic) which are meant explicitly for contact with new civilizations. The galactic community, in whole or in part, is probably aware of nascent civilizations, and if contact

---

[1] The reader is referred to these papers for further cogent argument on the likelihood of interstellar travel and colonization. The implications that Hart draws differ from ours, whereas Viewing reaches the same conclusions as those presented in this article without discussing the broad implications for possible tests of the presence of advanced civilizations.

[2] This is a rather arbitrary figure. However, increasing it to, say, 10,000 years does not significantly alter the argument [for example, see Viewing (1975)].

[3] Von Hoerner suggests that advanced civilizations may be rare or nonexistent because they tend either to annihilate themselves (or their technological capacity) or to undergo developmental stagnation crises at certain stages in their development. However, fear of annihilation or of irreversible stagnation of the home society may, in fact, prove a motivation for starting the colonization process: self-preservation on the species level.

with a new civilization is desired, it can be made directly (physically) or through a radio signal from somewhere in their local neighborhood. Since, in case (ii), nearby extraterrestrial beings are probably aware of the existence of our civilization, we might address ourselves to the implications of the fact that obvious overt contact has not yet been made. This certainly does not rule out case (ii) because there are plausible arguments in favor of the possibility that extraterrestrial beings have chosen not to reveal themselves until, perhaps, we reach a certain developmental threshold. As this involves anticipating the motives of an advanced civilization, it is somewhat more speculative than the main body of this article, and thus we offer a few of many possibilities in appendix A.

In conclusion, we envision four possible situations regarding SETI.

(1) There are no beacons, either because case (i) applies, or because estraterrestrial beings are concealing themselves or choose not to contact us at this time,

(2) A radio beacon meant for contact is being beamed toward the earth from the solar neighborhood (within, say, ~10 light years), and therefore great sensitivity is not required to detect it. In this case, extraterrestrial beings are revealing themselves to our society at its present technological level, although in a minimally spectacular manner which least perturbs our civilization.

(3) Beacons of a more sophisticated nature (beyond the capability of our technology to detect or interpret) are being sent to us from the solar neighborhood, so we must reach some relatively advanced technological level before we can establish contact.

(4) Signals are being beamed toward the solar system, but they are not intended for our earthbound civilization. Rather, they form a communications link between local and distant extraterrestrial beings. If these signals are coded in radio waves, then, as we shall show, the ability of a modestly sensitive system (that is, presently operating systems) to detect such a signal will depend on how it is coded.

These four situations form the set of reasonable alternatives that one can arrive at with the arguments above. In none of these alternatives do we see any advantage in the construction and use of a very-large-aperture radio telescope for SETI over the use of large single-dish antennas of the type that already exist. The arguments above cannot rule out the occurrence of signals for which a large collecting area would be required, but suggest that they are relatively improbable and that a search for them is premature at present.[4]

[4]An example of this might be "leakage radiation" from nearby inhabited systems (Oliver and Billingham, 1972, p. 59), but it seems likely that an advanced civilization would conduct their internal communications in an efficient manner. For example, accurate oscillators in the receivers would remove the need for strong carriers, the only part of a signal we might expect to detect.

Considerations for an Initial SETI

We have identified two ways in which we might, by passive electromagnetic means, detect the presence of ETC's. Both involve searching for transmissions beamed at the solar system. We consider here what would constitute an adequate search for such signals. We use the range equation in the form

$$\frac{\eta_T \eta_R d_t^2 d_r^2 P_t}{\lambda^2 R^2} = 2 \times 10^3 k_D k_R T \left(\frac{B}{\tau}\right)^{1/2} \quad (1)$$

where η_T and η_R are the aperture efficiencies of the transmitting and receiving antennas, whose diameters are d_t and d_r (meters); λ is the wavelength (meters); P_t is the radiated transmitter power (megawatts); R is the range (light years); k_D, the detection factor, is the minimum detectable signal divided by the root-mean-square noise in the receiver; k_R is a receiver efficiency factor; T is the system noise temperature (kelvins); B is the bandwidth (hertz); and τ is the integration time (seconds). Let us consider what beacons might be detected with a 26-m antenna, this aperture being selected because several antennas in the range 20 to 30 m are now underutilized and could be dedicated to a SETI. For the purpose of discussion, we set $\eta_T = \eta_R = 0.6$, $k_D = 5$, $k_R = 2$, $T = 20°K$, and $d_r = 26$ m, so that Equation 1 takes the form

$$\frac{d_t^2 P_t}{\lambda^2 R^2} = 1.64 \times 10^3 \left(\frac{B}{\tau}\right)^{1/2} \quad (2)$$

We have plotted Equation 2 in Figure 34.1 for (i) simple detection with $\tau = 2 \times 10^4$ seconds and (ii) demodulation of a simply encoded beacon for which $\tau = B^{-1}$. On the graph we have also noted the number of main sequence stars in the solar neighborhood as a function of distance, of which approximately 0.3 are in the spectral classes F0 to K7.

We can see that a directed beacon would be readily detectable. If the transmitting facility consisted of a 1-megawatt transmitter on a 100-m antenna radiating at a frequency of 12 centimeters (see appendix B) then we would be able to detect a 1-hertz signal out to a distance of 200 light years and demodulate it out to 20 light years. If we expect, by our previous arguments, that the transmitter is located on one of the nearest, say, ten stars, we could demodulate signals with bandwidths as high as 10 hertz. Even if the transmitter agency were operating on a low budget ($d_t = 26$ m and $P_t = 0.25$ Mw), we could still demodulate a 0.1-hertz signal from one of those stars and could detect (after 6 hours of integration) such a signal at a distance exceeding 40 light years. At shorter wavelengths, of course, the range would be greater.

The possibility of intercepting communications with an alien outpost in the solar system can be assessed by asking

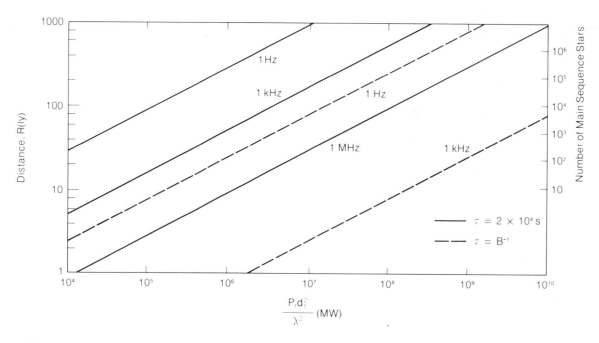

Figure 34.1. Distance over which a transmitter is detectable with a 26-m antenna as a function of its effective radiated power and the bandwidth of the signal. The assumed receiver parameters are specified in the text. Abbreviation: *ly*, light years.

how large an alien antenna in our solar system would have to be to be able to demodulate signals that we could not detect with, say, 6 hours of integration on a 26-m antenna. If we assume similar noise temperatures for the alien receiver and our receiver, we see by Equation 1 that

$$\left(\frac{B}{\tau}\right)^{1/2} \frac{1}{d_r^2} = \text{constant}$$

for both systems. If the bandwidth of the alien communication channel does not greatly exceed its capacity (see below), then $\tau \sim B^{-1}$. If our receiver has $\tau = 2 \times 10^4$, we find

$$d_r \text{ (alien)} > 309.2\, B^{1/4}\,\text{m}$$

(This is an underestimate since the extraterrestrials would presumably expect a higher signal-to-noise ratio in their communications than we would consider sufficient for detection.) In this case, we see that it is probable that alien communications would be detected by a 26-m antenna: a 1-km aperture receiving at a data rate of 100 hertz, or a 10-km aperture receiving at a data rate of 1 Mhz, would imply signals at the detection threshold. A 100-m antenna, or at most an array of several such antennas, should be sufficient to detect such a signal.

However, there is reason to believe that the detection of alien communications may be more difficult. The maximum capacity of a communications channel, obtained when the transmitted signal has white noise characteristics, may be expressed as (Shannon and Weaver, 1964, theorems 16–17)

$$C = B \ln\left(1 + \frac{P_r}{kTB}\right)$$

where k is the Boltzmann constant. For a given system noise temperature and received signal power level, the capacity is maximized at

$$C \simeq P_r / kT$$

when $B \geqslant 10C$. Thus, the alien signals may be spread so widely in frequency that they would be very difficult to detect with a modest antenna.

In searching for a directed beacon, we may feel we are able to restrict the frequencies to be examined (appendix B), but this cannot be the case for attempting to intercept communications. Nevertheless, a wide frequency search could be conducted within a reasonable time span. Using technology now planned or under development, we can modestly envision examining 100 Mhz at a time with 100-hertz resolution. Spending 6 hours on each of 100 candidate stars, we could scan the spectrum at the rate of 250 days per gigahertz or 35 years to cover the spectrum up to the oxygen absorption band above 55 Ghz. Less time or more stars would be covered in a search that was aimed only at restricted frequency intervals. Thus, a search for directed transmissions from nearby stars using a single 26-m antenna would not be of unreasonable duration.

Summary

We have argued that planning for a search for extraterrestrial intelligence should involve a minimum number of assumptions. In view of the feasibility (at our present level of understanding) of using nuclear fusion to effect interstellar travel at a speed of $0.1c$, it appears unwarranted (at this time) to assume that it would not occur for at least some technologically advanced civilizations. One cannot even conclude that humans would not attempt this within the next few centuries. On the contrary, the most likely future situation, given the maintenance of technological growth and the absence of extraterrestrial interference, is that our civilization will explore and colonize our galactic neighborhood.

A comparison of the time scales of galactic evolution and interstellar travel leads to the conclusion that the galaxy is either essentially empty with respect to technological civilizations or extensively colonized. In the former instance, a SETI would be unproductive. In the latter, a SETI could be fruitful if a signal has been deliberately directed at the earth or at an alien outpost, probe, or communication relay station in our solar system. In the former case, an existing antenna would probably be sufficient to detect the signal. In the latter case, success would depend on the way in which the communications were coded.

Failure to detect a signal could permit any of the following conclusions: (i) the galaxy is devoid of technological civilizations, advanced beyond our own, (ii) such civilizations exist, but cannot (for some reason which is presently beyond our ken) engage in interstellar colonization, or (iii) such civilizations are not attempting overt contact with terrestrial civilizations and their intercommunications, if present, are not coded in a simple way. To plan at this time for a high-cost, large-array SETI based on the last two possibilities appears to be rather premature.

Appendix A: On the Present Lack of Contact with Extraterrestrial Beings

If the cost and speed of interstellar travel are practicable, and the probability of the evolution of intelligent life in 10^{10} years is not small, then, following the arguments in this article, there should be some explanation why we are not in open contact with extraterrestrial beings. The most common explanation is that one of the two premises is

incorrect. However, we have argued that interstellar travel is practicable. Therefore, either the probability of the evolution of intelligent life is extremely small (that is, we are essentially unique in the galaxy), or we must seek an alternative explanation. Indeed, the possibilities that we are being ignored, avoided, or discreetly watched are logically possible and have been discussed previously (Sagan, Reading 29; Ball, Reading 45).

In considering these alternative possibilities, we must address ourselves to the reasons an advanced civilization might have for contacting us (we assume that alien beings from an advanced society are constrained to act primarily on reasoned principles). To do this we must identify the general class of resources that are most likely to be of value to an advanced society. Bell (1976) has pointed out that in preindustrial societies, the strategic resource is raw materials. The industrial society centers around capital and labor (which may in a general sense be equated). In the postindustrial society, knowledge is the most valued resource. Thus, an industrial society is able to adjust to the unavailability of a particular raw material by transforming some other material. (The World War II period offers a wealth of examples.) In a postindustrial society, the practical application of theoretical or empirical knowledge is able to compensate for a shortage of labor and, in many cases, can even replace labor. Thus we suggest that knowledge, in a general sense that encompasses science and culture, is likely to be most highly prized by an advanced civilization. This could be formal, codified knowledge, or experiences whose value we have not yet appreciated. Furthermore, this resource is one that grows with time. We believe that there is a critical phase in this. Before a certain threshold is reached, complete contact with a superior civilization (in which their store of knowledge is made available to us) would abort further development through a "culture shock" effect. If we were contacted before we reached this threshold, instead of enriching the galactic store of knowledge we would merely absorb it. Consider, for example, that the motivation a terrestrial researcher (or research funder) might have for pursuing new ideas would be considerably diminished, as the best human minds could be occupied for generations digesting the technology and cultural experiences of a society advanced far beyond our own. Thus, by intervening in our natural progress now, members of an extraterrestrial society could easily extinguish the only resource on this planet that could be of any value to them.

Opposed to the advantage of letting our civilization achieve some level of maturity may be the possibility that the earth has some other critical resource. In many cases, however, the needs of the aliens could be satisfied without undue impact on our civilization. The removal of rare elements or chemicals, of genetic material, or of samples for biological or psychological studies (including even an occasional human) could be effected with no more attention from us than a UFO article or a missing person's report. To establish that avoidance of open contact is not the most likely alien behavior, one would need to identify a resource that does not fall into this category.

There remains the possibility that members of an extraterrestrial society might choose limited contact without offering their store of knowledge. They might wish to do this (i) as part of an experiment to gauge the reaction of our society, (ii) in an attempt to stabilize terrestrial civilization to prevent an impending crisis of self-annihilation, or (iii) to plant selected information in order to stimulate our evolution in some preferred direction. In none of these cases can it be concluded that contact would necessarily occur in an overt way, so that we would immediately recognize it as such.

Finally, one must consider why the earth was not colonized by another civilization well before the advent of human or even animal life. Out of a wealth of possibilities, we offer three: (i) the earth has been a preserve for a long time, as the extraterrestrials would not want to halt natural evolution completely by occupying all suitable environments, (ii) extraterrestrial technology has advanced beyond the stage where a planetary base is required or even desired, or (iii) the biology on the earth is incompatible with or even hostile to that of the species which dominate our part of the galaxy.

In considering the subject matter of this appendix, we have ventured far beyond the realm proper to physics and astronomy. We hope that experts in relevant disciplines will address these questions and address their impact on the search for extraterrestrial intelligence.

Appendix B: Frequencies for Interstellar Beacons

Although it might ultimately be desirable to cover the entire range from 1 to 100 Ghz or more (perhaps simultaneously), the preliminary phases of SETI should involve choosing specific frequencies, or frequency bands, which there is some rationale for believing an extraterrestrial civilization would transmit at to attract our attention. At least, in a search intended eventually to cover a large portion of frequency space, these are the frequencies that would logically be chosen first.

The 21-cm hydrogen line, first suggested by Cocconi and Morrison (Reading 20), has received the most attention. Many cogent arguments in favor of a search at this frequency have been put forward, although it has been generalized into the "water hole" band lying between 1420 and 1720 Mhz (Oliver and Billingham, 1972, p. 63). This frequency band is attractive because (i) lying between the frequencies of the most fundamental or noticeable transitions of H and OH, the constituents of water, it might be used for a contact beacon if water is universally basic to the chemistry of life, and (ii) it minimizes the

combination of galactic background noise, receiver quantum noise, and atmospheric absorption. For similar reasons, the 22,235 Mhz line of H_2O, ubiquitous in the galaxy as maser radiation, has been suggested as an equally likely carrier of intelligent signals. (Although atmospheric absorption is higher at the frequency of the H_2O line, this cannot be considered to have a great impact on the sender's choice of frequency, especially since it may be more cost-effective in the near future to build large antenna systems in space.) Since we have argued that isotropic beacons of this kind envisioned by Project Cyclops (Oliver and Billingham, 1972, p. 60) are unlikely, Equations 1 and 2 and Figure 34.1 show that there is a distinct advantage in operating at the higher frequencies: the effective range is inversely proportional to the wavelength for particular transmitting and receiving systems. Given this consideration, the water line has an advantage over the lower frequencies that have been suggested.

We propose a frequency of a rather different nature, but which may be considered fundamental and universal. This is a frequency constructed from fundamental natural constants. Examples of frequencies constructed in this way are

$$(1) \quad \frac{c}{2\pi r_c} = 1.7 \times 10^{22} \text{ hertz}$$

where $r_c = e^2/m_e c^2$ is the classical electron radius; m_e and e are the mass and charge of the electron, respectively.

$$(2) \quad \frac{m_e c^2}{h} = \frac{c}{2\pi r_c} = 1.2 \times 10^{20} \text{ hertz}$$

where r_c is the compton wavelength of the electron. Note that $m_e c^2/h$ is the "intrinsic" frequency of the electron.

$$(3) \quad \frac{c}{2\pi r_B} = 9.0 \times 10^{17} \text{ hertz}$$

where $r_B = h^2/m_e e^2$ is the Bohr radius.

(Note: The 2π's are inserted to yield cyclical frequencies instead of angular frequencies when physical interpretations are applied to each expression.)

Two things are apparent from this compilation of frequencies.

(1) The frequencies are all very high (well above the radio regime).

(2) They differ only by integral powers of the (dimensionless) fine structure constant, $\alpha = 1/137$. This suggests that, by scaling with further powers of α, one can find a unique frequency that lies in the reasonable radio range. Thus we arrive at the "fundamental" frequency

$$2556.8 \text{ Mhz} = \alpha^4 \left(\frac{c}{2\pi\, r_{\text{B}}} \right)$$

$$= \alpha^5 \left(\frac{c}{2\pi\, r_{\text{c}}} \right) = \alpha^6 \left(\frac{c}{2\pi\, r_{\text{e}}} \right)$$

The following arguments support the unique character of this frequency.

(1) A different integral power of α applied to the sequence above would yield a frequency far from the range that one can presently consider appropriate for interstellar communication.

(2) Use of similar constants for the proton, for example, would not be nearly as physically meaningful as those in the sequence above, mostly because the proton is a relatively complex particle, and entities such as the "classical proton radius" have little relevance.

(3) One could consider using the gravitational constant, G, but frequencies derived from it [such as $(c^5/Gh)^{1/2} \approx 7.4 \times 10^{42}$ hertz, where h = Planck's constant] generally have no apparent physical meaning, and cannot be scaled to the radio range by some appropriate dimensionless number, such as $e^2/Gm_e^2 \approx 4.2 \times 10^{42}$.

The motivations that a transmitting civilization might have for choosing this frequency include (i) the lack of interference arising naturally from atomic or molecular species in the galaxy [for example, there are no recombination lines with a change in the principal quantum number $(\Delta n) < 6$ near this frequency, nor are there known molecular lines], (ii) the minimization of galactic background noise, receiver quantum noise, and atmospheric absorption, (iii) the presupposition of a modest breadth in the technical capacity of the receiving civilization, and (iv) the absence of any assumptions about the chemistry of life throughout the galaxy.

Acknowledgments

We gratefully acknowledge helpful criticisms from R. Carpenter, R. Edelson, B. Gary, S. Gulkis, M. Janssen, M. Jura, G. and S. Knapp, D. Kunth, R. McEliece, B. C. Murray, J. Rather, E. Rodriguez Kuiper, P. Swanson, and B. Zuckerman. This work has been carried out at the Jet Propulsion Laboratory (JPL) and the Owens Valley Radio Observatory, partially supported by a grant from the JPL Director's Discretionary Fund. Jet Propulsion Laboratory is operated by the California Institute of Technology under NASA contract NAS 7-100. Research at Owens Valley Radio Observatory is sponsored by the National Science Foundation under grant AST 73-04677AO3.

35

Bernard M. Oliver
Proximity of Galactic Civilizations

A galactic community of interacting planetary civilizations may have arisen in a variety of ways, but one plausible scenario is that communication began between fortuitously close civilizations and spread from these centers. It is therefore of considerable interest to estimate how many intelligent civilizations have existed in proximity. Obviously this depends upon the number of advanced cultures and upon how this number has varied with time.

Galactic Demography

The conventional approach to estimating the number of advanced races in the Galaxy is to assume (pessimistically) that such life needs reasonably Earth-like conditions. The problem is then reduced to estimating from astronomical evidence the number of "good earths"; from biological evidence the probability that, given a good earth, life will begin and evolve; from sociological evidence the probability that technology will develop; and from little or no evidence the lifetime of technological societies. Since the subject has been discussed at length in the literature (Cameron, 1963a; Ponnamperuma and Cameron, 1974; Sagan, 1973a; Shklovskii and Sagan, 1966; Dole, 1970) we will give only a brief outline here.

Terrestrial planets are presumed to exist only around Population I stars since only these would have had a sufficient abundance of heavy elements. Although supernovae explosions have continued to increase the heavy element abundance of the interstellar medium, the oldest Population I stars are as old as the estimated age of the Galaxy ($\sim 10^{10}$ yr). This suggests that considerable nuclear fusion took place in giant Population II stars before the galaxy had collapsed to its present form. Studies by Schmidt (1959, 1963) indicate that the rate of star formation goes as something between the first and second powers of the interstellar gas density and was therefore high in

the young galaxy and is less today, the present rate being about one-third of the average rate up to the present time. The present number of Population I stars is $\sim 2 \times 10^{11}$. Thus we can expect a buildup of Population I stars not too different from that shown in the curve marked N_* in Figure 35.1.

Not all Population I stars are appropriate for biology. Stars more massive than F stars are too short-lived for Darwin-Wallace evolution to be effective (if Earthly evolution is typical), while M stars have so low a luminosity that planets, to be warm enough for life, would be so close that tidal coupling would synchronize their rotation. Post main sequence stars will have no life about them now, but many giants and white dwarfs may formerly have warmed races of intelligent creatures. About one-fourth of all stars are F, G, K main sequence stars. Of these, about 15% are single. Thus, we take the fraction that are (or have been) "good suns" to be $f_g \approx 4 \times 10^{-2}$.

Most stars are now believed to have planetary systems, so we take the fraction with planets to be $f_p \sim 1$. The number of planets per planetary system that are in the stellar ecosphere is also on the order of unity, but not all these will be of the right size to outgas an atmosphere. Of the terrestrial planets in the solar system only Earth and Venus have done this to a sufficient extent. Mercury and the Moon have not, while Mars is a borderline case. Thus we take the number of good earths per planetary system as $10^{-1} < n_e < 1$.

Many but by no means all workers on the origin of life believe that, given a good earth, life will almost certainly start. Thus we take a fraction of the good earths on which life starts as $f_l \sim 1$. Likewise we assume that intelligence is so favored in natural selection that the fraction of life starts that develop an intelligent species is $f_i \sim 1$.

With us, technology began with attempts to improve our environment and our chances of survival. Stone weapons, fire, shelter, clothing, agriculture were all early technological triumphs. Technology was advanced by science which, in the last analysis, is the result of intellectual curiosity. Whether such curiosity is an inevitable aspect of

Reprinted from *Icarus*, 25 (1975), 360. Copyright © 1975 by Academic Press, Inc.

$$N(t) = s[N_*(t - G) - N_*(t - G - L)]. \quad (2)$$

Almost certainly G and L are not constants, but differ from civilization to civilization. This means that $N_*(t - G)$ should be convolved with the distribution for G, and $N_*(t - G - L)$ should be convolved with the distributions for G and L before taking their differences in (2) as proposed by Kreifeldt (1973). The effect would be to smooth out the curve for $N(t)$, reducing its peak height and extending it in time without appreciably affecting the area under the curve. For the time being we shall ignore this refinement.

Unless L is very long indeed ($\geqslant 10^9$ yr) the difference in (2) can be replaced by the differential to give

$$N(t) = R_*(t - \tau)sL, \quad (3)$$

where $\tau = G + L/2$, and $R_* = dN_*/dt$ is the rate of star formation. Equation (3) is essentially the formulation introduced by Drake in 1961.

If we let $F = sR_*(-\tau)$, the present number of galactic civilizations is

$$N(0) = FL. \quad (4)$$

Assuming the genesis time of intelligent life on Earth to be typical, we set τ equal to 5 billion years and find $R_*(-\tau) = 20$ yr^{-1}. Combining this with our estimate for s, we obtain

$$F \sim 0.07. \quad (5)$$

For certain purposes it is convenient to write $N(t)$ in terms of the present population $N(0)$ as

$$N(t) = FLf(t), \quad (6)$$

where

$$f(t) = R_*(t - \tau)/R_*(-\tau). \quad (7)$$

The approximate shapes of $N(t)$ and $f(t)$ are shown by the dashed curve in Figure 35.1.

Mean Separation

If we let $p = N/N_*$ be the probability that a given star, selected at random, now supports an advanced civilization, then the probability that it does not is $(1 - p)$ and the probability that n stars chosen at random do not is $(1 - p)^n$. Thus the probability that at least one out of n stars is the sun of another advanced civilization is

$$P = 1 - (1 - p)^n. \quad (8)$$

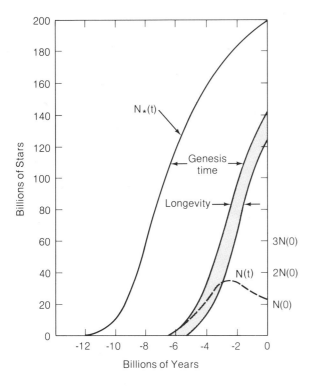

Figure 35.1. Buildup of stellar Population I in the Galaxy. Number of coexisting civilizations at any time is proportional to the difference between the number of such stars old enough to have evolved life and the number so old that life has perished. This difference is the height of the shaded band.

intelligence is moot. Other factors may militate against technology: a marine environment, or an absence of metals or fossil fuels, for example. We shall assume that the fraction of intelligent species that eventually develop a technology able to accomplish interstellar contact is given by $10^{-1} < f_t < 1$. If we let

$$s = f_g n_e f_l f_i f_t \quad (1)$$

be the product of all these selectivity factors, then under the above assumptions $4 \times 10^{-4} < s < 4 \times 10^{-2}$ with an expectation value $s \sim 4 \times 10^{-3}$. Other estimates (Sagan, 1973) cluster around $s \sim 10^{-2}$.

Let G be the average genesis time, and L be the average longevity of technologically advanced civilizations. If these times were constant for all cultures, then the number of evolved civilizations would be proportional to $N_*(t - G)$ while the number that have evolved and perished would be proportional to $N_*(t - G - L)$. These two curves are the left and right hand limits of the shaded band in Figure 35.1, and are simply the curve $N_*(t)$ shifted to the right by the times G and $G + L$. The number of civilizations coexisting at any time t would therefore be

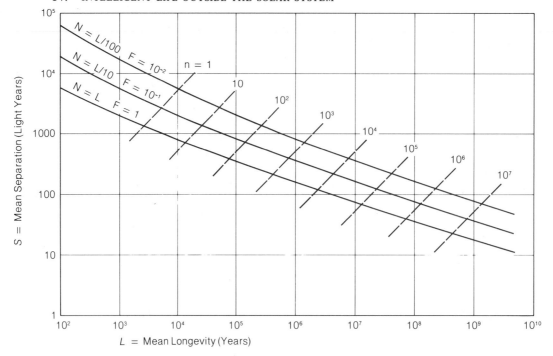

Figure 35.2. Relation between mean separation and longevity. Present estimates are that N is on the order of $L/10$. The dashed lines show the number of two-way exchanges that can occur at the speed of light.

For $p \ll 1$,

$$P \simeq 1 - e^{-np}. \tag{9}$$

The mean separation between civilizations is the range needed to encompass enough stars to make $np = 1$. Taking into account the actual number density of F, G, and K main sequence stars in the solar neighborhood and the falloff density with distance from the galactic midplane we obtain the solid lines of Figure 35.2. The dashed lines show the number of round trip radio message exchanges that can occur within the time L.

Clearly for $L \lesssim 3000$ yr interstellar communication, if it existed, would be a one-way affair. But if the longevity of civilizations is very long a great many exchanges can take place. Over the millenia these would profoundly modify the science, philosophy, and destinies of the races in contact. Perhaps, as von Hoerner (1961) has suggested, this prolonged discourse would encourage great longevity.

There are probably regions in the Galaxy nearer the center where the stellar densities are 10 to 30 times greater than in the solar neighborhood. Civilizations there, contemplating a search for interstellar life, would estimate mean separations one-half to one-third as great and would design antennas with one-quarter to one-tenth the area of ours. Interstellar communication is most likely to have begun (long ago) in these dense regions.

Neighboring Civilizations

Let us now estimate how many civilizations have existed in close proximity to one another over the history of the Galaxy. We will, for the moment, ignore star density variations and assume that the N civilizations are distributed uniformly, but randomly, throughout the volume $V_g \approx 2 \times 10^{13}$ ly³ of the Galaxy. The mean density is then $\rho = N/V_g$ and the probability of finding another civilization in a volume V surrounding a given one is

$$P = 1 - e^{-\rho V}. \tag{10}$$

For $\rho V \ll 1, P \approx \rho V$, and we may take V to be a sphere of radius R. The total number, n, of civilizations within a (small) distance R of one another at any time t is

$$\begin{aligned} n(R,t) &= NP = \tfrac{4}{3}\pi R^3 \, N^2(t)/V_g \\ &= \tfrac{4}{3}\pi R^3 V_g \, \rho^2(t). \end{aligned} \tag{11}$$

The velocity dispersion of stars is $\sim 10^{-4}c$. The duration of a close encounter is therefore $T \sim 10^4 R$, where R is measured in light years. If $L \ll T$, then for each generation of simultaneous cultures, the galaxy is frozen in place. If $L > T$, many pairs of stars will drift apart during the time L while a similar number will drift into proximity. This mixing can only increase the number of close encounters. Therefore the total number of civilizations that have been in proximity to others since intelligent life began is at least

$$v(R) = \frac{1}{L} \int n(R,t)\, dt, \qquad (12)$$

which can be expressed in terms of our earlier variables in several equivalent forms:

$$v(R) = \frac{4}{3}\pi R^3 \, (1/V_g L) \int N^2(t)\, dt, \qquad (13a)$$

$$= \frac{4}{3}\pi R^3 \, (V_g/L) \int \rho^2(t)\, dt, \qquad (13b)$$

$$= \frac{4}{3}\pi R^3 \, (F^2 L/V_g) \int f^2(t)\, dt. \qquad (13c)$$

From the integral of the square of the dashed curve in Figure 35.1 we estimate the integral in (13c) to be $\approx 10^{10}$, and so obtain

$$\nu(R) \approx 2 \times 10^{-3} F^2 L R^3. \qquad (14)$$

Using our earlier estimate of $F \sim 0.07$ this becomes

$$\nu(R) \sim 10^{-5} L R^3. \qquad (15)$$

The dispersions in G and L, which we have neglected, would tend to reduce the peak height of $f(t)$ and thus reduce $\int f^2(t)\,dt$ even though $\int f(t)\,dt$ were unchanged. On the other hand, our neglect of the large variations in stellar density means that ν should be increased by the factor 2. We suspect therefore that the value of ν given by (14) and (15) is conservative.

Multiple Stars

Although it is customary to eliminate multiple star components from the roster of "good suns", as we have done in the preceding sections, it is not at all clear that this is necessary. The arguments usually advanced are:

1. Most of the angular momentum in the original collapsing gas cloud went into the orbiting of the stars themselves, leaving too little in the protostar components to assure the formation of planets.
2. Planetary orbits in multiple star systems are likely to be unstable.

Neither of these objections is necessarily true. It is highly probable that multiple systems form from a gas globule large enough to form a single star with the total mass of the multiple system, but with *too much* angular momentum and turbulence to allow this to happen. Only by generating multiple protostars can this excess be disposed of. Further, the angular momentum of the *terrestrial* planets of the solar system is only 0.16% of the total.

A protostar could have far less angular momentum than the solar system and still form *terrestrial* planets, which are all we are interested in. Thus we are inclined to take $f_p \sim 1$ even for multiple stars.

In binary systems, stable orbits are possible around both components at distances greater than 2.5 times their separation (Dole, 1970). This case is probably of little interest since either the planet would not be in the ecosphere, or the components would be so close to as raise grave questions about the stability of their radiation. Stable orbits are also possible around the larger component at radii up to one-fourth to one-half the separation, and around the smaller component at radii up to zero to one-fourth the separation, depending on the mass ratio (Dole, 1970). For nearly equal stellar masses the figures of one-fourth apply.

If we regard Jupiter as a dark companion of the Sun, then the outer planets orbit both components, while the terrestrial planets and the satellites of Jupiter represent planets about each component. Thus the solar system itself exhibits both classes of stable orbit.

In multiple star systems the dynamics are quite complicated but in all systems in which the components remain well separated, stable orbits close to each component are certainly possible.

In about half of the observed binary systems the separation (major axis) is 60AU or more (Allen, 1973). Even with large eccentricities for the stellar orbit these could have stable planetary orbits out to ~ 2AU around each component. If we let f_m be the fraction of systems of multiplicity m that are widely spaced enough to have stable orbits around all components, then $f_2 > 1/2$. We will not introduce much error it we consider all more complex systems to be triple and take $f_3 = 1/3$.

Table 35.1 gives Allen's (1973) counts per 100 star systems. For systems with multiplicity greater than 2, the average component count is $81/23 = 3.5$.

On the basis of our assumption for f_m it appears that we should be able to increase our roster of "good suns" by 100/30 or 3.3-fold. The immediate effect is to raise N in the same ratio, giving

$$N \sim L/4. \qquad (16)$$

Since the number of proximities varies as N^2, ν will be increased by an order of magnitude, giving

$$\nu \sim 10^{-4} L R^3. \qquad (17)$$

Twin Civilizations

The components of multiple star systems are stars of nearly the same age. This tends to synchronize somewhat

Table 35.1.

Type of system	Number of systems	Number of components
Single	30	30
Binary	47	94
Multiple	23	81
Total	100	205

the emergence of civilizations in systems with two good suns and two good earths. Assuming no correlation between the spectral types of the components of widely separated systems, we can express the total number of such twin civilizations that have coexisted up to the present as

$$\nu_t = \sum_m \frac{m!}{(m-2)!} \, s^2 (1-s)^{m-2} \, N_{m*} f_m f_s , \qquad (18)$$

where $m \geq 2$ is the multiplicity, N_{m*} = number of star systems of multiplicity m that existed one average genesis time ago, s = product of the selectivity factors, f_m = fraction of multiple systems with widely spaced components, and f_s = fraction of civilizations that coexist. The coefficient $m!/(m-2)!$ is twice the number of combinations of m things taken two at a time; twice, because each combination involves two civilizations. It is of course possible to have triplet or n-tuplet civilizations, but the probability is very low.

Using our previous assumptions for f_m and taking all higher multiple systems to be ternary, (18) becomes

$$\nu_t \approx (N_{2*} + 2N_{m*})s^2 f_s, \qquad (19)$$

where we have neglected the factor $(1-s)^{m-2}$ since $s \ll 1$. Once again taking 5×10^9 yr as the average genesis time we find from Figure 35.1 that $N_* \approx 1.4 \times 10^{11}$. Out of 205 stars, there are 47 binary systems and 23 multiple systems, so $N_{2*} \approx 3.2 \times 10^{10}$ and $N_{m*} \approx 1.6 \times 10^{10}$. In the solar neighborhood 25% of all stars are F, G, and K main sequence stars. Assuming this proportion holds for wider-spread multiple systems,[1] we set $f_g = 1/4$ in s, leaving the other selectivity factors unchanged, and obtain $s \sim 2.5 \times 10^{-2}$. Making these substitutions in (19), we find

$$\nu_t \sim 4 \times 10^7 f_s. \qquad (20)$$

[1] Data in Allen (1973, p. 227) suggest that the proportion of F, G, and K stars in binaries may be as high as 60%. If so, then our estimates for ν_t are low by a factor of 5 and our estimate of ν for single stars only may be high.

Let us now assume that the genesis time is normally distributed with a standard deviation σ about some mean value $\bar{G}$. Then

$$p(G) = \frac{1}{(2\pi)^{1/2}\sigma} \, e^{-(G-\bar{G})^2 \, 2/\sigma^2} \qquad (21)$$

is the probability density function. If $L \ll \sigma$ we can write

$$f_s = L \int p^2 (G) \, dG = \frac{L}{2\pi^{1/2}\sigma} . \qquad (22)$$

Combining this result with (20) we obtain

$$\nu_t \sim 10^7 L/\sigma. \qquad (23)$$

We see no way at present to assign any firm values to L or σ. However, if $\bar{G} \approx 5 \times 10^9$ yr, a value of $\sigma = 10^9$ yr would give a ratio of 4 to 1 in G between the 3σ points, which does not seem unreasonable. This gives

$$\nu_t \sim 10^{-2}L. \qquad (24)$$

Summary of Results

Adding (17) and (24) we derive our best present estimate of the total number of civilizations that have coexisted within R light years of another,

$$\nu(R) \sim (10^{-2} + 10^{-4}R^3)L. \qquad (25)$$

Equations (24) and (25) begin to fail below about 0.1 light year as very wide-spaced binaries and multiple systems begin to be excluded. Figure 35.3 shows a graph of $\nu(R)$ in which this short-range correction has been included.

We are again confronted with the great uncertainty in L. If we solve our problems of planetary management we could flourish for 10^9 yr; if we do not we may well retrogress or vanish in 10^2 yr, as some doomsayers fear. It has often been suggested (Shklovskii and Sagan, 1966) that the distribution of longevities may be bimodal; a fraction of the civilizations, say 1%, succeed in solving their ecological and societal problems and live for 10^9 yr while the remaining 99% do not and vanish in, say, 10^3 yr, on the average. If this is true then $L \sim 10^7$ yr and the galaxy is teeming with life. Something on the order of 10^5 civilizations will have coexisted within several tens of AU of each other and 10^6 within 10 light years. Interstellar communications will have been established independently countless times.

But let us take the more pessimistic value of $L \sim 10^3$ yr,

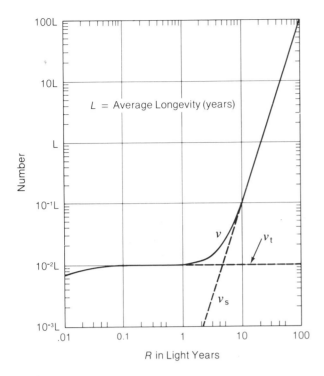

Figure 35.3. Number of civilizations that have coexisted within a range R of each other. The number is proportional to the square of the selectivity factors and so may easily be an order of magnitude greater or less than shown. Coexisting civilizations in the same planetary systems are not included.

beacon at 500 light years and thus reach 1000 times as many stars.

Referring back to (20), we note that the coefficient of f_s is 4×10^7. This means that there have been $\sim$40 million cases where a space voyage by an advanced culture around one component of a multiple star system to the planets of another component would have discovered either early life evolving or the remnants of an advanced culture long dead. (The former is more likely, since the other culture would then have made the voyage first.) A space voyage over only a few tens of AU is within our near-future capability.

Positive proof of life of independent origin obtained by any of the above methods would give great impetus to further search efforts. The number of close civilizations, though uncertain, seems large enough to support the hypothesis that an intercommunicating galactic community of advanced cultures already exists.

which assumes virtually no long-lived cultures. There will then have been only $\sim$10 twinned civilizations within a few tens of AU of each other. At this close range the planets of the other star could be resolved and leakage radiation could easily be detected.

For these early radio searches it is probably more significant to interpret L as the average length of time civilizations radiate, for their own purposes, signals that are detectable at the range R. For a cost less than that of space programs we have already undertaken, we could construct a receiving system that would detect the presence of signals like our own UHF-TV at a 50-light-year range (Oliver and Billingham, 1972). We have been radiating such signals for 15 years and will probably do so for at least another 35 yr. Taking $R = 50$ light years and $L = 50$ yr we see from Figure 35.3 that such a strategy could have succeeded $\sim$500 times.

Had we been one of these early "lucky" civilizations our success would almost certainly have encouraged us to search deeper into space and to radiate a beacon to attract others. Given a high-powered beacon the situation changes drastically. The same system that could detect leakage radiation at 50 light years could detect a 250-MW

V
Optimists and Pessimists

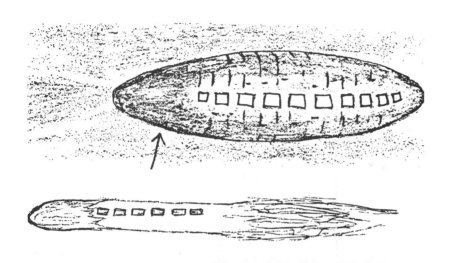

Almost four centuries ago, Johannes Kepler wrote an inspired dream, the *Somnium*, in which he imagined a flight to the moon. Today when travel to the moon has become a reality, most people vaguely imagine that travel to the stars must lie within the near future. In fact, since the distance from Earth to Alpha Centauri equals a *hundred million* times the distance from Earth to the moon, we have far to go before we reach the stars. To put things another way round, travel to the nearest stars compares with travel to the moon in the same ratio that a trip to the moon compares with a trip across the room.

The articles by Edward Purcell and Sebastian von Hoerner pursue the implications of this fact. Purcell's summary has often been misquoted as "interstellar travel belongs back on the cereal box," but a closer reading shows that the energy question predominates, and that Purcell does not rule out interstellar spaceflight entirely. Von Hoerner shows that *if* the energy problem could be overcome, then the time dilation effect, part of Einstein's relativity theory, would allow for incredibly long journeys within a single lifetime. When Von Hoerner wrote this article, he concluded that civilizations must communicate by photons, not spaceflight; since then, he has expressed a somewhat more positive attitude toward the possibilities of interstellar voyages, but his analysis of the general limits of space travel remains entirely valid.

Carl Sagan's "Direct Contact among Galactic Civilizations by Relativistic Interstellar Spaceflight" extends von Hoerner's analysis to consider the probability that Earth has been visited by other civilizations throughout its history. Of course, Sagan's considerations cover 4.6 billion years of time, so that any visits within, say, the past million years would be highly unlikely. This fact did not hinder the sales of millions of books suggesting precisely the reverse, namely that all of our recent history bears the stamp of visits from much more advanced civilizations.

George Gaylord Simpson's article on "The Nonprevalence of Humanoids" furnishes a good attack on the unproven assumption that life something like our own should be common. The conclusion of Simpson's article brings us back to a perennial problem for our civilization: how to allocate limited resources for research into unsolved problems. Alfred Adler's attempt to rip into shreds the scientific logic behind a SETI program provides a reminder that such searches must contend with an innate longing to remain unique. The first obstacle this longing has to overcome is the apparent tendency of scientists to reduce mystery to logic. This may be a serious problem (though in fact scientists love mystery as well as anyone), but an accusation that scientists are promoters barely deals with it.

The exchange between Michael Hart and Laurence Cox concerning the implications of the fact that extraterrestrials have not visited Earth sets in sharp perspective the difference between optimistic and pessimistic views on the likelihood of life throughout the universe. Hart's work aims to show that we can conclude from the absence of extraterrestrial visitors that life must be rare in the Milky Way; Cox attempts to rebut Hart's argument through a sociological analysis of other civilizations. As Hart's work on "Habitable Zones about Main Sequence Stars" shows rather clearly, he believes that only a near-unique set of circumstances has allowed Earth to develop life. Of course, Hart's personal attitude does not bear on the validity of his scientific arguments, which must be considered on their merits; and everyone who deals with the probability of extraterrestrial life has some bias for or against finding another civilization.

Many people would be shocked at Hart's statement that extraterrestrials are not visiting Earth. Quite the opposite appears to be the case from what you read in newspapers; however, the bald fact is that no reliable evidence does exist for such visits. Hence John Ball examines the idea that we live in a sort of cosmic zoo; Michael Papagiannis and David Stephenson examine the notion that interstellar interlopers may inhabit the asteroid belt; Sebastian von Hoerner reexamines the question of where "everybody" may be; William Markowitz and David Stephenson apply some scientific reasoning to UFOs; and David Schwartzman tries to find acceptable evidence for *some* UFOs as extraterrestrial spacecraft. The truth about UFOs will most likely remain obscure; the conclusions we can draw from the arguments of Hart, Ball, Papagiannis, and Stephenson remain mixed.

36

Edward Purcell
Radioastronomy and Communication through Space

It is a great privilege for me to open the series of Brookhaven Lectures. The principles on which these are conceived I heartily endorse, but I am just about to violate them by giving a talk which is really not, for the most part, a description of my own work. Indeed some of it will not be a description of any one's work, but instead some speculations about the future. In a way, you might regard this talk as a logical sequel to Dr. DuBridge's Pegram Lectures (DuBridge, 1960) of a year ago. It has three parts whose relation to one another will not be obvious until the end. The first part, at least, has to do with solid scientific matter, radioastronomy. Without revealing now the nature or motive of the last two parts, I would like to describe one branch of radioastronomy and what has come out of it in the last several years. I have not been active in this field myself in recent years, but I have been watching it develop.

Radioastronomy

Until 15 or 20 years ago, all man's information about the external world beyond the earth came to him in a small band of wavelengths of visible light. Everything the astronomers saw, all the images on their photographic plates, were collected by absorbing light within a range of wavelengths varying by no more than a factor of two from the shortest to the longest waves. It was the discovery, about two decades ago, that there were also radio waves coming through which started off radioastronomy.

These two great apertures, or windows, as they are often called, may be seen in Figure 36.1, which shows the absorption spectrum of the atmosphere of the earth on a scale of wavelengths, running from very short wavelengths in the ultraviolet region, through the visible, up into the range of radio wavelengths. Over nearly all of that range with two exceptions either the atmosphere, or the ionosphere just beyond it, absorbs 100 per cent of

Reprinted from U.S. Atomic Energy Comm. Report BNL-658, 1961.

incoming radiation. It is only in these two regions of the spectrum that our atmosphere will let anything come through. The radio "window" extends from a few centimeters to several meters wavelength. Electromagnetic waves in this band from any celestial source can reach our antennas on the earth. The branches of radioastronomy are many because radiation comes to us from all sorts of objects. A great deal of radio energy comes from the sun; radio waves come from stars and various odd astronomical objects. I shall discuss only one branch of radioastronomy, the study of the structure of the galaxy, that is, our Milky Way, by means of radio waves. My purpose in talking about it is to show how much information one can derive, from enormous distances, with little energy.

To begin with, let us place ourselves in the universe in the usual way by taking a look at a galaxy (Figure 36.2). No talk like this is complete without a picture of a spiral nebula. This is one of the most beautiful and, furthermore, is one which is probably rather like the galaxy in which we live. Of course it is not the one in which we live, or we could not have taken this picture. This is a large flat cluster of about 100 billion stars seen more or less on a slant. Observe its irregular shape with rather ill-defined arms spiraling off; it is a spiral nebula. There are hundreds and thousands of galaxies of this type. We happen to inhabit one of them. The one we inhabit is perhaps an ordinary one, but of course it is of special interest to us, and we would like to know what it looks like. It is very hard to find out because we cannot see it from the outside. Let me describe our galaxy by showing what it might look like in cross section, if we could examine a slice taken right down through the disk (Figure 36.3).

There are about 10^{11} stars in an object of this sort; the sun is one of these and happens to be out rather near the edge, about 25,000 light-years from the center. The thickness of this disk is only some 700 light-years on this scale. In addition to the stars, the galaxy has in it dust (small grains of matter) and hydrogen atoms. It has hydrogen atoms to the tune of about one per cubic centimeter through most of the spaces where there are no stars. In

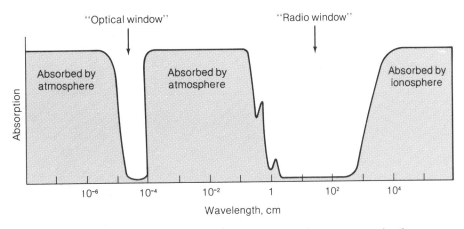

"Optical window"   "Radio window"

Figure 36.1. Absorption spectrum of the earth's atmosphere versus wavelength.

saying this I am getting ahead of my story, but it will make the story easier to follow. The stars make up most of the mass, but the hydrogen atoms are a nonnegligible part; they make up perhaps $\frac{1}{3}$ or $\frac{1}{4}$ of the mass of this whole assembly. The dust in itself doesn't amount to much—except as a nuisance; the dust makes this large collection of stars almost opaque to visible light. A telescope situated at the position of the sun or the earth can see only a little way into the galaxy, in most directions, because before long the path of vision is interrupted by a cloud of dust. One cannot see anything like the *whole* structure looking out with a telescope, or with the eye. Indeed if one could, the Milky Way, which is what we do see of the galaxy, from our vantage point, would present a very different spectacle. It would be a very narrow, very bright band, absolutely straight, going across the sky like a great circle. We are buried within this pancake, out near the edge, able to see with a telescope only part of the pancake in our vicinity. For this reason, until a tool became available to explore the greater depths of the system, one had rather little idea of the details of its structure. The dust grains, being very small, do not hinder the passage of radio waves at all. A 1-m-wavelength radio wave oozes around a tiny dust grain without the slightest trouble and goes on as if nothing were there. Thus the pancake is, by and large, completely transparent to radio waves, and, if there is a source of them, one can see that source no matter how far away it may be in the disk of stars and gas.

There is a radio wave that is emitted by the gas itself, and I will briefly describe this source before telling what it leads to (Figure 36.4). The hydrogen atom, which consists of an electron and a proton, happens to have in its structure a natural frequency which is in the radio range. The frequency is 1420 Mc/sec, corresponding to a wavelength of 21 cm. This frequency arises from the magnetic interaction between the electron and the proton. The cloud in Figure 36.4 represents the electron, and the

arrow represents the axis about which the electron spins. The proton spins around an axis, too. Because of the spin each particle acts like a bar magnet. The two little magnets try to set themselves parallel, but because they are spinning they don't achieve it. Instead, they precess around like gyroscopes. When the hydrogen atom is all by itself out in free space, with no perturbations or anything, and is in its lowest possible energy state, the electron spin axis quietly precesses around with a frequency of 1,420,405,750 cycles/sec.

I confess that the fact that this number is so long has no bearing whatever on the present subject, but I did want to write it out to show what kind of measurements are made nowadays in the branch of physics which measures these atoms in the laboratory. It is the branch carried on in Dr. Cohen's atomic beam laboratory at Brookhaven. This number is an actual experimental measurement, not a

Figure 36.2. Spiral nebula.

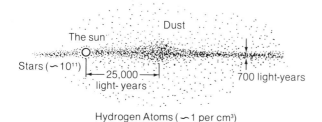

Figure 36.3. The galaxy, seen edge on.

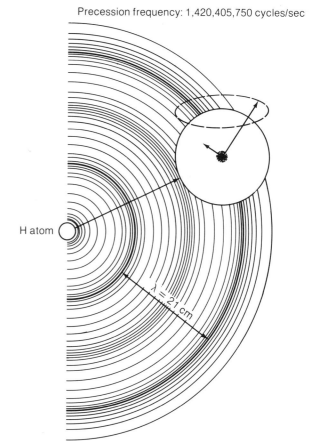

Precession frequency: 1,420,405,750 cycles/sec

Figure 36.4. The hydrogen atom emits radio waves.

theoretical number like π, and not a social security number, which it rather resembles. There is, at present, some argument among the fraternity about the last one or two digits. But there is also a recent development in atomic beams which makes it quite certain that within a year or two even more digits will be known. This is probably one of the most accurately known numbers in all of physics. As we shall see, that doesn't really do us very much good in the astronomical problem, but it does some good. From Figure 36.1, which shows the radio window in the spectrum, it may be seen that, fortunately, the wavelength of 21 cm falls right in the middle of the gap where there is practically no absorption in either the atmosphere or the ionosphere. Furthermore, the atom which emits this frequency is by all odds the most abundant atom in the universe. Hydrogen in the ground state makes up probably 65 per cent of the gas in the galaxy. The only difficulty is that the emission from any one atom is exceedingly feeble, so that we just about need that much hydrogen in order to get a result.

Nowadays, there is wide activity in this field. Many observatories are studying the emission that comes in from the hydrogen atoms in the galaxy. This is done with a standard kind of radio telescope....

If we look at the radiation that does come in from a hydrogen cloud or a concentration of hydrogen that was in the Milky Way, this is what we might see (Figure 36.5). Indeed this is what we do see in one particular direction. If one were looking at hydrogen in the laboratory, a frequency scan would give a single narrow line, the dotted peak in the figure, at the precise frequency I wrote down earlier. Instead of that, looking out into the galaxy, one sees quite a broad affair which often has a structure such as the three-humped curve I have drawn. The reason is very simple. It is the old business of the Doppler effect. The hydrogen which is emitting this "light" is not at rest with respect to us. It may, as a whole, be moving and streaming. We know that astronomical objects are commonly in motion. If the hydrogen cloud is coming toward us, the line will come in at a somewhat higher frequency, and if it is going away, at lower frequency, than if it were stationary with respect to our antenna. In this case we know,

and I will try to explain in a minute how we know, that the three humps are emissions from hydrogen located at three different places; at these different places the hydrogen is moving with different speeds. And that is about all one can say; something can be inferred about the temperature and density of the hydrogen, but we needn't go into that.

Despite its limitations, the astronomers, notably the Dutch astronomers Oort and Van de Hulst at Leiden, found how to exploit this kind of information. Van de Hulst, incidentally, was the first one to recognize the possiblity of detecting the galactic hydrogen emission. Oort and Van de Hulst discovered a way to extract, from records like Figure 36.5, by a kind of indirect argument, the actual *location* of the hydrogen along the line of sight. Remember, this is not like radar. We are not sending out a wave and getting it back; there is no "echo time" to tell us how far away the stuff is. We are just sitting here receiving, and the only thing we can tell directly is how fast the source is moving toward or away from us. To deduce the location of the source we need to know something else about the galaxy.

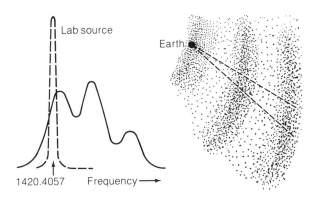

Figure 36.5. A hydrogen line from the galaxy.

1420.4057   Frequency ——>

Lab source

Earth

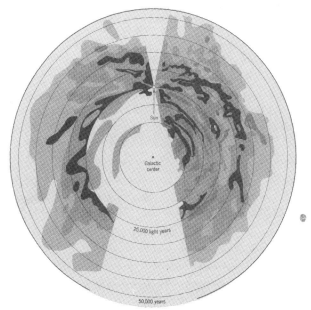

Figure 36.6. Westerhout's map of the galaxy.

... [The galaxy] is just a cloud of stars, and we know that it is not stationary; it is rotating. Astronomers knew that from their observations of the motion of stars. But it is not rotating like a phonograph record, all as one piece, because the stars are not rigidly connected to one another. Rather, it is rotating much more the way the planets revolve around the sun: The outer planets are moving relatively slowly, the inner planets, closer to the central mass, are moving with higher velocity. In fact, one can think of the galaxy as a sort of planetary system. It has no single, dominant body at its center, but it does have a general concentration of mass in the central portion of the disk. The rotational velocity must vary with distance from the center in a way that we can easily predict once we know how the mass is distributed. We begin by adopting a radial mass distribution—a galactic model—that is reasonable in the light of other astronomical evidence. For this distribution we work out the required speed of revolution for material at any given distance from the galactic center—the modified "Kepler's law" for the system.

If one looks out now in a certain direction and sees a source with a certain velocity, one can pin it to a certain position on the line of sight. Of course this involves the assumption about the radial distribution of mass. But one can work backward, and continue until the whole picture is consistent with itself. This is what has been done by the radioastronomers in both Holland and Australia, who have gradually built up a map of the hydrogen gas in the galaxy.

The chart (Figure 36.6) is one that was made and published by Westerhout in Leiden, who has been one of the leaders in this exploration. Westerhout, for reasons that we needn't go into here, left out the central part. There is a tremendous amount of hydrogen in the middle but not much is known about it.... There is no doubt whatever that this is a spiral nebula. In fact, we can even locate ourselves in one of the arms. It is also evident that this is still a self-centered view of the galaxy; there is no reason for the near half to look so different from the far

half except that we happen to have a better view of the former....

It is a pity that one cannot say for sure which way the spiral arms go. It is surprising to learn from astronomers that this question was not settled long ago. The naive assumption that because spiral nebulae look like pinwheels they must be moving like pinwheels is hard to defend without a convincing theory of galactic evolution. As for direct observation of another spiral nebula, Doppler shift of spectral lines reveals which side is approaching us, but there is no easy way to tell which edge is *nearer* to us, so the tantalizing ambiguity remains. I believe majority opinion favors the pinwheel sense. Further refinement of the hydrogen map of our own galaxy—where we *know* the absolute sense of rotation—may eventually settle the question beyond doubt.

This is what has been learned from this one branch of radioastronomy, and the point that I would like to make before I turn to the second part of my talk is that this has been learned by receiving a rather astonishingly small amount of energy, energy which has traveled a very long way to us. The total amount of power that comes to the earth in hydrogen radiation from everywhere in the universe, that is, the power falling on the entire earth, is about *one watt*. The radioastronomers at Leiden, Harvard, Sydney, Greenbank, and elsewhere have been picking up a tiny fraction of that with their antennas. A more astonishing figure is one that I had to compute three times before I was sure of my arithmetic: The total *energy* received by *all* 21-cm observatories over the past nine years, is less than 1 erg! From less than 1 erg of energy we have built this picture of our galaxy. Most of you know what an erg

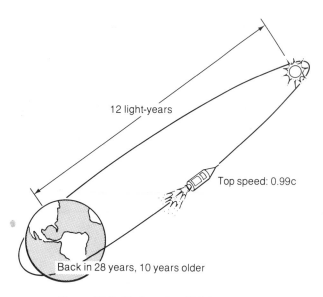

12 light-years

Top speed: 0.99c

Back in 28 years, 10 years older

Figure 36.7. Trip to a place 12 light-years away.

Final mass

ellant

V_{ex}

Propellant

Initial mass

Figure 36.8. Rocket.

is—you can't knock the ash off your cigarette with an erg. That point I want you to remember. It is germane to the thesis which I shall try to establish in the last two parts of this talk, which depart from sober science and go into other directions.

Space Travel

In the second part I shall talk briefly about space travel, and I want to say very distinctly that I am not going to argue the case, pro or con, for travel around the solar system—visiting the moon and Mars and so on. We shall look at wider horizons, as all the astonautical types do, and talk about travel *beyond* the solar system. A lot has been written about this. You are probably as tired of hearing about it as I am, but I hope that if we look at it in one particular way, it may present a fresh aspect. Of course, everything is very far away. The stars are very far away. The nearest star, Alpha Centauri, is 4 light-years distant. People have worried about this but they blandly say, "That's all right because we will travel at nearly the speed of light. Even without relativity we will get there fast and with relativity we will get there and be young anyway." That is perfectly correct, in my view, so far as it goes. Special relativity is reliable. The trouble is not, as we say, with the *kinematics* but with the *energetics*. I would like to develop that briefly, with a particular example. Figure 36.7 defines my example. Let us consider taking a trip to a place 12 light-years away, and back. Because we don't want to take many generations to do it, let us arbitrarily say we will go and come back in 28 years earth time. We will reach a top speed of 99 per cent speed

of light in the middle, and slow down and come back. The relativistic transformations show that we will come back in 28 years, only 10 years older. This I really believe. It would take 24 years for light to go out and come back; it takes the traveler 28 years as seen by the man on earth but the traveler is only 10 years older when he gets back. I don't want to stop and argue the "twin paradox" here because if one *does not* accept its implications then the conclusion that I am going to draw becomes even *stronger*. Personally, I believe in special relativity. If it were not reliable, some expensive machines around here would be in very deep trouble.

Now let us look at the problem of designing a rocket to perform this mission. Let us begin with a reminder of what a rocket is (Figure 36.8). It is a device that has some propellant which it burns and throws out the back. The mechanical reaction accelerates the rocket. When the propellant is all gone, the rocket has reached its final speed and only the payload remains. That is the *best* one can do—carrying along extra hardware only makes it worse. Staging of rockets, i.e., the use of four or five successively smaller stages, is merely a way of trying to *approach* this ideal. The performance of a rocket depends almost entirely on the velocity with which the propellant is exhausted, V_{ex}, as I have called it, ex for exhaust. The rocket people talk about specific impulse, but the impulse they talk about really has the dimensions of a velocity. Let us look at the role this velocity plays in rocket propulsion (Figure 36.8). Here is the rocket with its V_{ex} and we want to get it up to some final speed V_{max}. Then the elementary laws of mechanics—in this case relativistic mechanics, but still the elementary laws of mechanics—inexorably impose a certain relation between the initial mass and final mass of the rocket in the *ideal* case. This relation, shown in Figure 36.9, is relativistically exact. It follows very simply from conservation of momentum and energy, the mass-energy relation, and *nothing else*. In other words, the only thing that could possibly be wrong with this equation is that I made a mistake in deriving it. That is

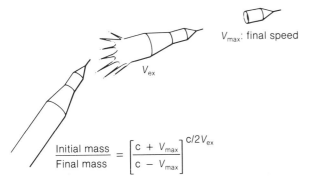

$$\frac{\text{Initial mass}}{\text{Final mass}} = \left[\frac{c + V_{\max}}{c - V_{\max}}\right]^{c/2V_{ex}}$$

Figure 36.9. Relation between the initial mass and the final mass of a rocket in the ideal case.

always possible, but I don't think I did. It checks all right at the limits.

You can plainly see the disadvantage of low exhaust velocity. If we demand a final speed $V_{\max}$ very near the velocity of light, this denominator is going to get awfully small, and the exponent will get large. This is not peculiar to the relativistic domain but occurs in ordinary rocketry, too, wherever the final speed required greatly exceeds the exhaust velocity—as it unfortunately does in the case of earth satellites launched with chemically fueled rockets.

For our vehicle we shall clearly want a propellant with a *very* high exhaust velocity. Putting all practical questions aside, I propose, in my first design, to use the *ideal nuclear fusion* propellant (Box 36.1). I am going to burn hydrogen to helium with 100 per cent efficiency; by means unspecified I shall throw the helium out the back with kinetic energy, as seen from the rocket, equivalent to the entire mass change. You can't beat that, with fusion. One can easily work out the exhaust velocity: it is about $\frac{1}{8}$ the velocity of light. The equation of Figure 36.9 tells us that to attain a speed of 0.99c we need an initial mass which is a little over a *billion* times the final mass. To put up a ton we have to start off with a million tons; there is no way to beat this if we can't find a better reaction.

There simply *are* no better fusion reactions in nature, except one. This is no place for timidity, so let us take the ultimate step and switch to the perfect matter-antimatter propellant (Box 36.2). Matter and anti-matter annihilate; the resulting energy leaves our rocket with an exhaust velocity of c or thereabouts. This makes the situation very much better. To go up to 99 per cent the velocity of light only a ratio of 14 is needed between the initial mass and the final mass. But remember, that isn't enough; we have only reached $V_{\max}$ and our mission is only one-quarter accomplished, so to speak. We have to slow down to a stop, turn around, get up to speed again, come home, and stop. That does not make the ratio 4×14, that makes it 14^4, which is 40,000. So to take a 10-ton payload over the trip described

in Figure 36.7 I see no way whatever to escape from the fact that at take-off we must have a 400,000-ton rocket, half matter and half antimatter.

Incidentally, there is one difficulty which I should have mentioned earlier, but at this stage it is comparatively trivial. If you are moving with 99 per cent the velocity of light through our galaxy, which contains one hydrogen atom per cubic centimeter even in the "empty spaces," each of these hydrogen atoms looks *to you* like a 6-billion-volt proton, and they are coming at you with a current which is roughly equivalent to 300 cosmotrons per square meter. So you have a minor shielding problem to get over before you start working on the shielding problem connected with the rocket engine. That problem is quite formidable as you will see from Figure 36.10, which shows our final design. We have 200,000 tons of matter, 200,000 tons of antimatter, and a 10-ton payload, preferably pretty far out. The accelerations required are of the order of 1g over the whole trip, and not merely in leaving the earth. It just happens that g times 1 year is about equal to the speed of light, so if we want to reach the speed of light in times of the order of years, we are going to be involved in accelerations of the order of 1g. (This is the *one* respect in which relativistic astronautics is simple. No space-medical research is needed to assure us that we can stand 1g. We have been doing it all our lives.) In order to achieve the required acceleration our rocket, near the beginning of its journey, will have to radiate about 10^{18} watts. That is only a little more than the total power the earth receives from the sun. But this isn't sunshine, it's gamma rays. So the problem is not to shield the *payload*, the problem is to shield the *earth*.

Well, this is preposterous, you are saying. That is exactly my point. It *is* preposterous. An remember, our conclusions are forced on us by the elementary laws of mechanics. All those people who have been seriously talking about *lebensraum* in space, and so on, simply haven't stopped to make this calculation and until they do, what they say is nonsense—no matter how highly placed they may be or how big a budget they may control.

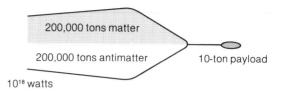

Figure 36.10.

Box 36.2

With perfect *antimatter* propellant:

$$V_{ex} \approx c$$

For $V_{max} = 0.99\, c$,

$$\frac{\text{initial mass}}{\text{final mass}} = 14$$

But to stop, return home, and stop,

$$\frac{\text{initial mass}}{\text{final mass}} = (14)^4 = 40,000$$

Communication through Space

Now I would like to turn to a quite different subject, one which is also speculative, but which involves an entirely different scale of magnitudes, the problem of communication through space. We have already seen how little energy was involved in the amount of information which revealed the structure of our galaxy. An example, in terms of practical communication of messages, is given in Figure 36.11. If I can transmit a message by point-to-point operation with a reasonably large antenna at each end, a 10-word telegram can be transmitted over the 12-light-year path discussed above with a dollar's worth of electrical energy. This is possible because we can detect, amplify, and identify in a radio circuit an amount of energy exceedingly small, and because the energy travels to us suffering no loss whatever except the "inverse square" diminution of intensity as it spreads.

Of course, the trouble is that there isn't anybody at the other end to communicate to. Or is there? What I would like to talk about now is not a new subject and I may not say anything new about it, but I have thought about it a good bit. It is the question of communicating with other people out there, if there are any.

Let us look at the galaxy again. There are some 10^{11} stars in the galaxy. *Double* stars are by no means uncommon, in fact there appear to be almost as many double stars as single stars. Astronomers take this as a hint that invisible companions in the form of planets may not be very uncommon either. Moreover, a large number of stars have lost their angular momentum and are not spinning. One good way for a star to lose its angular spin is by making planets; that is what probably happened in our own solar system. So the chance that there are hundreds of millions of planetary systems among these hundred billion stars seems pretty good. One can elaborate on this, but I am not going to try here to estimate the probability that a planet occurs at a suitable distance from a star, that it has an atmosphere in which life is possible, that life developed, and so on. Very soon in such a speculation the word probabilitiy loses operational meaning. On the other hand, one can scarcely escape the impression that it would be rather remarkable if only one planet in a billion, say, to speak only of our own galaxy, had become the home of intelligent life.

Since we can communicate so easily over such vast distances, it ought to be easy to establish communication with a society (let us use that word) in a remote spot. It would be even easier for them to initiate communication, if they were ahead of us. Shall we try to listen for such communications, or shall we broadcast a message and hope someone hears it? If you think about it a little, I think you will agree that we want to listen *before* we transmit. The time scale of the galaxy is very long. Wireless telegraphy is only 50 years old, and really sensitive receivers are much more recent. If we look for people who are able to receive our signals but have not surpassed us technologically, i.e., people who are not more than 20 years behind us but still not ahead, we are exploring a very thin slice of history. On the other hand, if we listen, we are looking for people who are *anywhere* ahead providing they happen to have the urge to send out signals. Also, being technologically advanced, they can transmit much better than we can. (For rather fundamental reasons, transmitting is harder than receiving in this game.)

So it would be silly to transmit before listening for a long time. This is an amusing game to play. I won't dwell on it long because you will have more fun trying it yourself, but let me suggest its nature. In the first place, it is essentially cryptography in reverse. Let me assume—this may not be true, but let me assume it—that there is somebody out there who is technologically ahead of us. He can transmit 10 Mw as easily as we can transmit a kilowatt, and he wants us to receive his signal. He surely knows more about us than we know about him, and moreover, he is a relatively close neighbor of ours in the galaxy. We share the same environment; he knows all about the hydrogen line—he learned it centuries ago. He knows that that line is the only prominent line in that window of the spectrum.

If you want to transmit to a fellow and you can't agree on a frequency, it's nearly hopeless. To search the entire

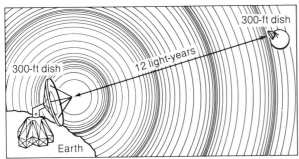

To transmit a 10-word telegram we have to radiate about $1 worth of electrical energy.

Figure 36.11.

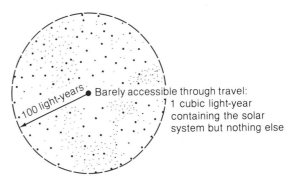

Easily spanned by radio communication: 3 million cubic light-years containing 500 stars like the sun

Figure 36.12.

radio spectrum for a feeble signal entails a vast, and calculable, waste of time. It is like trying to meet someone in New York when you have been unable to communicate and agree on a meeting place. Still, you know you want to meet him and he wants to meet you. Where do you end up? There are only two or three places: Grand Central Station, etc. Here, there is only one Grand Central Station, namely, the 1420-Mc line which is, by a factor of 1000 at least and probably more, the most prominent radio frequency in the whole galaxy. There is no question about where you transmit if you want the other fellow to hear, you pick out the frequency that he knows. Conversely, he will pick the frequency he knows we know, and that is the frequency to listen on. If you play this game carefully you will find the conclusion inescapable. We know what to do; we know where to listen. We don't know quite what his code will be but we know how to set up a computer program to search for various codes. Let us make some reasonable assumptions, for example, about power. Let us give the transmitter the capability of radiating a megawatt within a 1-cycle/sec band. This is something we could do next year if we had to; it is just a modest stretch of the present state of the art. Indeed, my information may be obsolete, there may be contracts out now calling for such performance. Suppose we receive with a 300-ft disk and he transmits with one as large. How we process the signals will affect the ultimate range, but, making very simple and conservative assumptions about that part of the problem, I find that we should be able to recognize his signal even if it comes from several hundred light-years away. With the new maser receivers which have just begun to be used in radioastronomy, 500 light-years ought to be easy. A sphere only 100 light-years in radius contains about 400 stars of roughly the same brightness (± 1 stellar magnitude) as the sun. And remember, the volume accessible by communication goes up as the cube of the range. I have argued that it is ridiculously difficult to travel even a few light-years, and ridiculously easy to communicate over a few hundred. I think these numbers actually under-

estimate the disparity. But even so, the ratio of the volumes is 1 million (Figure 36.12).

There are other interesting questions. When we get a signal, how do we know it is real and not just some accident of cosmic static? This I like to call the problem of the axe head. An archeologist finds a lump of stone that looks vaguely like an axe head, down in about the right layer. How does he know it is an axe head and not an oddly shaped lump of stone? Actually, they are usually *very* sure. An arrowhead can look rather like an elliptical pebble, and still there is no doubt that it is an arrowhead. Our axe-head problem can be solved in many ways. The neatest suggestion I know of originated with Cocconi and Morrison (Reading 20), who have published a discussion of this whole subject. Morrison would have the sender transmit a few prime numbers. That's all you need: 1, 3, 5, 7, 11, 13, 17—by then you *know*. There are no magnetic storms or anything on Venus making prime numbers.

What can we talk about with our remote friends? We have a lot in common. We have mathematics in common, and physics, and astronomy. We have the galaxy in which we are near neighbors. The Milky Way looks about the same to them; 400 light-years is only ⅛ inch on our model here. We have chemistry in common, inorganic chemistry, that is. Whether their organic chemistry has developed along the lines of ours is another question. So we can open our discourse from common ground before we move into the more exciting exploration of what is not common experience. Of course, the exchange, the conversation, has the peculiar feature of built-in delay. You get your answer back decades later. But you are sure to get it. It gives your children something to live for and look forward to. It is a conversation which is, in the deepest sense, utterly benign. No one can threaten anyone else with objects. We have seen what it takes to send *objects* around, but one can send information for practically nothing. Here one has the ultimate in philosophical discourse—all you can do is exchange ideas, but you do that to your heart's content.

I am not sure we are in a position to go about this yet. I am not advocating spending a lot of money setting up listening posts. Although, as a matter of fact, a listening program on a very modest scale is going on at Green Bank under Frank Drake, who has some very imaginative and, I think, sound ideas on how it should be done. They haven't heard anything yet.

But in my view, this is too adult an activity for our society to engage in, on a large scale, at the present time. We haven't grown up to it. It is a project which has to be funded by the *century,* not by the fiscal year. Furthermore, it is a project which is very likely to fail *completely.* If you spend a lot of money and go around every 10 years and say, "We haven't heard anything yet," you can imagine how you make out before a congressional committee. But I think it is not too soon to have the fun of thinking about it, and I think it is a much less childish subject to think about than astronautical space travel. In my view, most of the projects of the space cadets are not really imaginative. And the notion that you have to *go* there seems to me childish. Suppose you took a child into an art museum and he wanted to *feel* the pictures—you would say, "That isn't what we do, we stand back and look at the pictures and try to understand them. We can learn more about them that way." All this stuff about traveling around the universe in space suits—except for *local* exploration, which I have not discussed—belongs back where it came from, on the cereal box.

37

Sebastian von Hoerner
The General Limits of Space Travel

The goal of this discussion is to find out whether interstellar space travel (travel from star to star) might become possible for us in the far future, and might therefore already be possible for other, more advanced beings. Is there any hope of making direct interstellar visits, or will all communication between civilizations be confined to electromagnetic signals? Certainly, from present estimates we cannot give a direct and conclusive answer of the yes-no type, but we can point out the significant basic facts and get as close to an answer as is possible at our present state of knowledge, leaving the final conclusion to the reader.

I shall begin by summarizing the present limitations and problems of space flight, trying to pin down the few basic points, and trying to separate the general difficulties from the merely temporary ones. From this starting point we may then proceed to estimate the possibilities of future space flight.

The prime postulate in these estimates is a technology much more highly advanced than our present one. Thus, we completely neglect all technical problems, however serious they actually might be. Only such fundamental properties as time, acceleration, power, mass, and energy are considered.

The results are given in terms of the minimum travel times deriving from various assumptions. Furthermore, we calculate some basic requirements for reaching these travel times.

Chemical Binding Energy

The only propelling mechanism actually used at present is acceleration of exhaust material by combustion, where the relatively low binding energy between atoms sets a limit in two ways: in the energy content of fuels, and in the heat resistance of combustion-chamber and nozzle materials.

Reprinted from *Science*, 137 (1962), 18. Copyright 1962 by the American Association for the Advancement of Science.

In order to remove 1 kg of matter from the earth against gravity, we need an energy of 17.4 kwhr. But the best fuel burning hydrogen with oxygen yields only 3.2 kwhr/kg of fuel (explosives yield still less—for example, TNT yields 1.1 kwhr/kg). Thus we need 5.4 kg of fuel to remove 1 kg of matter, but the supply of fuel has to be accelerated, too, and this again requires much more fuel, and so on. Despite this difficulty of low-fuel-energy content, small payloads can still be removed, but with an extremely low efficiency.

The availability of more energetic fuels would not be of too much help. No nozzle material can stand temperatures above about 4000°C at the utmost limit. If a combustion gas of that temperature escapes through a nozzle, it will do so with an exhaust velocity of 4.0 km/sec if it is water vapor, and of less than that if substances other than hydrogen are burnt. But the rocket itself needs a velocity of 11.2 km/sec to leave the gravitational field of the earth, and far greater velocities if interstellar distances are to be covered within a reasonable time.

The velocity of a rocket after burnout of all its fuel, V, the exhaust velocity generated by the propellant, S, and the so-called mass ratio, $\mathbf{M} = M_i/M_0$, are connected by the well-known rocket formula

$$\frac{V}{S} = \ln \mathbf{M} \qquad (1)$$

where M_i is the total initial mass of the rocket (including fuel) and M_0 is the mass after burnout—that is, M_i minus the mass of fuel. Now, the logarithm is a function which increases very slowly with its argument; even if fuel constitutes 90 per cent of the initial mass, the rocket velocity will be only 2.3 times the exhaust velocity. And if a one-stage rocket with fuel constituting 99.9 per cent of its initial mass could be built, even then we would achieve a velocity of only $V = 6.9S$. We cannot at present build such a rocket, but we do imitate it through multistage rockets. With these the difficulty remains the same, because the mass ratios of all stages accumulate in a multiplicative way (a tiny payload in the last stage, as compared to a huge fuel mass in the first stage), but the stage velocities accumulate only additively. Thus, with combustion-powered rockets, even of many stages, we are just able to leave the earth, but we cannot reach very high velocities.

In order to see this more precisely, we define the efficiency of a rocket, Q, as the useful energy (the energy contained in the final velocity of the empty rocket) divided by the energy content of all fuel burnt. We then have

$$Q = \frac{(V/S)^2}{e^{V/S} - 1} \qquad (2)$$

This efficiency has its maximum value, $Q = 0.647$, at $V/S = 1.59$, but it drops off very fast and is only 1 per cent at $V/S = 9$. And in the case of many stages the efficiency gets still smaller by a large factor. The efficiency of an ideal multistage rocket is only 0.1 per cent for $V/S = 6$.

These difficulties connected with low binding energy are only temporary ones, because they are confined to combustion processes. In increasing the energy content of fuels we can make a huge step if we use atomic energy; the fission of uranium, for example, yields 20 million kwh/kg. And nozzles as well as high temperatures can be avoided completely when we learn to use ion thrust as a propelling mechanism (charged particles—ions—are accelerated by electrical fields to a high velocity, with which they leave the rocket). A 5000-volt acceleration, for example, could give S of the order of 100 km/sec. With this mechanism, S increases with the square root of the acceleration voltage.

In dealing with these difficulties we have discovered one fundamental principle:

> In order to avoid unreasonably low efficiencies, the exhaust velocity S should be about as large as the required final velocity of the rocket, V, or at least of the same order of magnitude.

Power-Mass Ratio

As soon as these two difficulties have been overcome, by obtaining a high energy content in fuel and a high exhaust velocity, one will immediately encounter the next fundamental difficulty. The acceleration of a rocket, b, is of course given by

$$b = \frac{\text{thrust of engine}}{\text{total mass of rocket}} \qquad (4)$$

Now, the thrust equals the exhaust mass-flow (mass/sec) times the exhaust velocity S; and the needed power of the engine equals the mass-flow times one-half the square of

the exhaust velocity. We thus can write [instead of Equation (4)]:

$$b = 2P/S \qquad (5)$$

where P is the ratio of power of engine to total mass of rocket.

This means that, if we are working with a high exhaust velocity S, we need a high power-mass ratio P, as otherwise we would get only a small acceleration b.

But nuclear reactors and all the equipment needed to give a strong ion thrust are so complicated and massive, as compared with the relatively simple combustion equipment, that there is no hope at present of reaching, with reactors, the value of P already attained with combustion rockets. The acceleration thus will be extremely small until we can find a way to increase the power-mass ratio of a reactor by many orders of magnitude.

Distances for Interstellar Space Travel

The only goal which may be important enough to justify the immense effort needed for interstellar space travel appears to be the search for other intelligent beings. In a recent article (von Hoerner, 1961) I tried to estimate the distances between neighboring technical civilizations in order to guide preparations and stimulate a search for possible electromagnetic signals. The points of interest for our present purpose may be summarized as follows:

1. It would be megalomania to think that we are the one intelligent civilization in the universe. On the contrary, we should assume from our present knowledge that life and intelligence will have developed, with about the same speed as on earth, wherever the proper surroundings and the needed time have been provided. From our present limited data, we judge that this might have been the case on planets of about 6 per cent of all stars. The nearest ten such stars are at an average distance from us of about 5.6 parsecs (1 parsec = 3.09×10^{13} km = 3.26 light-years).

2. It would be equally presumptuous to think that our present state of mind is the final goal of all evolution. On the contrary, we should assume that science and technology are just one link in a long chain and will be surpassed one day by completely new and unpredictable interests and activities (just as gods and demons unpredictably have been surpassed by science in offering explanations of many important phenomena). We should assume a finite longevity L of the technical state of mind; if we call T the age of the oldest stars and D the average distance to the nearest ten technical civilizations, we get

$$D = 5.6 \text{ parsecs} \left(\frac{T}{L}\right)^{1/3} \qquad (6)$$

We may take about 10 billion years for T, but it is extremely difficult to estimate the value of L. It is my personal opinion that we should take some ten thousands of years for L, but since many scientists regard this as being too pessimistic, we will take 100,000 years for our present purpose—a value which gives about 250 parsecs for D. Fortunately, the uncertainty of L enters the value of D only with the power $\frac{1}{3}$, and if we change L even by a factor of 8, D will change by only a factor of 2.

With respect to interstellar space travel we must clearly separate two questions: (a) We may want to know what the possibilities are for *our* future interstellar travel. In this case we are interested in locating *any* kind of intelligent life, and the distance we are required to reach is

$$5.6 \text{ parsecs } (= 18.6 \text{ light-years}) \qquad (7)$$

(b) We may examine the possibility of other beings visiting us. In this case the other civilization must be a *technical* one, and for the calculations that follow we will use, for the above-mentioned distance to be covered by these other beings

$$250 \text{ parsecs } (= 820 \text{ light-years}) \qquad (8)$$

In order to help visualize these astronomical distances, I will describe them with a model of scale 1:180 billion. The earth, then, is a tiny grain of desert sand, just visible to the naked eye, orbiting around its sun, which now is a cherrystone a little less than 3 ft away. Within approximately the same distance, some few feet, lies the goal of our present space travel: the other planets of our solar system, such as Mars and Venus. But the nearest star, Proxima Centauri, is another cherrystone 140 miles away; and the next stars with habitable planets, where we might look for intelligent life, are to be expected at a distance of 610 miles. The next technical civilizations, however, will be at a distance as great as the circumference of the earth. Just for fun one may add the distance to the Andromeda nebula, the next stellar system comparable to our own galaxy: In our model it is as far away as, in reality, the sun is from the earth. The most distant galaxies seen by astronomers with their best telescopes are 2000 times as far away, and here even our model fails to help.

Relativistic Treatment

One thing is now clear: In order to cover interstellar distances within reasonable times we ought to fly as close as possible to the velocity of light, the utmost limit of any velocity, according to the theory of relativity (and in accordance with all experiments with high-energy particles). But as we approach the velocity of light, the formulas of normal physics must be replaced by those of relativity theory.

This might be of some help, because one of the most striking statements of relativity theory is that time itself is not an absolute property but is shortened for systems approaching the velocity of light. If, for example, we are to move out and back a distance of 800 light years, then people remaining on earth will have to wait at least 1600 years for the return of the rocket. But if the speed of the rocket closely approaches the velocity of light, then the flow of time for this rocket and its crew becomes different from that on earth, and one may expect that the crew members will have to spend only a few years, perhaps, of their own lifetimes between start and return.

The equations that follow are derived under the assumption that the formulas of the special theory of relativity still hold under conditions of permanent acceleration and deceleration—a view which is generally assumed but not yet accepted. I will keep this part of the discussion as short as possible, and any reader who abominates formulas, relativistic or not, may skip to the next section.

We use the following definitions: τ is rocket time and t is earth time (both equal zero at the start of the rocket flight); v is the velocity of the rocket (as seen from the earth); c is the velocity of light; b is the acceleration of the rocket (as measured *within* the rocket by the pressure of a unit mass against a spring), assumed to be constant; and x is the distance between the rocket and the earth.

The differentials of τ and t are connected by

$$d\tau = dt(1 - [v/c]^2)^{1/2} \qquad (9)$$

and the differential equation for v reads

$$\frac{d}{dt}\left\{ \frac{v}{(1 - [v/c]^2)^{1/2}} \right\} = b \qquad (10)$$

which is easily integrated. Solving for v we get

$$v(t) = \frac{bt}{(1 + [bt/c]^2)^{1/2}} \qquad (11)$$

With the help of Equation (11) we can integrate Equation (9):

$$\tau(t) = \int_0^t (1 - [v/c]^2)^{1/2}\, dt = \frac{c}{b} \text{ arc sinh } \frac{bt}{c} \qquad (12)$$

while the distance is integrated according to

$$x(t) = \int_0^t v\, dt = \frac{c^2}{b}\left\{ (1 + [bt/c]^2)^{1/2} - 1 \right\} \qquad (13)$$

We realize that, in order to get an equivalent to our rocket formula [Equation (1)], thrust and acceleration, if both are measured within the rocket, should be connected as usual:

$$-\frac{dM}{d\tau} S = Mb \tag{14}$$

where M is the total mass of the rocket at any time and $-dM/d\tau$ is the exhaust mass flow. But in order to reach $v \approx c$, we need also $S \approx c$ according to relationship (3), and S in Equation (14) should be replaced by $S(1 - S^2/c^2)^{-\frac{1}{2}}$. Furthermore, $S \approx c$ demands so much energy that in integrating Equation (14) the mass loss due to mass defect should be considered, too. In the calculations that follow, however, we shall find that, even with atomic energy, both S and v still are much less than c, so that no relativistic treatment is needed, and Equation (1) may be used.

The only means of reaching $v \approx c$ turns out to be complete annihilation of matter. In this case one will use photon thrust, and Equation (14) should be written

$$-\frac{dM}{d\tau} c = Mb \tag{15}$$

We derive the following formulas for annihilation of matter as the energy source and photon thrust as the propelling mechanism (ignoring all doubts about the practical realization of either). We integrate Equation (15) from start ($M = M_i$) to burnout of all fuel ($M = M_0$). The durations of this period of acceleration for the crew, τ_0, and [from Equation (12)] for people on earth, t_0, then are

$$\tau_0 = \frac{c}{b} \ln \mathbf{M} \tag{16a}$$

and

$$t_0 = \frac{2c}{b} (\mathbf{M} - \mathbf{M}^{-1}) \tag{16b}$$

where, again, $\mathbf{M} = M_i/M_0$, and the distance traveled between start and burnout, x_0, is

$$x_0 = \frac{c^2}{2b} (\mathbf{M} + \mathbf{M}^{-1} - 2) \tag{17}$$

After burnout, the final velocity of the rocket, V, is given by

$$V = c \, \frac{1 - \mathbf{M}^{-2}}{1 + \mathbf{M}^{-2}} \tag{18}$$

—a velocity which causes a time dilatation, on the further unaccelerated) portion of the journey, of

$$\left(\frac{dt}{d\tau}\right)_0 = \frac{\mathbf{M} + \mathbf{M}^{-1}}{2} \tag{19}$$

We notice that the final velocity and time dilatation do not depend on b or t_0; it does not matter how quickly we burn our fuel.

The mass ratio should be, of course, as large as possible, and for $\mathbf{M} \gg 1$ the equations just given reduce to[1]

$$t_0 \approx 2 \, \frac{c}{b} \, \mathbf{M} \tag{20a}$$

$$x_0 \approx \frac{c^2}{2b} (\mathbf{M} - 2) \tag{20b}$$

$$V \approx c \, (1 - 2/\mathbf{M}^2) \tag{20c}$$

and

$$\left(\frac{dt}{d\tau}\right)_0 \approx \frac{1}{2} \, \mathbf{M} \tag{20d}$$

Finally, Equation (5) needs a slight modification, too; for photon thrust it reads simply:

$$b = P/c \tag{21}$$

Energy Content of Nuclear Fuels

Having seen, earlier in this article, that the energy per mass of fuel is one of the important considerations in space travel, we now ask for the most energetic fuels. The utmost possible limit is set by one of the fundamental laws of relativity:

$$E = mc^2 \tag{22}$$

which gives the energy E obtained by complete annihilation of matter of mass m. Or we might say it another way: Energy E has an inertial mass (its resistance against acceleration) of $m = E/c^2$. If we call ϵ the specific energy content (energy/mass), we have for complete annihilation, $\epsilon = c^2 = 9 \times 10^{20}$ ergs per gram.

Complete annihilation takes place only if matter and antimatter are brought together: when a proton combines with an antiproton, electron combines with positron, and so on. But the world we live in consists of matter only, and to store a large amount of antimatter with equipment consisting of matter seems quite impossible, from all we know. We thus have to look for some other source of energy.

---

[1] A different derivation of Equation (20), connecting V and M, has been given by J. R. Pierce (1959) together with a good explanation of the so-called clock paradox. Pierce also investigates interstellar matter as fuel, with the same negative result as that given in this article.

Table 37.1.

ENERGY PER MASS OF FUEL FOR NUCLEAR REACTIONS

Fuel	Final product	Energy/mass 10^{18} ergs/g
	Annihilation	
Matter plus antimatter	Radiation	900
	Fusion	
Hydrogen	Helium	6.3
Hydrogen	Iron	8.3
	Fission	
Uranium	Mixture, as produced in reactors	0.65
Uranium	Iron	1.1

Table 37.2.

TOTAL DURATION AND DISTANCE REACHED, WITH CONSTANT ACCELERATION AND DECELERATION AT 1G

Duration (out and back), years		Distance reached, parsec
For crew on board rocket	For people on Earth	
1	1.0	0.018
2	2.1	0.075
5	6.5	0.52
10	24	3.0
15	80	11.4
20	270	42
25	910	140
30	3,100	480
40	36,000	5,400
50	420,000	64,000
60	5,000,000	760,000

If antimatter is omitted, then according to another fundamental rule of nuclear physics the combined number of heavy elementary particles (protons plus neutrons) must stay constant, and the only thing left is to unite several light nuclei into a heavy one (fusion) or to split up a heavy nucleus into several light ones (fission), leaving the sum of protons and neutrons constant. In doing this we may gain or lose energy, according to the different amounts of nuclear binding energy of the various elements. The total energy content per nucleon (proton or neutron), in case of annihilation, would be 931.13 million electron volts for hydrogen; it drops quickly to 924.88 for helium, and then slowly to a flat minimum of 922.55 for iron. From then on it increases again, but very slowly, to 922.65 for uranium. The differences in these figures represent the energy available for nuclear reactions, and the most energy is gained if we start at either end and stop at the lowest point, at about iron. Since hydrogen has a higher value than uranium, we can gain more energy by fusion than by fission; and furthermore, since the minimum is an extremely flat one, it is not important to stop exactly at iron.

At present we use fission in reactors; the fission products are a mixture of elements of all masses, and the gain in energy is about half that which would result if all fission products were iron. Fusion of hydrogen into helium is the source of energy of our sun and of most other stars; it is used in hydrogen bombs only. Scientists in many countries have worked hard to produce controlled fusion, but without success so far.

The only fuel used at present for space travel releases energy by chemical reactions, where the burning of hydrogen to water yields only an energy-mass ratio of 1.15×10^{11} ergs per gram. If we learn to use uranium reactors instead, the energy-mass ratio will be increased by a factor of 5.6 million. If it ever became possible to use the fusion of hydrogen into helium as a power source for space travel, one would gain another factor of 10; and if complete annihilation were practicable, a further factor of 140 would be gained. Table 37.1 summarizes these facts.

Acceleration and Time

Thus equipped with an understanding of nuclear fuel, if not with the real thing, and with relativistic formulas, we proceed with estimating the general limits of future space travel.

As a first step we neglect even the requirements of energy and power. The only limitation then remaining will be the maximum amount of acceleration which a crew can stand. It has been estimated (U.S. Government, 1959) that a terrestrial crew can stand, for a period of *years*, approximately $b = 1g$. It seems likely that, over a long trip, any crew will stand only about as much acceleration as its members are used to experiencing on their home planet. This might differ from our case by a factor of, say 2 or 3 in either direction, but probably by less if the conditions for the development of life are carefully regarded. In the following discussion we will use a limit of $1g$.

If the acceleration is limited to this fixed value, the shortest travel time for a given distance will result if we accelerate with $1g$ half of the way and then decelerate with $1g$ over the second half of the trip, returning in the same way. On the basis of this assumption and of Equations (11) and (13), Table 37.2 has been calculated.

We see that the relativistic time dilation yields an effective gain for the crew only if the crew members spend more than about 10 years of their lives on the voyage. The further increase, however, is a very steep one (exponential); if the crew members spent 30 years of their lives on the voyage they would be able to fly to the Orion nebula and back, and 3000 years would have elapsed on earth between their departure and their return. For our goals of travel to distances of 5.6 and 250 parsecs, we obtain the values in Table 37.3.

With these results, many readers may already have lost

Table 37.3.

TOTAL DURATIONS, FOR ROCKET-CREW MEMBERS AND FOR
PEOPLE ON EARTH, OF ROUND-TRIP VOYAGES TO DISTANCES OF
5.6 AND 250 PARSECS, WITH CONSTANT ACCELERATION AND
DECELERATION AT 1G

Distance reached, parsec	Duration, years	
	For crew	On Earth
5.6	12.3	42
250	27.3	1550

hope of future interstellar space travel; others still may be optimistic. But so far we have neglected the requirements of energy and power.

Energy and Time

Acceleration and deceleration require a lot of energy, which has to come from somewhere. One might perhaps think of providing the rocket with a large funnel in order to sweep up the interstellar matter for fuel. But the interstellar matter has only a very low density (about 10^{-24} g/cm^3), and in order to collect 1000 tons of matter (10 times the fuel of one Atlas rocket) on a trip to a goal 5.6 parsecs away, one would need a funnel 100 km in diameter; we will rule out this possibility. We cannot refuel under way in this manner, or in any other way while traveling at high speed, and thus the rocket must be provided initially with all the energy it needs to reach its goal. But we might allow for refueling at the destination point, when the rocket will be at rest. For our estimate we will consider a three-stage rocket: Stages 1 and 2 are used for the trip to the destination, there stage 2 is refueled, and stages 2 and 3 are used for returning. There is thus one stage for each period of acceleration and deceleration.

The next thing to be fixed is the mass ratio **M** of a single stage. Our present values for **M** are around 10, but the values would become very low if any energy source other than combustion were to be used. Keeping in mind the extremely massive and complicated equipment needed for nuclear reactions, and for propelling mechanisms such as ion thrust, we think that a value of **M** = 10 could be used as an extreme upper limit, even for a much more advanced technology.

The only source of nuclear energy now in sight is the fission of heavy nuclei such as uranium, where 1 g yields 6.5×10^{17} ergs. The highest efficiency is achieved if the fission products themselves can be expelled for propulsion with their fission energy (although at present we have no idea how this can be accomplished). In this case, we will get an exhaust velocity of $S = 13,000$ km/sec $= c/23$—a value so small as compared to the velocity of light that

Equation (1) still may be used. With a mass ratio of 10 as a limit, the final velocity after burnout then is $V = 30,000$ km/sec $= c/10$. Relativistic effects, such as time dilatation, will not play a role of any importance. In order to reach greater distances we have to fly most of the time without acceleration (after burnout of the first stage) and decelerate shortly before reaching the goal with our second stage. The full travel time, out and back, is 380 years for 5.6 parsecs and 17,000 years for 250 parsecs. This certainly does not look very promising.

If one is optimistic enough to think that the fusion of hydrogen into helium might become usable for rocket propulsion, with a mass ratio of 10, even then only $V = c/5$ can be achieved, and time dilatation again will be unimportant. The full travel time is 180 years for 5.6 parsecs and 8000 years for 250 parsecs—not much better than before.

The utmost limit, which cannot be surpassed, is set by the mass equivalent of the needed energy itself (its resistance against acceleration), no matter how this energy is stored. Personally, I do not think that complete annihilation of matter, or some other means of storing "pure energy," ever will become practical for any purpose, let alone in rockets with a mass ratio of 10. But imagine that it does: Then 98 per cent of the velocity of light can be achieved, according to Equation (20c), and as a result of the time dilatation, the time for the crew will get shorter than the time on earth (after burnout) by a factor of 5.0. For 5.6 parsecs, the full travel time will be 14 years for the crew and 42 years on earth, and for a distance of 250 parsecs we get 300 years for the crew and 1500 years on earth. We still must spend 14 years within a rocket in order to search for intelligent beings, and only after 300 years in a rocket will the inhabitants of some alien planet have a fair chance of meeting other beings, like ourselves, who are in just the same state of science and technology as they are.

I should mention again that the final velocity after burnout does not depend on the amount of acceleration b, either in the classical treatment [Equation (1)] or in the relativistic one [Equation (18)]. Only the energy content of the fuel and the mass of the rocket are important, not the rate at which fuel is consumed. The latter rate will influence the duration of the acceleration period, of course, but not the final velocity. This means that if we should prepare the crew to resist very high acceleration (by freezing them in a solid block of ice, or the like), we could shorten the acceleration periods but not the duration of the unaccelerated flight in between.

In the case of fission or fusion, almost all of the travel time is spent in unaccelerated flight after burnout, and high acceleration will not help at all. In the case of annihilation, however, 9.5 years of the crew's time is spent in accelerated or decelerated flight, and this period could be shortened through greater acceleration, but we are still neglecting the power requirements. For a distance of 5.6 parsecs, 4.2 years of the crew's time is spent in unaccelerated flight, and this period cannot be shortened in any way; again, for a distance of 250 parsecs, almost all of the time is spent in unaccelerated flight.

Power-Mass Ratio and Acceleration

For the interstellar distances discussed earlier we need a travel velocity close to the velocity of light, and according to principle (3) we must have $V \approx S$ for reasonable efficiency. These criteria taken together then demand that $S \approx c$. Furthermore, we have seen that complete annihilation of matter is the only hope as a power source in interstellar space travel, and since we must not waste any matter by using it for propulsion, only photon thrust is left us. In that case Equation (21) applies, and $b = P/c$.

From Equation (16) we see that the acceleration must be as large as possible in order to hold τ_0 small, but we have argued that b must be limited to about $1g$. The two considerations then demand that $b \approx 1g$.

Now, if Equation (21) holds and $b \approx 1g$, then the power-mass ratio must have the extremely high value of

$$P = 3 \times 10^{13} \text{ cm}^2/\text{sec}^3 \qquad (23)$$

or, in the power units of watt or horsepower,

$$P = 3 \times 10^6 \text{ watts/g} = 4 \times 10^3 \text{ hp/g} \qquad (24)$$

In order to understand the full meaning of Equation (24) we might consider our present fission reactors—those with the highest power-mass ratios. Reactors for ship propulsion, with power output of 15 Mw and weight of 800 tons give $P = 0.02$ watt/gram—a value too low by a factor of 1.5×10^8 to fulfill Equation (24). If no shielding and no safety measures were needed, then the highest value theoretically possible would be $P = 100$ watts/gram, still too low, by a factor of 30,000. In fact, according to Equation (24), the whole power plant of 15-Mw output (enough for a small town) should weigh not more than 5 g (the weight of 10 paper clips). Or to express it another way, to fulfill Equation (24), the engine of a good car, producing 200 horsepower, could not weigh more than 50 mg—one-tenth the weight of a paper clip.

But that is not all. Not only do we need power, we have to get rid of it, too. Photons might be emitted in the optical or the radio range, and propulsion will result if all emission is in one direction. A large transmitting station of 100-kw power output can then give the tiny thrust of 30 milligrams, and so can an aggregate of searchlights with combined power of 100 kw. And all this should weigh not more than $^1/_{15}$ the weight of a paper clip. The power source and transmitter requirements must be combined,

and the mass entering Equation (24) must contain reactor as well as emitting stations.

So far we have neglected payloads and fuel, and the mass of these must be included in Equation (24), too. As an example we start with a "small" space ship of 10-ton payload, and we add another 10 tons for power plant plus emitters. If we want to reach a velocity within 2 per cent of that of light (with a dilatation factor of 5), we need a mass ratio $\mathbf{M} = 10$, according to Equation (20d), and the total mass of the rocket will be 200 tons. We find, from Equation (24), that in order to get an acceleration of $b = 1g$, we would need a power output of 600 million Mw. Thus,

> We would need 40 million annihilation power plants of 15 megawatts each, plus 6 billion transmitting stations of 100 kw each, altogether having no more mass than 10 tons, in order to approach the velocity of light to within 2 per cent within 2.3 years of the crew's time.

If requirement (25) is not fulfilled, we get equations for the periods of acceleration and deceleration as follows. From Equation (16a) we have

$$\tau_0 = \frac{c^2}{P} \ln \mathbf{M} \qquad (26)$$

and from Equations (20a) and (20b) we get

$$t_0 = 2 \frac{c^2}{P} \mathbf{M} \qquad (27a)$$

and

$$x_0 = \frac{c^3}{2P} (\mathbf{M} - 2) \qquad (27b)$$

If, for example, we fail to fulfill requirement (25) by a factor of 10^6 (if we have 40 power plants plus 6000 transmitters, weighing, in all, 10 tons—still a fantastic value), then $b = 10^{-6}g$, and it would take 2.3 millions of years for the crew to approach the velocity of light to within 2 per cent.

Or, to put it the other way round, if one wants to get an acceleration of, say, $b = 100g$, in order to take full advantage of having a deep-frozen crew, then 100 times the weight of the equipment mentioned in requirement (25) must not total more than 10 tons; this means that power plants plus transmitters should have an output of 6000 Mw/g. Purcell (Reading 36) has arrived at similar conclusions from a study of the requirements of relativistic rockets. There is no way of avoiding these demands, and definitely no hope of fulfilling them.

Conclusion

The various questions dealt with in this article have not led to the definitive answer that interstellar space travel is absolutely impossible. We have found simply the minimum travel times given by different assumptions, and we have found the requirements needed for reaching these limits. This is, at present, all we can do, and the final conclusion as to the feasibility of such ventures is up to the reader. The requirements, however, have turned out to be such extreme ones that I, personally, draw this conclusion: space travel, even in the most distant future, will be confined completely to our own planetary system, and a similar conclusion will hold for any other civilization, no matter how advanced it may be. The only means of communication between different civilizations thus seems to be electro-magnetic signals.

Acknowledgment

I wish to thank F. D. Drake for reading the manuscript.

38

Carl Sagan
Direct Contact among Galactic Civilizations by Relativistic Interstellar Spaceflight

In recent years there has been a resurgence of interest in the ancient speculation that civilizations exist on other worlds beyond the Earth. This question has retained a basic and widespread appeal from the beginnings of human history; but only in the past decade has it become even slightly tractable to serious scientific investigation. Work on stellar statistics and stellar evolution has suggested that a large fraction of the stars in the sky have planetary systems. Studies of the origin of the solar system and of the origin of the first terrestrial organisms have suggested that life readily arises early in the history of favorably-situated planets. The prospect occurs that life is a pervasive constituent of the universe. By terrestrial analogy it is not unreasonable to expect that, over astronomical timescales, intelligence and technical civilizations will evolve on many life-bearing planets. Under such circumstances the possibility then looms that contact with other galactic communities may somehow be established.

It has been argued that the natural channel for interstellar communication is radio emission near the 21 cm line of neutral hydrogen (Cocconi and Morrison, Reading 20); or between 3.2 and 8.1 cm (Drake, Reading 24 and private communication, 1961); or at 10.5 cm (von Hoerner, 1961). Alternatively, laser modulation of the intensity of core reversal in the Fraunhofer lines of late-type stars has been suggested (Schwartz and Townes, Reading 23); or automatic interstellar probe vehicles transmitting a precoded message to planetary sources of monochromatic radio emission which are randomly encountered (Bracewell, Reading 21).

The purpose of the present paper is to explore the likelihood and possible consequences of another communications channel: direct physical contact among galactic communities by relativistic interstellar spaceflight. Part of the impetus for publishing these remarks has been a paper by von Hoerner (1961) which arrives at very pessimistic estimates for the number of extraterrestrial civilizations; and three papers (Pierce, 1959; Purcell, Reading 36; von Hoerner, Reading 37) which reach distinctly negative conclusions on the ultimate prospect of relativistic interstellar spaceflight. I feel that the information now available permits rather different conclusions to be drawn.

The line of argument to be pursued involves a number of parameters which are only poorly known. The discussion is intended to stimulate further work in a number of disciplines. The reader is invited to adopt a skeptical frame of mind, and to modify the conclusions accordingly. Only through extensive discussion and experiment will the true outlines gradually emerge in this enigmatic but significant subject.

Distribution of Technical Civilizations in the Galaxy

We desire to compute the number of extant galactic communities which have attained a technical capability substantially in advance of our own. At the present rate of technological progress, we might picture this capability as several hundred years or more beyond our own stage of development. A simple method of computing this number is primarily due to F. D. Drake, and was discussed extensively at a Conference on Intelligent Extraterrestrial Life held at the National Radio Astronomy Observatory in November, 1961, and sponsored by the Space Science Board of the National Academy of Sciences.[1] While the details differ in several respects, the following discussion is in substantial agreement with the conclusions of the Conference.

The number of extant advanced technical civilizations possessing both the interest and the capability for interstellar communication can be expressed as

---

[1] Attending this meeting were D. W. Atchley, M. Calvin, G. Cocconi, F. D. Drake, S. S. Huang, J. C. Lilly, P. M. Morrison, B. M. Oliver, J. P. T. Pearman, C. Sagan and O. Struve.

$$N = R_* f_{\mathrm{p}} n_{\mathrm{e}} f_{\mathrm{l}} f_{\mathrm{i}} f_{\mathrm{c}} L. \qquad (1)$$

R_* is the mean rate of star formation averaged over the lifetime of the Galaxy, f_{p} is the fraction of stars with planetary systems, n_{e} is the mean number of planets in each planetary system with environments favorable for the origin of life, f_{l} is the fraction of such favorable planets on which life does develop, f_{i} is the fraction of such inhabited planets on which intelligent life with manipulative abilities arises during the lifetime of the local sun, f_{c} is the fraction of planets populated by intelligent beings on which an advanced technical civilization in the sense previously defined arises during the lifetime of the local sun, and L is the lifetime of the technical civilization. We now proceed to discuss each parameter in turn.

Since stars of solar mass or less have lifetimes on the main sequence comparable to the age of the Galaxy, it is not the present rate of star formation but the mean rate of star formation during the age of the Galaxy which concerns us here. The number of known stars in the Galaxy is $\sim 10^{11}$, most of which have mass equal to or less than the Sun. The age of the Galaxy is $\sim 10^{10}$ years. Consequently, a first estimate for the mean rate of star formation is ~ 10 stars/year. The present rate of star formation is at least an order of magnitude less than this figure, and the rate of star formation in early galactic history is possibly several orders of magnitude more (Schmidt, 1963b, and private communication, 1962). According to present views of stellar nucleogenesis (Burbidge et al., 1957), stars (and by implication, planets) formed in the early history of the Galaxy are extremely poor in heavy elements. Technical civilizations developed on such ancient planets would of necessity be extremely different from our own. But in the flurry of early star formation when the Galaxy was young, heavy elements must have been generated rapidly and later generations of stars and planets would have had adequate endowments of high mass number nuclides. These very early systems should be subtracted in our estimate of R_*. On the other hand, a suspicion exists that large numbers of low mass stars may exist to the right of the main sequence in the Hertzsprung-Russell diagram (Huang, 1961). Inclusion of these objects will tend to increase our estimate of R_*. For present purposes we adopt $R_* \sim 10$/yr.

There is a discontinuity in stellar rotational velocities near spectral type F5V; stars of later spectral type have very slow equatorial rotation rates. This circumstance is generally attributed to the transfer of angular momentum from the star to a surrounding solar nebula by magnetic coupling (Struve, 1950; Huang, 1957; Alfven, 1954; Hoyle, 1961). The solar nebula is then expected to condense into a planetary system (Hoyle, 1961; Kuiper, 1951; Urey, 1951; Cameron, 1962). The fraction of stars of later type than F5V is greater than 0.98; well over 60 per cent of these are dwarf M stars (Oort, 1958). It is not known what

influence the luminosity of the star has on the subsequent condensation and dissipation of the surrounding solar nebula. We might expect that stars of much earlier type than the Sun readily dissipate their solar nebulae; and that stars of much later type than the Sun dissipate very little of their solar nebulae, thereby forming large numbers of massive planets of the Jovian type. There is good evidence that many of the chemical processes in the early history of the solar system occurred at low temperature (Urey, 1951), and the low luminosity of late type stars is unlikely to impede condensation processes in the solar nebula. We therefore adopt $f_{\mathrm{p}} \sim 1$.

Planets of double and multiple star systems are expected in general to have—over astronomical time-scales—such erratic orbits that the evolution of life on them is deemed unlikely (Huang, 1960). I fail to find this argument entirely convincing; but for conservative reasons it will be included in the discussion. The fraction of stars which are not members of double or multiple systems is ~ 0.5 (Worley, 1962). In our own solar system the number of planets which are favorably situated for the origin of life is at least two (Earth and Mars), and the possibility that life arose at some time on the Jovian planets (Sagan, 1961a) has recently been raised. It is sometimes argued that life cannot develop on planets of M dwarfs, because the luminosity of the local sun is too small. However, especially for Jovian type planets of M dwarfs, the greenhouse effect in a methane-ammonia-water atmosphere should produce quite reasonable temperatures. We adopt $n_{\mathrm{e}} \sim 0.5 \times 2 = 1$.

The most recent work on the origin of life strongly suggests that life arose very rapidly during the early history of the Earth (Sagan, 1961a; Oparin et al., 1959; Oró, 1962; Sagan, Ponnamperuma, and Mariner, 1963). It appears that the production of self-replicating molecular systems is a forced process which is bound to occur because of the physics and chemistry of primitive planetary environments. Such self-replicating systems, situated in a medium filled with replication precursors, satisfy all the requirements for natural selection and biological evolution. Given sufficient time and an environment which is not entirely static, the evolution of complex organisms is apparently inevitable. In our own solar system, the origin of life has probably occurred at least twice. We adopt $f_{\mathrm{l}} \sim 1$.

The question of the evolution of intelligence is a difficult one. This is not a field which lends itself to laboratory experimentation, and the number of intelligent species available for study on Earth is limited. Intelligent hominids have inhabited the Earth for $< 10^{-3}$ of Earth history.

It is clear that the evolution of intelligence and manipulative ability has resulted from the product of a large number of individually unlikely events. If the history of the Earth were started again, it is highly improbable that the same sequence of events would recur and that intelligence would evolve in the identical manner. On the other hand, the adaptive value of intelligence and manipulative ability is so great—at least until technical civilizations are developed—that, if it is genetically feasible, natural selection is very likely to bring it forth. There is some evidence that surprisingly high levels of intelligence have evolved in the Cetacea (Lilly, 1961, and private communication). Phylogenetically, these are rather close to hominids; the neuroanatomy of Cetacea brains is remarkably similar to that of the primates, although the most recent common ancestor of the two groups lived more than 10^8 years ago (Lilly, 1961, and private communication). The Cetacea have very limited manipulative abilities.

Comparison of the rates of stellar and of biological evolution provides some perspective on the probability that intelligence will arise on an otherwise suitable planet. Terrestrial intelligence and civilization have emerged roughly midway in the Sun's residence time on the main sequence. The overwhelming majority of stars in the sky have longer lifetimes than the Sun. With the expectation that the Earth is not extraordinary in its recent evolution but allowing for the fact that apparently only one intelligent phylogenetic order with manipulative abilities has developed, and this only recently, we adopt $f_i \sim 10^{-1}$.

Whether there is one, or several, foci for the line of cultural development which has led to the present technological civilization on Earth is still an open question, depending in part on the extent of cultural diffusion over large distances some five or six thousand years in the past. It appears that little can be gained from speculation on, e.g. whether Aztec civilization would have developed a technical phase had there been no *Conquistadores*.

Recorded history—even in mythological guise—covers $\lesssim 10^{-2}$ of the period in which the Earth has been inhabited by hominids, and $\lesssim 10^{-5}$ of geological time. The same considerations are involved as in the determination of f_i. The development of a technical civilization has high survival value at least up to a point; but in any given case it depends on the concatenation of many improbable events; and it has occurred only recently in terrestrial history. It is unlikely that the Earth is very extraordinary in possessing a technical civilization among planets inhabited by intelligent beings. As before, over stellar evolutionary timescales, we adopt $f_c \sim 10^{-1}$.

The multiplication of the preceding factors gives

$$N = 10 \times 1 \times 1 \times 1 \times 10^{-1} \times 10^{-1} \times L = 10^{-1} L.$$

L is the mean lifetime in years of a technical civilization possessing both the interest and the capability for interstellar communication. For the evaluation of L there is—fortunately for us, but unfortunately for the discussion—not even one known terrestrial example. The present technical civilization on Earth has reached the communicative

phase (in the sense of high-gain directional antennas for the reception of extraterrestrial radio signals) only within the last few years. There is a sober possibility that L for Earth will be measured in decades. It is also possible that international political differences will be permanently settled, and that L may be measured in geological time. It is conceivable that, on other worlds, the resolution of national conflicts and the establishment of planetary governments are accomplished before weapons of mass destruction become available. We can imagine two extreme alternatives for the evaluation of L: (a) a technical civilization destroys itself soon after reaching the communicative phase ($L < 10^2$ years); or (b) a technical civilization learns to live with itself soon after reaching the communicative phase. If it survives $>10^2$ years, it will be unlikely to destroy itself afterwards. In the latter case its lifetime may be measured on a stellar evolutionary timescale ($L \gg 10^8$) years. Such a society will exercise self-selection on its members; genetic changes will be unable to move the species off the adaptive peak of the technical civilization. The technology will certainly be adequate to cope with tectonic and orogenic changes. Even the evolution of the local sun through the red giant and white dwarf evolutionary stages may not pose insuperable problems for the survival of an extremely advanced community.

It seems improbable that, surrounded by large numbers of flourishing and diverse galactic communities, a given planetary civilization will retreat from the communicative phase. This is one reason that L is itself a function of N. Von Hoerner (1961) has suggested another reason: he feels that the means of avoiding self-destruction will be among the primary contents of initial interstellar communications.

Gold (private communication, 1962) has talked of the possibility that interstellar space voyagers accidentally may biologically contaminate lifeless planets, and thereby initiate the origin of life. There is also some prospect that such initiation might be purposefully performed. In these cases $f_l = f_l(N)$. Below we will discuss the possibility that $f_c = f_c(N)$. For these reasons it should be remembered that equation (1) is in reality an integral equation.

The two choices for L ($<10^2$ years and $\gg 10^8$ years) lead to two values of N: <10 communicative technical civilizations per galaxy, or $\gg 10^7$. Thus the evaluation of N depends quite critically on our expectation for the lifetime of the average advanced community. Von Hoerner (1961) has made very pessimistic estimates for L, and his values of N are correspondingly small. It seems more reasonable to me that at least a few per cent of the advanced technical civilizations in the Galaxy do not destroy themselves, nor lose interest in interstellar communication, nor suffer insuperable biological or geological catastrophes, and that their lifetimes, therefore, are measured on stellar evolutionary timescales. Averaged over all technical civilizations, we therefore take $L \sim 10^7$

years. For the purposes of the following discussion then, we adopt as the steady-state number of extant advanced technical civilizations in the Galaxy:

$$N \sim 10^6.$$

Thus, approximately 0.001 per cent of the stars in the sky will have a planet upon which an advanced civilization resides. The most probable distance to the nearest such community is then several hundred light years.[2]

Feasibility of Interstellar Spaceflight

The difficulties of electromagnetic communication over such interstellar distances are serious. A simple query and response to the nearest technical civilization requires periods approaching 1000 years. An extended conversation—or direct communication with a particularly interesting community on the other side of the Galaxy—will occupy much greater time intervals, 10^4 to 10^5 years.

Electromagnetic communication assumes that the choice of signal frequency will be obvious to all communities. But there has been considerable disagreement about interstellar transmission frequency assignment even on our own planet (Cocconi and Morrison, Reading 20; Drake, Reading 24; von Hoerner, 1961; Schwartz and Townes, Reading 23; Bracewell, Reading 21); among galactic communities, we can expect much more sizable differences of opinion about what is obvious and what is not. No matter how ingenious the method, there are certain limitations on the character of the communication effected with an alien civilization by electromagnetic signalling. With billions of years of independent biological and social evolution, the thought processes and habit patterns of any two communities must differ greatly; electromagnetic communication of programmed learning between two such communities would seem to be a very difficult undertaking indeed. The learning is vicarious. Finally, electromagnetic communication does not permit two of the most exciting categories of interstellar contact—namely, contact between an advanced civilization and an intelligent but pre-technical society, and the exchange of artifacts and biological specimens among the various communities.

Interstellar space flight sweeps away these difficulties. It reopens the arena of action for civilizations where local exploration has been completed; it provides access beyond the planetary frontiers.

There are two basic methods of achieving interstellar spaceflight within characteristic human lifetimes. One involves the slowing down of human metabolic activities during very long flight times. In the remainder of the paper, we will discuss relativistic interstellar spaceflight, which, in effect accomplishes the identical function, and further, permits the voyager to return to his home planet in much shorter periods of time, as measured on the home planet.

If relativistic velocities can be achieved, time dilation will permit very long journeys within a human lifetime. Consider a starship capable of uniform acceleration to the midpoint of the journey, and uniform deceleration thereafter. The relativistic equations of motion have been solved by Peschka (1956) and by Sänger (1957). For our flight plan, their results are readily modified, and yield for the time t, as measured on the space vehicle, to travel a distance S, with a uniform acceleration a to $S/2$ and a uniform deceleration $-a$ thereafter:

$$t = (2\,c/a)\ \text{arc cosh}\ (1 + aS/2c^2) \qquad (2)$$

where c is the velocity of light. The results for such an acceleration-deceleration flight plan are shown in Figure 38.1.

At an acceleration of 1 g—as would be appropriate for inhabitants of a planet of terrestrial mass and radius—it takes only a few years shiptime to reach the nearest stars, 21 years to reach the Galactic Center, and 28 years to reach the nearest spiral galaxy beyond the Milky Way. With accelerations of 2 or 3 g—as would be appropriate for inhabitants of a planet of Jovian mass and radius—these distances can be negotiated in about half the time. Of course there is no time dilation on the home planet; the elapsed time in years approximately equals the distance of the destination in light years plus twice the time to reach relativistic velocities. For distances beyond about ten light years, the elapsed time on the home planet in years roughly equals the distance of the destination in light years. Thus, for a round trip with a several-year stopover to the nearest stars, the elapsed time on Earth will be a few decades; to Deneb, a few centuries; to the Vela Cloud Complex, a few millenia; to the Galactic Center, a few tens of thousands of years; to M31, a few million years; to the Virgo Cluster of Galaxies, a few tens of millions of years; and to the Coma Cluster, a few hundreds of millions of years. Nevertheless, each of these immense journeys could be performed within the lifetimes of a human crew. For transgalactic and intergalactic distances, equation (2) reduces to

$$t = 2\,c/a\ \ln\ (aS/c^2) \qquad (3)$$

and in this range the curves of Figure 38.1 are straight lines on a semi-logarithmic plot.

---

[2]In the Space Science Board Conference previously mentioned, the conclusions for N spanned 10^4–10^9, and the distance to the nearest advanced community ranged from ten to several thousand light years.

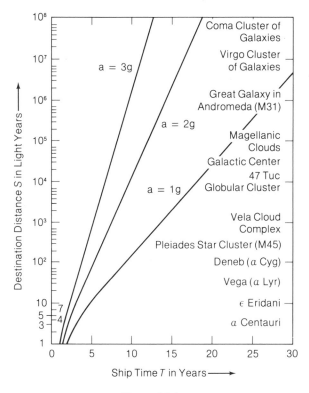

Figure 38.1.

A number of difficulties have been presented by early authors on the technical aspects of relativistic interstellar spaceflight. Even with complete conversion of mass into energy, extreme mass ratios are required if all the fuel is carried at launch. For relativistic velocities and the above flight plan, the mass ratio is approximately equal to $2/\phi$, where $\phi = 1 - (v/c)$, and v is the maximum vehicle velocity (Purcell, Reading 36; Ackeret, 1946, 1956). For example, to reach $v = 0.999c$, the liftoff weight must be some 2000 times the payload, and it is clear that enormous initial vehicle masses are required. For the round trip with no refueling, the mass ratio is $(2/\phi)^2$. Thus, Ackeret (1946, 1956) concluded that "even with daring assumptions," interstellar spaceflight at relativistic velocities would be feasible only for travel to the nearest stars. Furthermore, baryon charge conservation prevents the complete conversion of matter to energy, except if half the working fuel is antimatter (Marx, 1960); the containment of the antimatter—to say nothing of its production in the quantities required—is clearly a very serious problem. An additional difficulty with such an antimatter starship drive has been emphasized by Purcell (Reading 36); the gamma ray exhaust would be lethal for the inhabitants of the launch planet if the drive were turned on near the planet (and if atmospheric absorption is neglected). Staging of fusion rockets (Spencer and Jaffe, 1962) provides some

relaxation of the required mass ratios, but it appears that relativistic velocities cannot be obtained by such staging alone.

A way out of these difficulties has been provided by Bussard (1960) in a most stimulating paper. Bussard describes an interstellar ramjet which uses the interstellar medium both as a working fluid (to provide reaction mass), and as an energy source (by thermonuclear fusion). There is no complete conversion of matter into energy; the existing mass deficits and low reaction cross-sections for the conversion of hydrogen to deuterium are used. The reactor is certainly not available today; but it violates no physical principles, it is currently being very actively pursued, and there is no reason to expect it to be more than a few centuries away from realization on this planet. The Bussard interstellar ramjet requires very large frontal area loading densities: $\sim 10^{-8}$ g cm^{-2} per nucleon cm^{-3} in the interstellar medium. Thus, if the payload is 10^9 gm, the intake area must have a radius of ~ 60 km in regions where the interstellar density is as high as 10^3 nucleons cm^{-3}. In ordinary interstellar space, where the density is ~ 1 nucleon cm^{-3}, the intake area radius must be ~ 2000 km. If the latter radius seems absurdly large, even projecting for the progress of future technology, we can easily imagine the vehicle to seek trajectories through clouds of interstellar material, and vary its acceleration with the density of the medium within which it finds itself. Pierce (1959) had earlier considered and rejected interstellar ramjets, but the rejection was based on much smaller intake areas than Bussard proposes.

Of course the intake area may not necessarily be material; to the extent that the ramjet sweeps up ionized interstellar material magnetic fields could be used for collection. Starships would then seek trajectories through H II regions. The Bussard interstellar ramjet also requires moderate liftoff velocities; but even presently-achievable liftoff velocities as low as 1–10 km/sec would be adequate.

Bussard does not discuss the method of funnelling the interstellar matter so it can be collected and utilized for propulsion. Indeed this is one fundamental problem which must be faced by any relativistic interstellar vehicle; otherwise the structural and biological damage from the induced cosmic ray flux will prevent any useful application of the extreme velocities achieved. The maximum velocity of the vehicle in the rest frame, after covering $S/2$, half the distance to the destination, at uniform acceleration a, is given (Sänger, 1957; Ackeret, 1946) by

$$v = c[1 - (1 + aS/2c^2)^{-2}]^{1/2} \qquad (4)$$

Equation (4) is illustrated in Figure 38.2 for the same three choices of a which have already been used. The abscissa gives the maximum velocity reached, expressed as $\phi = 1 - v/c$, during a half acceleration–half deceleration flight plan to a destination at distance S. For example, for a trip to Galactic Center, maximum velocities within 10^{-7} to 10^{-8} per cent of the velocity of light are required. Also shown in Figure 38.2 are the velocities at which relative kinetic energies of 1 MeV, 1 BeV, and 1 erg are imparted to interstellar protons by the motion of the vehicle. For travel to even the nearest stars within a human shipboard lifetime, protection from the induced cosmic ray flux is mandatory. It is evident from the large mass ratios already required for boosted interstellar flight, and from the low frontal loading area surface densities required for an interstellar ramjet, that material shielding is probably not a feasible solution.

If some means of ionizing the impacting interstellar material could be found, the ions can be deflected and captured by a magnetic field. In the case that trajectories through H II regions are sought, the interstellar medium will be already largely ionized, and magnetic funnelling would be practicable. The configuration of the field would have to be designed very ingeniously, but the average field strengths required could be as low as a few hundred gauss even for very long voyages. Much higher field strengths would be required, at least in the propulsion module, for a fusion ramjet; or alternatively for a contained plasma driving a photon rocket (Sänger, 1962). It appears likely that superconducting flux pumps (Kash and Tooper, 1962) can provide the magnetic field strengths required for deflection of the induced cosmic ray flux.

Bussard's (1960) concluding remarks on the size of the frontal loading area and the magnitude of the effort involved in relativistic interstellar spaceflight are worth quoting: "This is very large by ordinary standards, but then, on any account, interstellar travel is inherently a rather grand undertaking, certainly many magnitudes broader in scope and likewise more difficult than interplanetary travel in the solar system, for example. The engineering effort required for the achievement of successful short-time interstellar flight will likely be as much greater than that involved in interplanetary flight as the latter is more difficult than travel on the surface of the Earth. However, the expansion of man's horizons will be proportionately greater, and nothing worthwhile is ever achieved easily."

The purpose of this Section is to lend credence to the proposition that a combination of staged fusion boosters, large mass-ratios, ramjets working on the interstellar medium and trajectories through H II regions is capable of travel certainly to the nearest stars within a human shipboard lifetime, without appeal to as yet undiscovered principles. Especially allowing for a modicum of scientific and technological progress within the next few centuries, I believe that interstellar spaceflight at relativistic velocities to the farthest reaches of our Galaxy is a feasible objective for humanity. And if this is the case, other civilizations, aeons more advanced than ours, must today

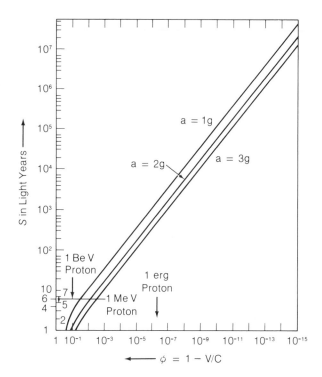

Figure 38.2.

be plying the spaces between the stars.

Frequency of Contact Among Galactic Communities

We can expect that if interstellar spaceflight is technically feasible—even though an exceedingly expensive and difficult undertaking, from our point of view—it will be developed. Even beyond the exchanges of information and ideas with other intelligent communities, the scientific advantages of interstellar spaceflight stagger the imagination. There are direct astronomical samplings—of stars in all evolutionary stages, of distant planetary systems, of the interstellar medium, of very ancient globular clusters. There are cooperative astronomical ventures, such as the trigonometric parallaxes of extremely distant objects. There is the observation and sampling of a multitude of independent biologies and societies. These are undertakings which could challenge and inspire even a very long-lived civilization.

For the civilization lifetimes, L, previously adopted we see that interstellar space flight to all points within the Galaxy, and even to other galaxies, is possible in principle. The voyagers will return far in the future of their departure, but we have already anticipated that the civilization will be stable over these immense periods of time. There will still be a record of the departure, a repository for the information collected, and a community interested in the results. To avoid unneccessary duplication in interstellar exploration, the communicative societies will pool information and act in concert, as Bracewell (Reading 21) has already pointed out. Direct contacts and exchange of information and artifacts will exist among most spacefaring societies possessing relativistic starships. In fact, over large distances, starship communication will occur very nearly as rapidly as, and much more reliably than, communication by electromagnetic radiation. The situation bears some similarity to the post-Renaissance seafaring communities of Europe and their colonies before the development of clipper and steam ships. If relativistic interstellar space flight is feasible, the technical civilizations of the Galaxy will be an intercommunicating whole; but the communication will be sluggish.

It is of some interest to estimate the mean time interval between contacts for a given planetary system. Although the shipboard transit times at relativistic velocities are very roughly the same to any place in the Galaxy, the elapsed time on the home planet is of course approximately proportional to the distance of the voyage. Consequently contact should be greatest among neighboring communities, although we can anticipate that occasional very long journeys will be attempted.

Let each of the N planets in the communicative phase launch q relativistic starships per year. These vehicles each effect at least one contact per journey, and are most often gone some 10^3–10^4 years from the home planet per mission. In the steady state, there are then q contacts effected by each starship-launching civilization per year, and $\sim qN$ contacts per year for the Galaxy as a whole. Relative to the economic capacity of such advanced civilizations, a value of $q = 1$ yr^{-1} seems modest. (Other choices of q will modify the results in an obvious manner.) Each civilization then makes ~ 1 contact per year, and an average of 10^7 contacts during its lifetime. The number of contacts per year for the Galaxy as a whole is then 10^6; a sizable fraction of these should be between two advanced communities. The mean number of starships on patrol from each technical civilization at any given time is $\sim 10^3$–10^4.[3]

If contacts are made on a purely random basis, each star should be visited about once each 10^5 years. Even the most massive stars will then be examined at least once while they are on the main sequence. Especially with a central galactic information repository, these advanced civilizations should have an excellent idea of which planetary environments are most likely to develop intelligent life. With average contact frequency per planet of 10^{-5} yr^{-1},

the origin and evolution of life on every planet in the Galaxy can be monitored efficiently. The successive development of metazoa, of cooperative behavior, of the use of tools, and of primitive intraspecific communication schemes would each be noted, and would each be followed by an increase in the interstellar sampling frequency. If $f_i \sim 10^{-1}$, then, on a purely random basis, the frequency of contact with intelligent pretechnical planetary communities should be $\sim 10^{-4}$ yr^{-1}. Once technical civilization has been established, and especially after the communicative phase has come into being, the contact frequency should again increase; if $f_c \sim 10^{-1}$, to some 10^{-3} yr^{-1}. Planets of extraordinary interest will be visited even more frequently. Under the preceding assumptions, each communicative technical civilization should be visited by another such civilization about once every thousand years. The survey vehicles of each civilization should return to the home planet at a rate of about one a year, and a sizable fraction of these will have had contact with other communities. The wealth, diversity and brilliance of this commerce, the exchange of goods and information, of arguments and artifacts, of concepts and conflicts, must continuously sharpen the curiosity and enhance the vitality of the participating societies.

The preceding discussion has a curious application to our own planet. On the basis of the assumptions made, some one or two million years ago, with the emergence of *Proconsul* and *Zinjanthropus*, the rate of sampling of our planet should have increased to about once every ten thousand years. At the beginning of the most recent postglacial epoch, the development of social structure, art, religion, and elementary technical skills should have increased the contact frequency still further. But if the interval between samplings is only several thousand years, there is then a possibility that contact with an extraterrestrial civilization has occurred within historical times.[4]

Possibility of Extraterrestrial Contact with Earth During Historical Times

There are no reliable reports of contacts during the last few centuries, when critical scholarship and nonsuperstitious reasoning have been fairly widespread. Any earlier contact story must be encumbered with some degree of fanciful embellishment, due simply to the views prevailing at the time of the contact. The extent to which subsequent variation and embellishment alters the basic fabric of the account varies with time and circumstance.

[3]It is easily shown that with the adopted values of N and q, and with even very large ramjet frontal loading areas, the exhaust from such interstellar vehicles makes a negligible contribution to the background galactic cosmic ray flux.

[4]This possibility has been seriously raised before; for example, by Enrico Fermi, in a now rather well-known dinner table discussion at Los Alamos during the Second World War, when he introduced the problem with the words "Where are they?"

Brailoiu (Eliade, 1959, p. 44) records an incident in Rumanian folklore, where, but forty years after a romantic tragedy, the story became elaborately embellished with mythological material and supernatural beings. At the time as the ballad was being sung and attributed to remote antiquity, the actual heroine was still alive.

Another incident, which is more relevant to the topic at hand, is the native account of the first contact with European civilization by the Tlingit people of the Northeast Coast of North America (Emmons, 1911). The contact occurred in 1786 with an expedition led by the French navigator La Perouse. The Tlingit kept no written records. One century after the contact the verbal narrative of the encounter was related to Emmons by a principal Tlingit chief. The story is overladen with the mythological framework in which the French sailing vessels were initially interpreted. But what is very striking is that the true nature of the encounter had been faithfully preserved. One blind old warrior had mastered his fears at the time of the encounter, had boarded one of the French ships, and exchanged goods with the Europeans. Despite his blindness, he reasoned that the occupants of the vessels were men. His interpretation led to active trade between the expedition of La Perouse and the Tlingit. The oral tradition contained sufficient information for later reconstruction of the true nature of the encounter, although many of the incidents were disguised in a mythological framework: e.g., the sailing ships were described as immense black birds with white wings.

The encounter between the Tlingit and La Perouse suggests that under certain circumstances a brief contact with an alien civilization will be recorded in a reconstructable manner. The reconstruction will be greatly aided if (1) the account is committed to written record soon after the event, (2) a major change is effected in the contacted society by the encounter, and (3) no attempt is made by the contacting civilization to disguise its exogenous nature.

On the other hand, it is obvious that the reconstruction of a contact with an extraterrestrial civilization is fraught with difficulties. What guise may we expect such a contact myth to wear? A simple account of the apparition of a strange being who performs marvelous works and resides in the heavens is not quite adequate. All peoples have a need to understand their environment, and the attribution of the incompletely understood to nonhuman deities is at least mildly satisfying. When interaction occurs among peoples supporting different deities, it is inevitable that each group will claim extraordinary powers for its god. Residence of the gods in the sky is not even approximately suggestive of extraterrestrial origin. After all, where can the gods reside? Obviously not over in the next county; it would be too easy to disprove their existence by taking a walk. Until very subtle metaphysical constructs are developed—possibly in desperation—the gods can only live beneath the ground, in the waters, or in the sky. And

except perhaps for seafaring peoples, the sky offers the widest range of opportunities for theological speculation.

Accordingly, we require more of a legend than the apparition of a strange being who does extraordinary works and lives in the sky. It would certainly add credibility if no obvious supernatural adumbration were attached to the story. A description of the morphology of an intelligent non-human, a clear account of astronomical realities for a primitive people, or a transparent presentation of the purpose of the contact would increase the credibility of the legend.

In the Soviet Union, Agrest (private communication from Shklovskii, 1962; Shklovskii and Sagan, 1966) and others have called attention to several biblical incidents which they suspect to reflect contact with extraterrestrial civilizations. For example, Agrest considers the incidents related in the apocryphal book, the "Slavonic Enoch," to be in reality an account of the visitation of Earth by extraterrestrial cosmonauts, and the reciprocal visitation of several galactic communities by a rather befuddled inhabitant of Earth. However, the Slavonic Enoch fails to satisfy several of the criteria for a genuine contact myth mentioned above: it has been molded into several different standardized supernatural frameworks; there is no transparent extraterrestrial motivation for the events described; and the astronomy is largely wrong. The interested reader may wish to consult standard versions of the manuscript (Charles, 1896).

There are other legends which more nearly satisfy the foregoing contact criteria, and which deserve serious study in the present context. As one example, we may mention the Babylonian account of the origin of Sumerian civilization by the *Apkallu,* representatives of an advanced, nonhuman and possibly extraterrestrial society (Hodges, 1876; Schnabel, 1923).

A completely convincing demonstration of past contact with an extraterrestrial civilization may never be provided on textural and iconographic grounds alone. But there are other possible sources of information.

The statistics presented earlier in this paper suggest that the Earth has been visited by various galactic civilizations many times (possibly $\sim 10^4$) during geological time. It is not out of the question that artifacts of these visits still exist, or even that some kind of base is maintained (possibly automatically) within the solar system to provide continuity for successive expeditions. Because of weathering and the possibility of detection and interference by the inhabitants of the Earth, it would be preferable not to erect such a base on the Earth's surface. The Moon seems one reasonable alternative. Forthcoming high resolution photographic reconnaissance of the Moon from space vehicles—particularly of the back side—might bear these possibilities in mind. There are also other locales in the solar system which might prove of interest in this context. Contact with such a base would, of course, provide the most direct check on the conclusions of the present paper.

Otherwise the abundance of advanced civilizations in the Galaxy could be tested by successful detection of intelligible electromagnetic signals of interstellar origin. In the next few decades mankind will have the capability of transmitting electromagnetic signals over distances of several hundreds of light years. The receipt and return of such a signal would announce our presence as a technical civilization, and, if the conclusions of the present paper are valid, would be followed by a special contact mission. Even if an intelligible interstellar signal were received and returned today, it would be several hundreds of years before the contact mission could arrive on Earth. Hopefully, there will then still be a thriving terrestrial civilization to greet the visitors from the far distant stars.

Acknowledgments
This research was supported in part by grant NsG-126-61 from the National Aeronautics and Space Administration while I was at the University of California, Berkeley, where much of the work was performed; the views presented, however, are entirely the responsibility of the author. I am indebted to J. Finkelstein, J. Lederberg, L. Sagan, M. Schmidt, C. Seeger, I. S. Shklovskii, C. Stern and W. Talbert for stimulating and encouraging discussions of various aspects of this paper, and to my fellow members of the Order of the Dolphins, especially F. D. Drake and P. M. Morrison.

39

George Gaylord Simpson
The Nonprevalence of Humanoids

The possibility that life exists elsewhere than on earth has excited human imagination since antiquity. In our own days it has become the principal basis for a whole school of writing: science fiction, which remains mere entertainment even though some of its devotees do make an unjustified claim that it should be taken more seriously. There has also long been discussion that was scientific, at least in the sense that it was by professional scientists who did not intend to write fiction. Even in the nineteenth century there was serious, if not invariably sober, discussion of the view that life exists not only elsewhere but even everywhere in the cosmos.

There is, then, nothing new in the fact that this subject is being widely discussed and publicized.[1] What is new is that the usual speculation and philosophizing are now accompanied by extensive (and incidentally expensive) research programs, by concrete plans for exploration, and by development of pertinent instrumentation. Although the interested scientists have by no means stopped talking, they are now, and for the first time in history, also acting. Our major space agency, NASA, has a "space bioscience" program. Biologists meeting under the auspices of the National Academy of Sciences have agreed that their "first and . . . foremost [task in space science] is the search for extraterrestrial life" (Hess et al., 1962). The existence of this movement is as familiar to the reader of the newspapers as to those of technical publications. There is even increasing recognition of a new science of extraterrestrial life, sometimes called *exobiology*—a curious development in view of the fact that this "science" has yet to demonstrate that its subject matter exists!

Another curious fact is that a large proportion of those now discussing this biological subject are not biologists. Even when biochemists and biophysicists are involved, the accent is usually on chemistry and physics and not on biology, strictly speaking. It would seem obvious that organic evolution has a crucial bearing on the subject, which is essentially a problem in evolutionary systematics. Surely, then, it is odd that evolutionary biologists and systematists have rarely been consulted and have volunteered little to the discussion. A possible reason for this blatant omission was suggested long ago by an evolutionary systematist, W. D. Matthew (1921), who wrote that, "[Physical scientists] are accustomed to hold a more receptive attitude . . . toward hypotheses that cannot be definitely disproved . . . [while] the [evolutionary and systematic] biologist . . . is compelled . . . to leave out of consideration all factors that have not something in the way of positive evidence for their existence."

Matthew also remarked that, "To admit the probability of extra-mundane life opens the way to all sorts of fascinating speculation in which a man of imaginative temperament may revel free from the checks and barriers of earthly realities." Both of his points are illustrated delightfully and without conscious humor by a contemporary leader in exobiology who wrote in 1962, "We do not really know [what the atmosphere of Venus is like], and we are thus not severely limited in our conclusions"! (exclamation point mine).

As an evolutionary biologist and systematist, I believe that we should make ourselves heard in this field. Since part of our role must be to point out "the checks and

[1]In the spring of 1963 I gave lectures on this subject (but entitled "Life on Other Worlds") at six member institutions of the University Center of Virginia and at the University of Colorado. This essay is based on those lectures but has been extensively revised.

I have stressed that "there are no direct observational data whatever" on any planetary systems but our own. On April 19, 1963, the *New York Times* announced that Dr. van de Kamp of Swarthmore had discovered the third such planetary (or "solar") system. The apparent contradiction is a matter of definition of "direct observation" and "solar system" and really calls for no correction of my text. Three stars are inferred to have *unobserved* dark companions on the basis of perturbations of the stars' motions interpreted as due to gravitational influence of the companion. Whether or in what sense the dark companions are to be considered planets is not clear. Inferences as to size, radiation, and so on make them unlike any planets of our system and entirely unsuited for life.

barriers of earthly realities," we may at times seem merely to be spoilsports, but we do have other contributions as well.

Three Major Questions

Exobiology has three major questions: "What kind of life?" "Where?" "How may it have evolved?" Each question in turn involves two complex, distinct fields of inquiry. Confusion of these fields frequently distorts judgment and confuses argument.

The alternative fields as to the kind of life are "life as we know it" and "life as we do not know it." Life as we know it obviously cannot be confined in this context to actual terrestrial species, but implies only a more general similarity. It must, at least, involve a carbon chemistry reacting in aqueous media and with such fundamental organic compounds as amino acids, carbohydrates, purine-pyrimidine bases, fatty acids, and others. It must almost certainly also involve the combination and polymerization of those or similar fundamental molecules into such larger molecules or macromolecules as proteins, polysaccharides, nucleic acids, and lipids. Life as we do not know it might be based on some multivalent element other than carbon, on some medium (perhaps even solid or gaseous) other than liquid water, and then necessarily on quite different kinds of compounds.

If we did encounter such systems or organisms, we might well fail to recognize them as living or might have to revise our conception of what life is. Here on earth, in spite of a border zone between, and enormous diversities within, each realm, we can recognize two kinds of configurations of matter, one living and one not. (Under "configuration" I mean to include not only chemical composition but also organization or anatomy in the fullest sense and energy states and transactions.) "Life as we do not know it," if recognized at all, might have to be recognized as a third fundamental kind of configuration and not, strictly speaking, as life. There has been considerable speculation along such lines, some of it diverting in a science-fictional sort of way. Yet there is not a scrap of evidence that "life as we do not know it" actually exists or even that it *could* exist—evidence, for example, in the form of detailed specifications for a natural system that might exhibit attributes of life without the basis of life as we do know it. (Computers and other artifacts that mimic some features of the life of their makers are not really pertinent to this question.) Here, at least, further consideration will be given only to life as we know it, to the minimal extent of depending on similar biophysical and biochemical substrates.

The dichotomy in discussing the "Where?" of possible extraterrestrial life is between our own solar system and presumed similar planetary systems anywhere else in the universe. We have extensive observational data on the planets of our system and a reasonable expectation of learning much more. Many facts have been learned over the years by earth-based astronomical methods. Recently rocketry and telemetry have given us closer looks at the moon and at Venus and promise to give us many additional facts. Human visits to the moon and the closer planets, at least, make no evident further demands on our theoretical knowledge and require only a reasonable extrapolation of our technical potentialities into the near future. Here, then, we have actual observational data to work with, and the promise of many more.

Not so for any planetary system that may exist outside our own. Statements in both the scientific and the popular literature that there are millions of such systems suitable for life and probably inhabited may give the impression that we know that they do exist. In fact we know no such thing in any way acceptable as sober science. There are no direct observational data whatever. It is inherent in any acceptable definition of science that statements that cannot be checked by observation are not really *about* anything—or at the very least they are not science. As long as we do not confuse what we are saying with reality, there is no reason why we should not discuss what we hope or expect to observe, but it is all too easy to take conjecture and extrapolation too seriously. It is not impossible that our descendants may some day make pertinent direct observations on other planetary systems, but that is far beyond our present capabilities or any reasonable extrapolation from them. With our present techniques, the only way we could obtain direct knowledge of life outside our solar system would be by receiving signals from someone or something out there. That point is involved in the third question, the directly evolutionary one, and its two major fields of enquiry: the origin of life and its subsequent history. Here is my main topic, to which I will return at length.

Within Our Solar System

First it is necessary to refer briefly to the environmental conditions and possible evidence of life on the only planets for which we have any actual data, the planets of our own solar system. Apart from a few eccentrics, astronomers have long since agreed that life as we know it is now quite impossible on any extraterrestrial body in our solar system except Venus and Mars (see, for example, Jackson and Moore, 1962). Opinion regarding Venus has been divided, but telemetry from the recent Venus probe seems to confirm beyond doubt the previous view that Venus is far too hot for life as we know it (Barath et al., 1963). Although somewhat equivocal, such evidence as we have on the composition of the Venusian atmosphere also seems to be unfavorable on balance (see, for example, Sagan, 1961a). It would appear, then, that Venus can now be ruled out as a possible abode of recognizable life.

The evidence for Mars is also highly equivocal, but it does not at present entirely exclude the possibility of life there. Temperatures are rigorous and there is little or no free oxygen. Obviously neither man nor any of our familiar animals and plants could possibly live in the open on Mars. Simple microorganisms have, however, been grown in conditions possibly similar to those that just might exist on Mars (Hawrylewicz, Gowdy, and Erlich, 1962). This possibility depends in part on the usual belief that the so-called ice caps of Mars are indeed composed of water and that the atmosphere is mainly nitrogen with some carbon dioxide. Both beliefs have been authoritatively challenged by Kiess, Karrer, and Kiess (1960), who maintain that the caps are N_2O_4. That and the accompanying concentrations of oxides of nitrogen in the atmosphere would make Mars lethal to life as we know it. In any case, there is increasing doubt that enough water exists on Mars to sustain any form of life.

Direct evidence for life on Mars has also been claimed. The old idea that the so-called canals of Mars were made by intelligent beings no longer merits sober consideration. It is, however, well known that there are dark areas on Mars that show seasonal changes in position and in apparent color. It has been claimed repeatedly that these areas must be covered with some form of plant life, and that idea received significant support when it was discovered that their infrared spectrum has a band similar to that of some organic compounds (Sinton, 1959). However, similar absorption can also be caused by oxides of nitrogen and by a variety of inorganic carbonates (partly unpublished work cited by Calvin, 1962). The question remains open, and plans to make direct observations by space probe are going forward (see Levin et al., 1962). These plans depend on the further doubtful proposition that there may be microorganisms on Mars that can be grown by the same methods used here to grow microorganisms in laboratories.

The only other direct evidence for extraterrestrial life worthy of serious consideration is derived from meteorites. It has been claimed that some of these contain hydrocarbons of organic origin and even actual fossils of microorganisms (see Nagy, Meinschein, and Hennessy, 1961). If confirmed, these observations would indicate that life (now extinct) had occurred on a planet of our system that has since been disrupted. However, further investigation strongly suggests that the materials observed are in part inorganic and in the remaining part terrestrial contaminants (Anders and Fitch, 1962). The most favorable possible verdict is "Not proven."

There is, then, no clear evidence of life anywhere else in our solar system. Wishful thinking, to which scientists are not immune, has obviously played a part here. The possibility is not excluded, but, on what real evidence we have, the chance of finding life on other planets of our system is slim.

Outside Our Solar System

It bears repeating that there are no observational data whatever on the existence, still less on the possible environmental conditions, of planets suitable for life outside our solar system. Any judgment on this subject depends on extrapolations from what we know of the earth and its life and from astronomical data that do not include direct observation. There are, indeed, considerable grounds for such extrapolations, but they still contain a large subjective element and have a strong tendency to go over into sheer fantasy.

There are four successive probabilities to be judged: the probability that suitable planets do exist; the probability that life has arisen on them; the probability that such life has evolved in a predictable way; and the probability that such evolution would lead eventually to humanoids (as defined in the next paragraph). The thesis I shall now develop, admittedly subjective and speculative but extrapolated from evidence, is that the first probability is fair, the second far lower but appreciable, the third exceedingly small, and the fourth almost negligible. Each of these probabilities depends on that preceding it, so that they must be multiplied together to obtain the over-all probability of the final event, the emergence of humanoids. The product of these probabilities, each a fraction, is probably not significantly greater than zero.

(Before proceeding, I should define "humanoid" for those not as addicted as I am to science fiction. A humanoid, in science-fiction terminology adaptable to the present also somewhat fanciful subject, is a natural, living organism with intelligence comparable to man's in quantity and quality, hence with the possibility of rational communication with us. Its anatomy and indeed its means of communication are not defined as identical with ours. An android, on the other hand, is a nonliving machine, servomechanism, or robot constructed in more or less human external shape and capable of performing some manlike actions.)

The first point, as to the existence of earthlike planets, need not detain us long. The astronomers seem to be in complete agreement that planets that are or have been similar to the earth when life arose here probably exist in large numbers (see Jackson and Moore, 1962; Hoyle, 1955; Shapley, 1958). Indeed the number of stars in the accessible universe (discernible by light or radio telescopy) is so incredibly enormous that even if the chances of any one of them having such a planet were exceedingly small, the probability that *some* of them do would be considerable. As a basis for further consideration, we may, then, reasonably postulate that conditions such as proved propitious to the origin of life on earth may have existed also outside our solar system.

The Origin of Life

The next question is: "How did life arise on earth, and is it probable or perhaps inevitable that it would arise elsewhere under similar conditions? This is largely in the field of the biochemists, and they certainly have not neglected it. The literature is enormous. Enough of it for our purposes is summarized or cited in the recent works of Oparin (1957; et al., 1959; Academy of Sciences of the U.S.S.R., 1957), Florkin (1961), Calvin (1961), and Ehrensvärd (1962). There are wide differences of opinion as to the particular course followed, but here again there is near unanimity on the essential points. Virtually all biochemists agree that life on earth arose spontaneously from nonliving matter and that it would almost inevitably arise on sufficiently similar young planets elsewhere.

That confidence is based on chemical experience. If atoms of hydrogen and oxygen come together under certain simple and common conditions of energy, they always deterministically combine to form water. Formation of more complex molecules requires correspondingly more complex concatenations of circumstances but is still deterministic in what seems to be a comparatively simple way. That has, indeed, been demonstrated in the laboratory. If energy such as would be available on a young planet is put into a mixture of the simplest possible compounds of hydrogen, oxygen, carbon, and nitrogen, such as also could well occur on a young planet, amino acids and other building blocks of the essential complex organic molecules are formed. The crucial experiment was that of Miller (1955). A large amount of later work, mostly noted in the books cited above, has confirmed and extended those results. The further synthesis of the building blocks into the macromolecules, especially nucleic acids and proteins, essential for life has not yet been accomplished under realistically primitive conditions. Nevertheless it is reasonable to assume that those steps, too, would occur deterministically, inevitably, if given enough time under conditions likely to hold on some primitive planets. It is also clear that there has indeed been enough time, for the earth is now definitely known to be more than three billion years old, and planets still older could well exist in this and other galaxies.

It is still a far cry from the essential preliminary formation of proteins, nucleic acids, and other large organic molecules to their organization into a system alive in the full sense of the word. This is the step, or rather the great series of steps, about which we now know the least even by inference and extrapolation. A fully living system must be capable of energy conversion in such a way as to accumulate negentropy, that is, it must produce a less probable, less random organization of matter and must cause the increase of available energy in the local system rather than the decrease demanded in closed systems by the second law of thermodynamics. It must also be capable of storing and replicating information, and the replicated information must eventually enter into the development of a new individual system like that from which it came. The living system must further be enclosed in such a way as to prevent dispersal of the interacting molecular structures and to permit negentropy accumulation. At the same time selective transfer of materials and energy in both directions between organism and environment must be possible. Systems evolving toward life must become cellular individuals bounded by membranes.

The simplest true organisms have all those characteristics and more, but they are very far from being simple in microscopic and submicroscopic organization. Less organized associations of organic macromolecules, such as are seen today in the viruses, cannot perform all those feats on their own and therefore cannot be meaningfully viewed as primitive and true forms of life.

If evolution is to occur and organisms are to progress and diversify, still more is necessary. Living things must be capable of acquiring new information, of alteration in their stored information, and of its combination into new but still integrated genetic systems. Indeed it now seems that these processes, summed up as mutation, recombination, and selection, must already be invoked in order to get from the stage of loose macromolecules to that of true organisms, or cellular systems. There must be some kind of feedback and encoding leading to increased and diversified adaptation of the nascent organisms to the available environments. Basically such adaptation is the ability to reproduce and to maintain or increase continuous populations of individuals by acquiring, converting, and organizing materials and energy available from existing environments. These processes of adaptation in populations are decidedly different in degree from any involved in the prior inorganic synthesis of macromolecules. They also seem to be quite different in kind, but that is partly a matter of definition and is also obscured by the fact that they must have arisen gradually on the basis of properties already present in the inorganic precursors. In any case, something new has definitely been added in these stages of the origin of life. It requires an attitude of hope if not of faith to assume that the acquisition of organic adaptability was deterministic or inevitable to the same degree or even in the same sense in which that was probably true of the preceding, more simply chemical origin of the necessary macromolecules.

By that I do not mean to say that material causality has been left behind or that some mysterious vitalistic element has been breathed into the evolving systems. All must still be proceeding without violation of physical and chemical principles. Those principles must, however, now be acting in different ways because they are involved in holistic, organic, increasingly complex, multimolecular systems that far transcend simple chemical bonding. It is here that one must stop taking for granted the expectations and extrapolations of the chemist and can obtain further enlightenment only from the biologist as such, that is, the student of whole organisms as they exist in reproducing populations and in communities adapted to environments.

Given ample time and rather simple circumstances not likely to be unique in the universe, there does seem to be considerable probability, perhaps even inevitability, in the progression from dissociated atoms to macromolecules. The further organization of those molecules into cellular life would seem, on the face of it, to have a far different, very much lower order of probability. It is not impossible, because we know it did happen at least once. Nevertheless that event is so improbable that even if macromolecules have arisen many times in many places, it would seem that evolution must frequently or usually have ended at that preorganismal stage. Only the astronomical assurance that there may be many millions of earthlike planets permits us to assume that the origin of true, that is of cellular, life may have happened more than once. In the observable universe the lowest recent estimate for earthlike planets by a competent astronomer is, as far as I know, that of Shapley (1958), who considers 100 million a highly conservative figure. On that basis it is reasonable to speculate that life has arisen repeatedly in the universe, even though we do not know and perhaps will never know whether that is a fact.

Here brief consideration may be given to the idea that once life had arisen somewhere, organisms in a state of cryptobiosis (Keilin, 1955) might have spread by "cosmozoan" transport from one planet to another. That possibility was especially urged by Arrhenius (1908), following the still earlier, curious speculation of Richter and others that life may be coextensive with the whole cosmos both in space and in time. It now appears extremely improbable but not quite impossible that any organism, even encapsulated and in a cryptobiotic state of entirely suspended metabolism, could survive the radiation hazards in space without artificial shielding (Becquerel, 1950). Furthermore, passage from one solar system to another at any speed attainable by natural means (e.g., by the pressure of light) would require vastly more time than any established or probable duration of the cryptobiotic state, which is not know to have lasted longer than about fifty years in microorganisms or about a thousand years in any organisms (Keilin, 1955). A conservative conclusion would be that it is extremely improbable, almost to the point of impossibility, that any form of life has ever traveled by natural means from one planetary system to another. Such travel between earth and Mars, within the same planetary system, is still improbable, but the possibility is not absolutely ruled out.

Subsequent Evolution of Postulated Life Forms

We now turn to the subsequent evolution of postulated life forms once life has appeared on a planet, and we again move to a different order of probability. We have only a single sample on which to base judgment. Paleobiologists have shown us the general course followed by evolution on this planet. Neobiologists have shown in great, although still incomplete, detail the outcome of that process at one point in time, the present. Although these are far from being the only accomplishments of systematists, they are in themselves so important for current problems as to justify intensified research on this enormous subject.

The problem of extrapolating from this unique sample is to decide whether it is inevitable, probable, improbable, or impossible for life of independent origin to have followed a similar or identical course. Opinions have indeed varied from one end to the other of that scale. I believe that a reasonable choice among opinions is possible, and furthermore that many, even most, of those who have recently considered the subject have made a wrong choice. Review of recent literature on exobiology, almost all of it by physical scientists and biochemists (or molecular biologists), shows that most of them have *assumed,* usually without even raising the question, that once life arose anywhere its subsequent course would be much as it has been on earth. Now, the only really sound basis for such an assumption would be the opinion that the course followed by evolution on earth is its only possible course, that life cannot evolve in any other way. In a review of two books in which that assumption is made, Blum (1963) has called this the "deterministic" point of view as contrasted with an "opportunistic" one. The choice of terms is not a happy one, if only because it is demonstrable that evolution fully deterministic in the philosophical sense would not necessarily, indeed would almost surely not, follow similar courses on different planets. Nevertheless, the two schools of thought do exist and what Blum calls the deterministic one is more commonly followed in current exobiological speculations.

There are here underlying problems of philsophy and indeed also theology. . . . The pertinent *scientific* questions are: If the processes of evolution are the same everywhere as they are here on earth, will they elsewhere lead to the same material results, including men or humanoids? Just how inevitable is that outcome?

Those questions can be followed up in two different but related ways. First, we can examine the course of evolution here on earth to see whether in fact it has proceeded as if directed toward a goal or an inevitable outcome. Second, we can investigate the mechanisms or processes of evolution in order to judge whether and under what conditions their outcome was limited to a course eventuating in some kind of humanoid, that is, in ourselves in the terrestrial example. . . .

The fossil record shows very clearly that there is no central line leading steadily, in a goal-directed way, from a protozoan to man. Instead there has been continual and extremely intricate branching, and whatever course we follow through the branches there are repeated changes both in the rate and in the direction of evolution. Man is the end of one ultimate twig. The housefly, the dog flea, the apple tree, and millions of other kinds of organisms are similarly the ends of others. Moreover, we do not find that life has simply expanded, branching into increasing diversity, until the organisms now living had evolved. On the contrary, the vast majority of earlier forms of life have become extinct without issue. Usually their places in the economy of nature have then been taken by other organisms of quite different origin. In some cases, their places seem simply to have remained empty for shorter or longer periods.

Neither in its over-all pattern nor in its intricate detail can that record be interpreted in any simply finalistic way. If evolution is God's plan of creation—a proposition that a scientist as such should neither affirm nor deny—then God is not a finalist. But this still does not fully answer the particular question we are pursuing here. The whole nonfinalistic pattern *might* have been followed nearly enough on a planet of some other star to produce humanoids there also. We must turn then to the causal elements and limitations inherent in the process for further judgment of the probability of such an outcome.

Each new organism develops in accordance with a figurative message, coded information, received from its one or two parents. Evolution occurs only if there are changes in that information in the course of generations. Such changes in individuals occur for the most part in two ways, although each takes numerous and sometimes complicated forms: mutations, which introduce new elements into the message, and recombinations, which put these elements into new associations and sequences. In a stricter' sense mutations are any changes within the code carried by a nucleic acid. Recombinations involve rearrangements of the various code units and particularly new associations of units from different sources. The latter sources of variation are sexual, and sexlike processes occur in even the most primitive living organisms although they have been secondarily lost in a relatively small number of both plants and animals.

In themselves, these processes are not adaptive; they have no direct relevance to fitting organisms into the economy of nature, permitting their survival and further evolution. Since most (but not all) evolutionary changes are adaptive and progressive evolution does occur, these processes alone cannot be the whole story. They are necessary for evolution, but something else must also be involved. There must be some interaction between organisms and environment and from this there must be some kind of feedback into the genetic code. The feedback is by natural selection and it occurs in populations through successive generations, not in individuals in their lifetimes. That is the whole point of natural selection: that it does feed back from environment to genetic code in such a way as to maintain or change the message in adaptive ways. It does this because, by and large, the better adapted organisms have more offspring. The more adaptive genetic messages thus tend to spread through the population in the course of generations. Also, in more complex ways that I need not go into here, new code combinations adaptive for the population as a whole are thus brought into being.

This feedback is basic for our present enquiry because it places definite limitations on the possible course of evolution. We can be quite sure that if the environments of their ancestors had been very different from what they were, the organisms of today would also be very different. It is also clear that evolution must be opportunistic in the sense that it can work only with what is there. Mutations can occur only in quite definite ways depending on the existing nature of the coded message. Recombination can recombine only the code elements that do exist in given organisms. Selection can work only on variations actually present in a population. The cause of evolution thus includes all the genetic, structural, physiological, and behavioral states of populations right back to the origin of life.

Even slight changes in earlier parts of the history would have profound cumulative effects on all descendent organisms through the succeeding millions of generations. In spite of the enormous diversity of life, with many millions of species through the years, it represents only a minute fraction of the possible forms of life. The existing species would surely have been different if the start had been different and if any stage of the histories of organisms and their environments had been different. Thus the existence of our present species depends on a very precise sequence of causative events through some two billion years or more. Man cannot be an exception to this rule. If the causal chain had been different, *Homo sapiens* would not exist (Simpson, 1949, 1960, 1962).

Not Repeatable

Both the course followed by evolution and its processes clearly show that evolution is not repeatable. No species or any larger group has ever evolved, or can ever evolve, twice. Dinosaurs are gone forever. Nothing very like them occurred before them or will occur after them. That is so not only because of the action of selection through long chains of nonrepetitive circumstances, as I have just briefly noted. It is also true because in addition to those adaptive circumstances there is a more or less random element in evolution involved in mutation and recombination, which are stochastic, technically speaking. Repetition is virtually impossible for nonrandom actions of selection on what is there in populations. It becomes still less probable when one considers that duplication of what are, in a manner of speaking, accidents is also required. This essential nonrepeatability of evolution on earth obviously has a decisive bearing on the chances that it has been repeated or closely paralleled on any other planet.

The assumption, so freely made by astronomers, physicists, and some biochemists, that once life gets started anywhere, humanoids will eventually and inevitably appear is plainly false. The chance of duplicating man on any other planet is the same as the chance that the planet and its organisms have had a history identical in all essentials with that of the earth through some billions of years. Let us grant the unsubstantiated claim of millions or billions of possible planetary abodes of life; the chances of such historical duplication are still vanishingly small.

Even if, as I believe, any close approximation of *Homo sapiens* elsewhere in the accessible universe is effectively ruled out, the question is not quite closed. Manlike intelligence is, after all, a marvelous adaptation, especially in its breadth. It has survival value in a wide range of environmental conditions, and therefore, if it became possible at all, might be favored by natural selection even under conditions different from those on earth. There is, to be sure, another serious hitch here. Man may be going to use one wild aspect of his intelligence to wipe himself out. I do not believe that will occur, but no realist can now deny it as a possibility. If it did happen, the adaptiveness of human intelligence would have been short-lived indeed, and the argument from its apparent broad adaptiveness would be negatived.

Apart from this point, is there not some play, so to speak, in the causations of history? Even in planetary histories different from ours might not some quite different and yet comparably intelligent beings—humanoids in a broader sense—have evolved? Obviously these are questions that cannot be answered categorically. I can only express an opinion. Evolution is indeed a deterministic process to a high degree. The factors that have determined the appearance of man have been so extremely special, so very long continued, so incredibly intricate that I have been able hardly to hint at them here. Indeed they are far from all being known, and everything we learn seems to make them even more appallingly unique. If human origins were indeed inevitable under the precise conditions of our actual history, that makes the more nearly impossible such an occurrence anywhere else. I therefore think it extremely unlikely that anything enough like us for real communication of thought exists anywhere in our accessible universe.

"Extremely unlikely" is not "impossible," and those who like to dream may still dream that mankind is not alone in the universe. But here another point comes up to trouble us. What is the nature and value of that dream? Unless we know or can seriously hope to learn in fact of other humanoids, the dream remains a dream, a fantasy, a science-fiction *divertissement,* a poetic consolation with no substance of reality. Suppose the near-impossible were to be true. What are the chances that we could in fact learn of the existence of extraterrestrial humanoids and eventually communicate with them? With a feeling almost of sorrow, I must conclude that the chances are vanishingly small.

Communication

In the present or any foreseeable state of our technology, the only way we could learn of other humanoids would be by their sending us a message or actually visiting us. That requires, in the first instance, that they must have developed manlike technology, which by no means follows automatically from the mere development of intelligence. (They *might* be intelligent enough to use their brains in better ways!) They must also have done so at just the right time, which involves another tricky point. Out of the billions of years of life on earth, there has been only an infinitesimal length of time, some sixty years, since man has been in a position either to send or to receive messages through outer space. How small the chance of coincidence that any other humanoid reached just this stage at just the right time!

Theoretically, the improbability of humanoids becomes a little less the farther out in space. If humanoids were on a planet a million light years away—and that is a very small distance in the vastness of the galaxies—a message to reach us now would have had to be sent precisely a million years ago. Improbability piled on improbability approaches impossibility. If again the apparently impossible happened, it would certainly be one of the most exciting events in history, but to what avail? The senders of the message would obviously be dead when we received it; their whole species might well be extinct. If, finally stretching the barest possibility to the utmost, we received a message from the relatively nearby stars, it would take years or more likely generations to send a message and receive a reply. Under those conditions the establishment of useful, intelligible intercommunication would still be impossible.

An actual visit to earth by extraterrestrial humanoids would require a technology extremely far advanced beyond ours. We do not, at present, even know that such a stage of technology is possible. All the difficulties previously noted, and more, here pile up. If such a feat is remotely possible and if humanoids are at all prevalent in the universe—the if's do tend to pile up, too, in this subject!—then one would think that we would have been visited by now. In spite of reports of flying saucers and little green men, which belong only in science fiction, the fact is that none have visited us. That would seem, indeed, a logical added reason to believe that humanoids are, to say the least, nonprevalent.

Conclusions

I cannot share the euphoria current among so many, even among certain biologists (some of them now ex-biologists converted to exobiologists). The reasons for my pessimism are given here only in barest suggestion. They will not, I know, convince all or indeed many. There are too many emotional factors and, to put it bluntly, selfish interests opposed to these conclusions. In fact I myself would like to be proved wrong, but a rational view of the evidence seems now to make the following conclusions logically inescapable:

1. There are certainly no humanoids elsewhere in our solar system.

2. There is probably no extraterrestrial life in our solar system, but the possibility is not wholly excluded as regards Mars.

3. There probably are forms of life on other planetary systems somewhere in the universe, but if so it is unlikely that we can learn anything whatever about them, even as to the bare fact of their real existence.

4. It is extremely improbable that such forms of life include humanoids, and apparently as near impossible as does not matter that we could ever communicate with them in a meaningful and useful way if they did exist.

I shall close this chapter with a plea. We are now spending billions of dollars a year and an enormously disproportionate part of our badly needed engineering and scientific manpower on space programs. The prospective discovery of extraterrestrial life is advanced as one of the major reasons, or excuses, for this. Let us face the fact that this is a gamble at the most adverse odds in history. Then if we want to go on gambling, we will at least recognize that what we are doing resembles a wild spree more than a sober scientific program.

To some it seems that the reward could be so great that facing any odds whatever is justified. The biological reward, if any, would be a little more knowledge of life. But we already have life, known, real, and present right here in ourselves and all around us. We are only beginning to understand it. We can learn more from it than from any number of hypothetical Martian microbes. We can, indeed, learn more about possible extraterrestrial life by studying the systematics and evolution of earthly organisms. Knowledge from enlarged programs in those fields is not a gamble because profit is sure.

My plea then is simply this: that we invest just a bit more of our money and manpower, say one-tenth of that now being gambled on the expanding space program, for this sure profit.

40

Sidney W. Fox
Humanoids and Proteinoids

In "The nonprevalence of humanoids," Simpson makes several points with which I agree. Terms which presuppose the existence of extraterrestrial life are understatements of our ignorance. I am glad that Simpson said this emphatically. However, . . . we can be sure that organic chemicals will be found on Mars, and knowledge of the state of molecular evolution there may be fundamentally informative.

Part of Simpson's section on "The origin of life" might have been titled "The nonprevalence of proteinoids" in view of his statement that the "synthesis of the building blocks into macromolecules, especially nucleic acids and proteins, essential for life has not yet been accomplished under realistically primitive conditions." Two papers which appeared in *Nature* (Fox, 1964a; Harada and Fox, 1964) too recently to have been cited in Simpson's article offer evidence of the existence of terrestrial thermal conditions under which proteinoids would form. Even at this stage in the history of Earth, the thermal ranges which exist in the regions of over 450 active volcanoes provide an extensive number of dry and moderately hot zones having conditions such as are employed in laboratory polymerization of amino acids.

Simpson says, "It is still a far cry from the essential preliminary formation of proteins . . . to their organization into a system alive in the full sense of the word. This is the step, or rather the great series of steps, about which we now know the least even by inference and extrapolation." Continued research with the organized units which form spontaneously from proteinoid in the presence of water (Fox, 1960) permits inferences through experiments simulating natural conditions, inferences which are consistent with the concept of the self-organizing properties of macromolecules as enunciated by George Wald in the *Scientific American* in 1954, and by others. The similarities to biocells of the structural details of the ex-

perimental units are best judged from pictures, which are to be found in a number of recent publications.

Whatever coefficient of validity one places upon such an experimental model, its present development makes at least one demonstration. This is the possibility of naturalistically and experimentally accounting for the nonbiological emergence of complex macromolecules and complex protocellular forms from simple systems by processes which are terrestrially realistic. This experimental model, as yet unique in its span of an evolutionary continuum, is also unique in having emerged from theoretical studies of (protein) systematics. This investigative history is consistent with Simpson's emphasis on the value of studying facts in terrestrial organisms.

[Additional comments made by Sidney Fox in 1980]

In 1964, most scientists thought that organic compounds would be found on Mars. The Viking landings, however, showed that not even that optimism was justified, since no organic molecules are present, as measured down to a very low limit [R. S. Young, *BioScience* 28(8), 502 (1978); H. P. Klein, *J. Mol. Evol. 14,* 161 (1979)].

Furthermore, the emphasis on volcanic conditions in my letter of 1964 proved unneeded, in the light of studies of heating in polyphosphoric acid [see K. Harada and S. W. Fox, *Origins of Prebiological Systems* (1965), p. 289] and of heating without such a catalytic agent [see "Thermal Polyamino Acids: Synthesis at less than 100°C," by D. L. Rohlfing, *Science 193,* 68 (1976)].

Simpson's emphasis on "alive in the full sense of the word" was well justified. Perhaps the greatest advance in laboratory studies of the origin of life has occurred in what Simpson called a "great series of steps" [see Chapter 11 on Extraterrestrial Molecular Evolution in S. W. Fox and K. Dose, *Molecular Evolution and the Origin of Life*

(1977)]. In a modern view of the origin of life, an organism that is "fully" alive could appear only through a series of steps, each of which was less than fully alive. This became evident through the laboratory studies made from 1965 to 1980. Steps that go beyond even the laboratory work described in a 1978 reference [S. W. Fox in *The Nature of Life,* W. H. Heidcamp, ed., University Park Press, Baltimore (1978)] have now been modelled [S. W. Fox and T. Nakashima, *BioSystems 12,* 155 (1980]. Peptides formed geologically assemble to cell-like structures that can catalyze the production of other peptides.

In addition, two other pools of data-derived inferences now exist. One consists of the amino acid precursors found in two extraterrestrial sources, the lunar surface and meteorites. Although the meteorites inevitably suffer from contamination, the lunar samples do not. Analyses of proteinous amino acids from these two sources show good agreement when normalized to the percentage of carbon. This indicates that early-stage chemical precursors of life exist in cosmic locales, but does *not* mean that life has ever existed in such sources [see S. W. Fox and K. Dose, *Molecular Evolution and the Origin of Life,* rev. ed. (1977), pp. 342–348]. The same reference assessed the prospects for finding organic molecules on Mars, made before the Viking data had been acquired but after the Apollo samples had been returned from the Moon. The latter, as indicated, showed a little organic matter when correct methods of sampling and analysis were used. These consisted of a standard set of five or six amino acids.

The second pool of inferences may be found in my paper on "The Origins of Behavior in Macromolecules and Protocells," in *Comparative Biochemistry and Physiology* (1980). This documents protocommunication in simulated protocells, which would be a primitive kind of intelligence. In other words, primordia of intelligence existed at the moment of origin of the protocell. I think this is central to questions of the quest for extraterrestrial intelligence.

I might add that, as a result of an analysis I have been performing, I have come to the conclusion that the conditions for the origin of humanoid life are extremely special, and that earlier optimism about extraterrestrial life is far less justified than it has appeared to be, mainly on the basis of data that no one seems to have paid enough attention to.

41

Alfred Adler
Behold the Stars

In September, 1971, the first international conference on extraterrestrial civilizations was held in Soviet Armenia. Jointly organized by the US and USSR Academies of Sciences, it brought together eminent physicists (such as James Dyson), astronomers, biologists (Francis Crick), engineers, and some social scientists. Both the importance of the subject and the scientific credentials of the participants invited attention and respect, even a certain amount of hopeful anticipation. It is, after all, a reasonable hypothesis that life may exist in other parts of the universe, though nothing is yet understood of where it resides, or of how to approach it, or of what forms it might take. To those of us who believe that man's future resides in the whole universe rather than on earth alone, the conference held the promise of revelation and inspiration. Yet such confidence was totally misplaced. For, the mountain labored, and brought forth a

CONFERENCE RESOLUTION
List of Possible Research Directions
It would be useful to concentrate efforts in two directions, both of which seem promising:
I. Searches for civilizations at a technical level comparable with our own.
II. Searches for civilizations at a technological level greatly surpassing our own.
A wide circle of specialists, from astrophysicists to historians, should participate in the planning of this research.

But this was only the last of a collection of blows designed to make those who love science weep, and those who do not, laugh.

The full account is available in an instructive book called *Communication With Extraterrestrial Intelligence (CETI)* (MIT Press, $10.00). The primary occupation of the conference was explained by Carl Sagan, a young astronomer and an organizer of the event. Sagan proposed a formula for the number N of other civilizations within

Reprinted from the *Atlantic Monthly,* 234 (October 1974), 109.

our galaxy:

$$N = R_* f_p\, n_e f_l f_i f_c\, L.$$

Since the formula appears scientific (mathematical), but is in fact totally nonscientific and even meaningless, and since the formula lies at the heart of almost all the work and interest and significance of the conference, I will have to try the reader's patience by identifying each of its terms. The definitions are all quotations from the book.

R_* is the rate of star formation averaged over the lifetime of the Galaxy, in units of numbers of stars per year . . . the province of astrophysics.
f_p is the fraction of stars which have planetary systems.
n_e is the mean number of planets within such planetary systems which are ecologically suitable for life . . . n_e is determined at the boundary between astronomy and biology.
f_l is the fraction of such planets on which life actually occurs.
f_i is the fraction of such planets on which, after the origin of life, intelligence in some form arises.
f_e is the fraction of such planets in which the intelligent beings develop to a communicative phase . . . a topic in anthropology, archaeology, and history.
L is the mean lifetime of such technical civilizations . . . it involves psychology and psychopathology, history, politics, sociology, and many other fields.

And how are these numbers to be determined? The answers, fortunately, are provided by the distinguished conferees.

[FRANCIS] CRICK: The point that I am making adds up to the following conclusion: It is not possible at the moment, with our knowledge of biochemistry, to make any reasonable estimate whatsoever of the factor f_l . . . until we have further information, we cannot really guess about the matter.
[L.M.] MUKHIN: I do not quite understand how we can estimate f_i . . . when we cannot choose any rational approach for assessing f_l. I think it is correct to say that a reliable estimate cannot be given.
[LESLIE] ORGEL: It is our opinion, as a result of experimental

and theoretical work on the subject, that our science has not yet progressed to the point where a meaningful estimate [of f_1] can be made.

[Carl] Sagan: We are faced . . . with very difficult problems of extrapolating . . . in the case of L, from no examples at all.

[R.B.] Lee: The first tool [for determining f_c] is the modern synthetic theory of evolution. . . . The second tool is the theory and method of historical materialism . . . pioneered by Marx and Engels. . . . The third tool is the commitment shared by most of us to search for the broadest, most comprehensive generalizations that can be drawn from available facts.

The "second tool," with its lunatic assertion that historical materialism has direct bearing on the nature of f_c (the fraction of planets in which intelligence develops to a communicative phase), causes one almost to forget that the conference was dealing with questions of extraterrestrial life. The "third tool" is, if possible, an even greater intellectual pollutant. And finally, the major substantive purpose of the conference, a determination of estimates for the number N, is quite clearly a total fraud.

And yet: *"Mon derrière est divisé en deux parts,"* said Gladstone of his own past, thus providing a fair description of the conference as well. A (lesser) part was indeed devoted to useful pursuits, to questions of message transmission, message reception, message decoding, and, finally, to the possible consequences of contact with other civilizations. On the whole these were technical exchanges of theory and experiment, with two expositions of particular general interest. One, by Philip Morrison, speculated on the possible impact on Earth of contact with extraterrestrial civilizations. The other, by Dyson, was a compassionate essay on the human dilemma, beginning with a discussion of Bernal's three enemies of man's rational nature (the cruelty of nature, the frailty of physical man, and the power of man's irrational components), and closing with an invocation which almost totally defied the spirit and content of the conference that preceded it:

> If we are wise, we shall preserve intact these qualities of the human species [toughness, courage, unselfishness, foresight, common sense, and good humor] through the centuries to come, and they will see us safely through the many crises of destiny that surely await us.

A moderately intelligent and humane note. Almost at the conference's end, yet precisely the point from which the conference should have proceeded. For the human qualities most displayed by the conferees were of another nature: those of cupidity, inanity, and triviality:

Cupidity: "Their [the components of the number N] *only* value is in assessing how much effort, time, and money we are willing to devote to the problem. . . . If it turns out that there is some rigorous argument to exclude extraterrestrial intelligence, a convincing demonstration of a small value of N, then a search would not be a useful allocation of resources." (Sagan)

We have already seen that a demonstration of any value of N, let alone of a small value of N by a rigorous argument of any kind, is out of the question. The disclaimer, in short, is an exercise in cynicism.

Inanity: "I believe that the state of society characteristic of Polynesia is one that we can look forward to in [all] the industrial nations. . . . And if their brains resembled ours . . . then all those advanced [extraterrestrial] societies would make up a galactic Polynesian archipelago." (Gunther Stent)

Trivia: "The strategy [in our quest for other civilizations] should be organized roughly as follows: First, the energy for transmitting one bit of information should be minimized. Second, interference in the vicinity of the sending side should be minimized. Third, the cost of the receiving apparatus should be minimized. Fourth, a signal-to-noise ratio greater than unity . . . is desirable. And lastly, the point I should particularly like to stress, is that the ultimate time must be minimized." (N.S. Kardashev).

Is there absolutely nothing that these pedants could not simply take for granted?

It is almost incredible that the truly distinguished scientists among the conferees (and there were indeed several of these) could be willing, almost eager, participants in a travesty of all that is taken seriously by men and women who love and value science and intellect. Possibly some of the scientists were not quite so willing as they appeared to be. Dyson in particular was terse and belligerent almost from beginning to end. But that does not explain the behavior of those who were indeed willing participants. Plato would have understood them well:

> In a state of democratic anarchy, the master fears and flatters his scholars . . . the old men condescend to the young and are full of pleasantry and gaiety; they are loath to be thought morose and authoritative, and therefore they adopt the manners of the young. . . . *(The Republic)*

So now, who are these young? Fortunately, they are not hard to find, nor shy or secretive. Carl Sagan, for one, has written a short, quite personal book of essays in which he has disclosed much more of himself and of his peers than perhaps even he or the reader would like. Called *The Cosmic Connection* (Anchor/Doubleday, $7.95), it is part autobiography, part personal philosophy and speculation, part self-aggrandizement, direct and indirect. (Sagan's wife, who provided an aesthetically offensive drawing for an intellectually offensive metal plaque sent off by NASA to greet extraterrestrial beings, is said to have created art "based on the classical models of Greek sculpture and the drawings of Leonardo da Vinci.") This book has much to teach about the nature of the technological mind.

The modern technologist is first of all a promoter. "It [the subject of extraterrestrial life] has now reached a practical stage where it can be pursued by rigorous scientific techniques where it has achieved scientific respectability," writes Sagan, fully aware of the total absurdity of every part of his assertion, and having in fact participated in a conference which exhibited the emperor's nakedness for even the most obtuse to see. The technologist, too, is gracefully immodest about his own accomplishments: "The greatest significance of the . . . plaque is not as a message to out there; it is as a message to back here." He is the vacant and specious authority on intellect: "The deflation of some of our more common conceits is one of the practical applications of astronomy." Whose conceits? What conceits? Today, in the second half of the twentieth century? His wit is arch and flat (chapter headings such as "Hello, Central Casting? Send Me Twenty Extraterrestrials," or "Some of My Best Friends Are Dolphins," or "The Cosmic Cheshire Cats," are a numskull's delight). He believes himself to be Renaissance Man, with a profound understanding of man's creative sources: "As the results of space exploration . . . permeate our society, they must, I believe, have consequences in literature and poetry, in the visual arts and music"—a platitude that in fact provides another perfect illustration of Sherlock Holmes's dog that did not bark. And he apparently regards himself as a master of politics and economics:

> Old economic assumptions, old methods of determining political leaders . . . may once have been valid . . . but today may no longer have survival value at all. . . . At the same time, there are vested interests opposed to change. These include individuals . . . who are unable in middle years to change the attitudes inculcated in their youth.

The modern technologist is a gifted, highly trained, opportunistic, humorless, and unimaginative ass. Not a barbarian, certainly, and not to be feared; but not to be flattered, pampered, or praised, either. None of his fatuous pseudo-science is science; all of it is empty of intellectual content, inflated with self-importance, and held accountable for nothing. He charges through subtleties and profundities where wise men hesitate to walk on tiptoe; he usurps domains about which he knows nothing and then proceeds to pre-empt them. He tells us that the way to roast a pig is first to find a cow, and expects to be rewarded, applauded, and honored for this genial advice.

The fundamental issues of a search for extraterrestrial civilizations are after all not so arcane or so inaccessible to the nonscientist. Reasonable hypotheses can be advanced and a few promising steps taken, even though at present, since we do not really know where we are going, almost any road will get us there. Already something is understood of how best to attempt the reception of messages from space, of how best to attempt to decode them, of where to seek them. These methods are highly technical, but not unacceptably expensive. Modest conjectures about what to expect of these efforts, and what some of their consequences might be, are also in order. Even though there is no a priori certainty that there exist any civilizations at all in the universe beyond Earth, it is reasonable to postulate that other civilizations exist, even that they exist in large numbers, and it does no harm, in any case, to proceed from such assumptions.

If, in fact there exist civilizations in the galaxy that have attained a level of technology far beyond ours, and if there also exist those whose technologies are approximately equivalent to ours, then it is also reasonable to conjecture that the second of these will contact us long before the first. Civilizations whose development stands to ours as does ours to the grasshopper's will either ignore us entirely or else observe us without bothering to communicate with us in any way. We would not understand most of what they had to tell us in any case. Thus the cultural shock of the first intragalactic messages to earth might well be far less severe than we now expect. We could not even be certain that any message from space would be a call from a living civilization. The speed of light (186,000 miles per second) is a universal upper limit for the speed of messages through space, so that signals would reach us years, or centuries, or more likely yet, millennia after their transmission, possibly arriving here long after the death of the civilization that had sent them.

Morrison, in one of the conference's few bright spots, proposed the thesis that contrary to all superficial expectations, contact with extraterrestrial civilizations would in time have a slow and soberly meditated impact on Earth, much like that of the Western world on nineteenth-century Japan, rather than an overpowering, even cataclysmic effect. Morrison's argument revolves around the observation that other civilizations would be so far removed from us, both in space and (because of the great distance any message must travel) in time, that there could be no question of military dominance or technical economic competition.

A similar argument suggests that the step from reception of the first message to a two-way communication would be a long and difficult one. Each of a series of questions and replies would take (at best) a prohibitive amount of time to travel from sender to receiver, the interval between successive messages spanning many human generations. A reasonable corollary to this conclusion is the suggestion that the matter of communication with other civilizations is not now, and will not be in the future, an urgent one, requiring vast expenditures and intensive exploratory programs. The whole subject, though of great interest and importance, can be approached in patient, measured fashion. It might well, in fact, be postponed entirely for another century (although there is no objective reason for such a postponement).

Human technology has only recently approached a level at which it can begin to liberate man from some of his terrestrial constraints; and even this accomplishment, if regarded from a slightly different perspective, indicates that human technology is still so much more primitive than that of civilizations capable today of contacting us that we could barely understand the other civilization's messages, or benefit from their advice.

The reception of a galactic message would be significant most of all because it would replace a human expectation with a certainty. It would prove to humanity that we are not alone in the universe, and in the proving might diminish man's self-consciousness and self-centeredness. The consequences could only be beneficial. And that is about all the profit that can be expected. Something tangible might indeed be learned from an extraterrestrial civilization, but this would have to be within the realm of our present knowledge and capabilities, or else we could not comprehend its meaning. Being in this realm, it would be something we would sooner or later have discovered for ourselves, without outside help. It would slightly accelerate our progress, no more. Any scientist, any rational person, would reach conclusions more or less like these if ever he decided to think seriously about the matter.

And yet most of the participants at the conference on extraterrestrial life, playing by the rules of their patrons, their new technologists, chose to avoid such rational discourse. Rational, civilized man appears to have become very tired, no longer able to withstand the onslaughts of the manic young masters who promote large grants of money and influence, leave him breathless at conferences, lavish the currency of vague new ideas upon all those around them, and exhaust their weakened elders with pure, assured, unself-conscious power. The behavior of scientists is but one small indication of this syndrome. The fact, for example, that our entire culture turns to M.D.'s for an understanding of the meanings and possibilities of conjugal and erotic love (M.D.'s! Is this to be believed?) suggests that our whole humanist and intellectual tradition has become tired too. It knows better, but goes along all the same, paying obeisance to its new masters in the spirit of the Arab proverb: If the King at noonday says it is night, behold the stars.

42

Michael H. Hart
An Explanation for
the Absence of Extraterrestrials on Earth

Are there intelligent beings elsewhere in our Galaxy? This is the question which astronomers are most frequently asked by laymen. The question is not a foolish one; indeed, it is perhaps the most significant of all questions in astronomy. In investigating the problem we must therefore do our best to include all relevant observational data.

Because of our training, most scientists have a tendency to disregard all information which is not the result of measurements. This is, in most matters, a sensible precaution against the intrusion of metaphysical arguments. In the present matter, however, that policy has caused many of us to disregard a clearly empirical fact of great importance, to wit: *There are no intelligent beings from outer space on Earth now.* (There may have been visitors in the past, but none of them has remained to settle or colonize here.) Since frequent reference will be made to the foregoing piece of data, in what follows we shall refer to it as 'Fact A'.

Fact A, like all facts, requires an explanation. Once this is recognized, an argument is suggested which indicates an answer to our original question. If, the argument goes, there were intelligent beings elsewhere in our Galaxy, then they would eventually have achieved space travel, and would have explored and colonized the Galaxy, as we have explored and colonized the Earth. However, (Fact A), they are not here; therefore they do not exist.

The author believes that the above argument is basically correct; however, in the rather loose form stated above it is clearly incomplete. After all, might there not be some other explanation of Fact A? Indeed, many other explanations of Fact A have been proposed; however, none of them appears to be adequate.

The other proposed explanations of Fact A might be grouped as follows:

(1) All explanations which claim that extraterrestrial visitors have never arrived on Earth because some physical, astronomical, biological or engineering difficulty makes space travel infeasible. We shall refer to these as 'physical explanations'.

(2) Explanations based on the view that extraterrestrials have not arrived on Earth because they have chosen not to. This category is also intended to include any explanation based on their supposed lack of interest, motivation or organization, as well as political explanations. We shall refer to these as 'sociological explanations'.

(3) Explanations based on the possibility that advanced civilizations have arisen so recently that, although capable and willing to visit us, they have not had time to reach us yet. We shall call these 'temporal explanations'.

(4) Those explanations which take the view that the Earth *has* been visited by extraterrestrials, though we do not observe them here at present.

These four categories are intended to be exhaustive of the plausible alternatives to the explanation we suggest. Therefore, if the reasoning in the next four sections should prove persuasive, it would seem very likely that we are the only intelligent beings in our Galaxy.

Physical Explanations

After the success of Apollo 11 it seems strange to hear people claim that space travel is impossible. Still, the problems involved in interstellar travel are admittedly greater than those involved in a trip to the Moon, so it is reasonable to consider just how serious the problems are, and how they might be overcome.

The most obvious obstacle to interstellar travel is the enormity of the distances between the stars, and the consequently large travel times involved. A brief computation should make the difficulty clear: The greatest speeds which manned aircraft, or even spacecraft, have yet attained is only a few thousand km hr^{-1}. Yet travelling at 10

Reprinted with permission from the *Quarterly Journal of the Royal Astronomical Society*, 16 (1975), 128.

per cent of the speed of light (~one billion km hr⁻¹) a one-way trip to Sirius, which is one of the nearest stars, would take 88 years. Plainly, the problem presented is not trivial; however, there are several possible means of dealing with it:

(1) If it is considered essential that those who start on the voyage should still be reasonably youthful upon arrival, this could be accomplished by having the voyagers spend most of the trip in some form of 'suspended animation'. For example, a suitable combination of drugs might not only put a traveller to sleep, but also slow his metabolism down by a factor of 100 or more. The same result might be effected by freezing the space voyagers near the beginning of the trip, and thawing them out shortly before arrival. It is true that we do not yet know how to freeze and revive warm-blooded animals but: (a) future biologists on Earth (or biologists in advanced civilizations elsewhere) may learn how to do so; (b) intelligent beings arising in other solar systems are not necessarily warm-blooded.

(2) There is no reason to assume that all intelligent extraterrestrials have life spans similar to ours. (In fact, future medical advances may result in human beings having life expectancies of several millenia, or even perhaps much longer.) For a being with a life span of 3000 years a voyage of 200 years might seem not a dreary waste of most of one's life, but rather a diverting interlude.

(3) Various highly speculative methods of overcoming the problem have been proposed. For example, utlilization of the relativistic time-dilation effect has been suggested (though the difficulties in this approach seem extremely great to me). Or the spaceship might be 'manned' by robots, perhaps with a supplementary population of frozen zygotes which, after arrival at the destination, could be thawed out and used to produce a population of living beings.

(4) The most direct manner of handling the problem, and the one which makes the fewest demands on future scientific advances, is the straightforward one of planning each space voyage, from the beginning, as one that will take more than one generation to complete. If the spaceship is large and comfortable, and the social structure and arrangements are planned carefully, there is no reason why this need be impracticable.

Another frequently mentioned obstacle to interstellar travel is the magnitude of the energy requirements. This problem might be insurmountable if only chemical fuels were available, but if nuclear energy is used the fuel requirements do not appear to be extreme. For example, the kinetic energy of a spaceship travelling at one-tenth the speed of light is:

$$KE = (\gamma - 1) Mc^2 = ([1.0 - 0.01]^{-1/2} - 1) Mc^2 = 0.005 \, Mc^2. \quad (1)$$

Now the energy released in the fusion of a mass F of hydrogen into helium is approximately $0.007 \, Fc^2$. In principal, the mechanical efficiency of a nuclear-powered rocket can be more than 60 per cent (von Hoerner, Reading 37; Marx, 1963). However, let us assume that in practice only one-third of the nuclear energy could actually be released and converted into kinetic energy of the spacecraft. Then the fuel needed to accelerate the spaceship to 0.10 c is given by:

$$0.005 \, Mc^2 = 0.007 \, Fc^2/3. \quad (2)$$

This gives: $F = 2.14 \, M$, and $T = 3.14 \, M$, where T is the combined mass of spaceship and fuel. The necessity of starting out with enough fuel first to accelerate the ship, and later to decelerate, introduces another factor of 3.14; so initially we must have $T = 9.88 \, M$. In other words, the ship must start its voyage carrying about nine times its own weight in fuel. This is a rather modest requirement, particularly in view of the cheapness and abundance of the fuel. (The enormous fuel-to-payload ratios computed by Purcell (Reading 36) are a result of his considering only relativistic space flight; a travel speed of 0.1 c seems more realistic.) Furthermore, there are several possible ways of reducing the fuel-to-payload ratio, including (a) refuelling from auxiliary craft; (b) scooping up H atoms while travelling through interstellar space; (c) greater engine efficiencies; (d) travelling at slightly lower speeds (travelling at 0.09 c instead of 0.10 c would reduce the fuel-to-payload ratio to 6.5:1); and (e) using methods of propulsion other than rockets. For some interesting possibilities see Marx (1966) and other papers listed by Mallove & Forward (1972).

It can be seen that neither the time of travel nor the energy requirements create an insuperable obstacle to space travel. However, in the past, it was sometimes suggested that one or more of the following would make space travel unreasonably hazardous: (a) the effects of cosmic rays; (b) the danger of collisions with meteoroids; (c) the biological effects of prolonged weightlessness; and (d) unpredictable or unspecified dangers. With the success of the Apollo and Skylab missions it appears that none of these hazards is so great as to prohibit space travel.

Sociological Explanations

Most proposed explanations of Fact A fall into this category. A few typical examples are:

(a) Why take the anthropomorphic view that extraterrestrials are just like us? Perhaps most advanced civilizations are primarily concerned with spiritual contemplation and have no interest in space exploration. (The Contemplation Hypothesis.)

(b) Perhaps most technologically advanced species

destroy themselves in nuclear warfare not long after they discover atomic energy. (The Self-Destruction Hypothesis.)

(c) Perhaps an advanced civilization has set the Earth aside as their version of a national forest, or wildlife preserve. (The Zoo Hypothesis [Ball, Reading 45].)

In addition to variations on these themes (for example, extraterrestrials might be primarily concerned with artistic values rather than spiritual contemplation) many quite different explanations have been suggested. Plainly, it is not possible to consider each of these individually. There is, however, a weak spot which is common to all of these theories.

Consider, for example, the Contemplation Hypothesis. This might be a perfectly adequate explanation of why, in the year 600,000 BC, the inhabitants of Vega III chose not to visit the Earth. However, as we well know, civilizations and cultures change. The Vegans of 599,000 BC could well be less interested in spiritual matters than their ancestors were, and more interested in space travel. A similar possibility would exist in 598,000 BC, and so forth. Even if we assume that the Vegans' social and political structure is so rigid that no changes occur even over hundreds of thousands of years, or that their basic psychological makeup is such that they always remain uninterested in space travel, there is still a problem. With such an additional assumption the Contemplation Hypothesis might explain why the Vegans have never visited the Earth, but it still would not explain why the civilizations which developed on Procyon VI, Sirius II, and Altair IV have also failed to come here. The Contemplation Hypothesis is not sufficient to explain Fact A unless we assume that it will hold for *every* race of extraterrestrials—regardless of its biological, psychological, social or political structure—and at *every* stage in their history after they achieve the ability to engage in space travel. That assumption is not plausible, however, so the Contemplation Hypothesis must be rejected as insufficient.

The same objection, however, applies to any other proposed sociological explanation. No such hypothesis is sufficient to explain Fact A unless we can show that it will apply to every race in the Galaxy, and at every time.

The foregoing objection would hold even if there *were* some established sociological theory which predicted that most technologically advanced civilizations will be spiritually oriented, or will blow themselves up, or will refrain from exploring and colonizing. In point of fact, however, there is no such theory which has been generally accepted by political scientists, or sociologists, or psychologists. Furthermore, it is safe to say that no such theory will be accepted. For any scientific theory must be based upon evidence, and the only evidence concerning the behaviour of technologically advanced civilizations which political scientists, sociologists and psychologists

have comes from the human species—a species which has neither blown itself up, nor confined itself exclusively to spiritual contemplation, but which has explored and colonized every portion of the globe it could. (This is *not* intended as proof that all extraterrestrials must behave as we have; it *is* intended to show that we cannot expect a scientific theory to be developed which predicts that most extraterrestrials will behave in the reverse way.)

Another objection to any sociological explanation of Fact A is methodological. Faced with a clear physical fact astronomers should attempt to find a scientific explanation for it—one based on known physical laws and subject to observational or experimental tests. No scientific procedure has ever been suggested for testing the validity of the Zoo Hypothesis, the Self-Destruction Hypothesis, or any other suggested sociological explanation of Fact A; therefore to accept any such explanation would be to abandon our scientific approach to the question.

Temporal Explanations

Even if one rejects the physical and sociological explanations of Fact A, the possibility exists that the reason no extraterrestrials are here is simply because none have yet had the time to reach us. To judge how plausible this explanation is, one needs some estimate of how long it might take a civilization to reach us once it had embarked upon a programme of space exploration. To obtain such an estimate, let us reverse the question and ask how long it will be, assuming that we are indeed the first species in our Galaxy to achieve interstellar travel, before we visit a given planet in the Galaxy?

Assume that we eventually send expeditions to each of the 100 nearest stars. (These are all within 20 light-years of the Sun.) Each of these colonies has the potential of eventually sending out their own expeditions, and their colonies in turn can colonize, and so forth. If there were no pause between trips, the frontier of space exploration would then lie roughly on the surface of a sphere whose radius was increasing at a speed of $0.10\,c$. At that rate, most of our Galaxy would be traversed within 650,000 years. If we assume that the time between voyages is of the same order as the length of a single voyage, then the time needed to span the Galaxy will be roughly doubled.

We see that if there were other advanced civilizations in our Galaxy they would have had ample time to reach us, unless they commenced space exploration less than 2 million years ago. (There is no real chance of the Sun being accidentally overlooked. Even if the residents of one nearby planetary system ignored us, within a few thousand years an expedition from one of their colonies, or from some other nearby planetary system, would visit the solar system.)

Now the age of our Galaxy is $\sim 10^{10}$ years. To accept the temporal explanation of Fact A we must therefore hypothesize that (a) it took roughly 5000 time-units (choosing one time-unit $\equiv 2 \times 10^6$ years) for the first species to arise in our Galaxy which had the inclination and ability to engage in interstellar travel; but (b) the second such species (i.e. us) arose less than 1 time-unit later.

Plainly, this would involve a quite remarkable coincidence. We conclude that, though the temporal explanation is theoretically possible, it should be considered highly unlikely.

Perhaps They Have Come

There are several versions of this theory. Perhaps the most common one is the hypothesis that visitors from space arrived here in the fairly recent past (within, say, the last 5000 years) but did not settle here permanently. There are various interesting archaeological finds which proponents of this hypothesis often suggest are relics of the aliens' visit to Earth.

The weak spot of that hypothesis is that it fails to explain why the Earth was not visited earlier:

(a) If it is assumed that extraterrestrials have been able to visit us for a long time, then a sociological theory is required to explain why they all postponed the voyage to Earth for so long. However, any such sociological explanation runs into the same difficulties described earlier.

(b) On the other hand, suppose it is assumed that extraterrestrials visited us as soon as they were able to. That this occurred within 5000 years (which is only 1/400 of a time-unit) of the advent of our own space age would involve an even more remarkable coincidence than that discussed in the previous section.

Another version of the theory is that the Earth was visited from space a very long time ago, say 50 million years ago. This version involves no temporal coincidence. However, once again, a sociological theory is required to explain why, in all the intervening years, no other extraterrestrials have chosen to come to Earth, and remain. Of course, any suggested mechanism which is effective only 50 per cent (or even 90 per cent) of the time would be insufficient to explain Fact A. (For example, the hypothesis that *most* extraterrestrials wished only to visit, but not to colonize, is inadequate. For colonization not to have occurred requires that *every* single civilization which had the opportunity to colonize chose not to.)

A third version, which we may call 'the UFO Hypothesis', is that extraterrestrials have not only arrived on Earth, but are still here. This version is not really an explanation of Fact A, but rather a denial of it. Since very few astronomers believe the UFO Hypothesis it seems unnecessary to discuss my own reasons for rejecting it.

Conclusions and Discussion

In recent years several astronomers have suggested that intelligent life in our Galaxy is very common. It has been argued (Shklovskii and Sagan, 1966) that (a) a high percentage of stars have planetary systems; (b) most of these systems contain an Earth-like planet; (c) life has developed on most of such planets; and (d) intelligent life has evolved on a considerable number of such planets. These optimistic conclusions have perhaps led many persons to believe that (1) our starfaring descendants are almost certain, sooner or later, to encounter other advanced cultures in our Galaxy; and (2) radio contact with other civilizations may be just around the corner.

These are very exciting prospects indeed; so much so that wishful thinking may lead us to overestimate the chances that the conjecture is correct. Unfortunately, though, the idea that thousands of advanced civilizations are scattered throughout the Galaxy is quite implausible in the light of Fact A. Though it is possible that one or two civilizations have evolved and have destroyed themselves in a nuclear war, it is implausible that every one of 10,000 alien civilizations had done so. Our descendants might eventually encounter a few advanced civilizations which never chose to engage in interstellar travel; but their number should be small, and could well be zero.

If the basic thesis of this paper is correct there are two corollary conclusions: (1) an extensive search for radio messages from other civilizations is probably a waste of time and money; and (2) in the long run, cultures descended directly from ours will probably occupy most of the habitable planets in our Galaxy.

In view of the enormous number of stars in our Galaxy, the conclusions reached in this paper may be rather surprising. It is natural to inquire how it has come about that intelligent life has evolved on Earth in advance of its appearance on other planets. Future research in such fields as biochemistry; the dynamics of planetary formation; and the formation and evolution of atmospheres, may well provide a convincing answer to this question. In the meantime, Fact A provides strong evidence that we are the first civilization in our Galaxy, even though the cause of our priority is not yet known.

43

Laurence Cox

An Explanation for
the Absence of Extraterrestrials on Earth

Hart (Reading 42) has argued that other advanced civilizations do not exist in the Galaxy. He starts his argument by what he calls his empirical Fact A, namely: There are no intelligent beings from outer space on Earth now. Although this paper is concerned with putting forward alternative explanations for Fact A, it should be mentioned at this point that it is not an empirical fact as Hart asserts, but a theory (the empirical fact would be: No intelligent beings from outer space are observed on Earth now). With Fact A as a theory I am not in disagreement and it will be taken as the starting point for this paper also.

Having established Fact A, Hart's argument runs as follows: Since the time to cross the Galaxy is small compared to its age, even for non-relativistic spaceflight ($<0.1c$), any civilization as little as 10^6 yr in advance of our own will have had sufficient time to explore and colonize the entire Galaxy. However (Fact A) they are not here; therefore unless a physical or sociological explanation prevents them from visiting Earth, they do not exist. Two further categories of explanations for Fact A are considered by Hart: temporal explanations, which assume that other civilisations simply have not had time to reach us yet; and the assumption that Earth has been visited by extraterrestrials in the past although we do not observe them here at present. Implicit in this fourth explanation is that Earth may have been visited by extraterrestrials with a desire to colonize but were unable to do so. Since my paper, like Hart's, is indirectly concerned with the prospects of our exploration and colonization of planetary systems of other stars this restriction will be considered first under the heading of physical explanations.

One further point needs to be made at this stage: our knowledge of intelligent life in the Universe is limited to one dominant species on one planet. *Homo sapiens* (and his precursors) provide us with all the observational data which we have. But in making use of this data we must be

Reprinted with permission from the *Quarterly Journal of the Royal Astronomical Society*, 17 (1976), 201.

particularly careful not to ascribe to all other intelligent beings any characteristic of ourselves which may have come about by chance rather than necessarily during our biological and social evolution. This caveat is particularly relevant to the class of sociological explanations of Fact A. On the other hand the synthesis of organic compounds prior to the appearance of life on Earth is well understood now and it is from this viewpoint that the physical explanations of Fact A will be considered.

Physical Explanations

The purely physical obstacles to interstellar travel have already been investigated in depth (Reading 42 and references therein) and these will not be considered further here. Of the four methods of countering the enormity of the travel times involved, that of planning the voyages to last for more than one generation requires fewest assumptions about future advances in technology and can be assumed to be applicable to all intelligent species. The biological obstacles to interstellar flight, for example, cosmic rays and weightlessness, have already been discounted by Hart and neither need be a problem in a very large spaceship designed to maintain a colony for the several generations needed to reach the nearest stars. Instead a more fundamental obstacle is proposed.

The abiological synthesis of amino acids and monosaccharides such as glucose necessarily produces equal quantities of the D- and L-forms (enantiomorphic states) unless some asymmetric force, such as circularly polarized light, favours the production of one of the enantiomorphs (Keosian, 1964). In living organisms on Earth one finds almost exclusively L- but not D-amino acids and D- but not L-sugars. Since the D- and L- forms of a molecule are mirror images of one another, it is apparent that an enzyme can only interact with one of the two forms. Three possibilities may be recognized:

1. L-amino acids are predominant in carbon-based life on all planets where it is found.

2. L-amino acids are predominant on about half of these planets and D-amino acids on the remainder.

3. Any amino acid may be of either the L- or of the D-type without fixing the enantiomorphic state of other amino acids on that planet.

Similar statements may also be made about the two enantiomorphic states of each sugar. Now Hart implicitly assumes that the first of these statements applies and hence that any space-travelling race could colonize all available planets with similar biospheres to its own. If the second statement represents the true case then this would halve the number of colonizable planets; whereas if the third possibility were chosen, to find a planet where just the same 20 amino acids appeared in the same form as on Earth would require the searching of about 2^{20} (10^6) other planetary systems. Fortunately the establishment of one of these three possibilities can be achieved fairly rapidly once exploration is started.

A further factor that could reduce the number of colonizable planets is that we do not know whether the same 20 amino acids are metabolized by living cells elsewhere in the Galaxy. For example the abiological synthesis of alanine produces not only the L- and D- forms of α-alanine but also the optically inactive β-alanine, which is not metabolized by living cells on Earth. If the use of one isomer rather than another is the result of a chance process in our evolution then it may be that the number of Earthlike planets which we could colonize might be small or even zero. This is not an absolute bar to colonization for we could still create suitable biospheres on sterile planets, as Sagan has suggested for Venus (Shklovskii and Sagan, 1966), or transform the asteroids into space cities after the fashion of Tsiolkovskii & Cole (in Shklovskii and Sagan, 1966). It seems reasonable to expect asteroids to be a common feature of other planetary systems and no doubt these would be used by any exploring civilization as a convenient source of raw materials if for nothing else.

Evidently the biological obstacles to colonization cannot be dismissed lightly. Future research in biochemistry and molecular biology should reveal to what extent our own evolution from non-living matter to the single-celled stage is constrained by the shape of the molecules. To decide between the three possibilities above, however, is likely to be impossible without further data obtained either by interstellar exploration or by communication with other civilizations. Since the latter course is considerably less expensive this is a strong argument in favour of an extensive search for radio messages from other civilzations before undertaking interstellar travel.

Sociological Explanations

Hart takes three of these as examples: The Contemplation Hypothesis, The Self-Destruction Hypothesis and the Zoo Hypothesis (Ball, Reading 45). His objection to all of these is that it is necessary that one of them or a similar explanation applies to all races of extraterrestrials at all times after they achieve interstellar spaceflight. If we are to overcome this objection to all sociological explanations it is necessary to put forward an explanation which is applicable to all civilizations, including our own. It may be that this explanation has not applied to us in the past, it need not apply even now, but it must apply by the time that we are capable of interstellar travel and it must continue to apply indefinitely thereafter. Furthermore it must be demonstrable that the explanation will require a fundamental change in our own society which will render our past history valueless for the purposes of predicting the future course of our society. This is necessary to counter Shklovskii & Sagan's arguments (1966).

There is one sociological explanation that satisfies these requirements: it is that any society undertaking interstellar travel must have a stable population. To illustrate this Table 43.1 shows the consequences if it were possible for the population of the Earth to grow at its present rate for the next 3000 years.

Comparing this time scale with Hart's estimated time for an advanced civilization to cross the Galaxy (650,000 years) it becomes apparent that any civilization with a population growth rate approaching our own would find its exponential population growth outstripping the rate at which new planets could be colonized. To populate 1 Earthlike planet around all the 10^{11} stars in our Galaxy with a population density equal to that of the present day would require that our population must double in not less than 20,000 years, compared with the 30 years doubling period at present. Even taking Tsiolkovskii & Dyson's (in Shklovskii and Sagan, 1966) extreme predictions of the order of 10^{20} people to each system would do no more than halve this minimum doubling time.

The possibility that another civilization could undertake interstellar travel with a population much smaller than our present population is germane to this argument. In fact with a population growth rate equal to our present rate no other civilization could avoid reaching a similar population to our own within at most a few centuries, whatever its initial population. Thus the stable population requirement is in itself a sufficient sociological explanation and its acceptance leads to further consequences which render the class of temporal explanations more probable.

Table 43.1.

Assumptions: Population at 2000 AD = 7 × 10⁹ people
Population doubling time = 30 yr, equivalent to a factor of 10 per century (for ease of computation)
Average weight of person = 60 kg

Year AD	*No. of people*	*Mass of people (g)*	*Comments*
2100	7×10^{10}	4×10^{15}	
2500	7×10^{14}	4×10^{19}	
3000	7×10^{19}	4×10^{24}	Exceeds mass of largest asteroid
4000	7×10^{29}	4×10^{34}	20 solar masses
5000	7×10^{39}	4×10^{44}	Exceeds mass of Galaxy

Temporal Explanations

It is in the possibility that extraterrestrials are not on Earth because they have not had time to reach us that the most promising explanation of Fact A lies. It has been shown above that any civilization undertaking interstellar travel must have a stable population and the consequences of this for temporal explanations of Fact A may most easily be demonstrated by re-examining Hart's argument.

Let us suppose, as Hart does, that we eventually send expeditions to each of the nearest 100 stars, all of which are within 20 light years. If there is no pause between trips, and if after exploring a planetary system each starship continues travelling away from the Earth, then after 200 years the starships would lie within a sphere of radius 20 light years (or on its surface if they were all launched simultaneously). This does not mean, however, that the sphere of explored space would continue to expand at 0.1c (the starship speed). If the density of stars is constant over this volume then there will be 7 times as many stars in the annular region between the two spheres of radii 20 and 40 light years as there are within the inner sphere. Furthermore, none of the stars in the larger region can be reached by a starship launched from Earth in less than a further 200 years, making a total of 400 years. Hence all the exploration of this outer region must be done by the original starships if the rate of expansion of explored space is to be maintained. This is clearly impossible, since if progressively larger spheres are considered the required starship speed exceeds that of light; the time dilation effect does not save the situation as we are considering Earth-time rather than starship time.

On the other hand, it might seem possible to circumvent this difficulty by allowing the occupants of the starships to create new starships from material in the planetary systems which they explore. Each starship would need to create a further six vessels within rather less than the 200 years which it would take another starship from Earth to reach it. The physical difficulties involved in this seem almost insurmountable but let us assume that this is possible. The real difficulty lies in the crew of these additional starships which must come from the population of the original starship. This population, like that of Earth, must be stable during the 200 year spaceflight since the starship is a limited size. Upon reaching the planetary system which will provide the materials for these extra starships the original starship population must expand at a rate comparable with present-day Earth and then return to a zero growth rate for the next 200-yr journey and so on *ad infinitum*. Although in principle this problem could be avoided if the population of the first starship included a large proportion of hibernating (or frozen) humans, sooner or later there would be insufficient crew for all the starships needed and it would cease to be possible to maintain the rate of expansion of explored space. We may therefore conclude that Hart's assumption of:

$$r_{\text{explored space}}(\text{ly}) = 0.1 \times t(\text{y})$$

will not be valid. Fortunately there are a number of possibilities based on a variety of assumptions. For example:

Assumption 1: All starships are produced on Earth; the number is limited by the number of suitably qualified personnel to man them.
Assumption 2: The lifetime of starships is limited
Assumption 3: Starships travel in random directions—not always outwards (this permits interstellar traffic from colony to colony or from colony to Earth).

These assumptions lead to relations of the form:

$$r_{\text{exp}} \propto t^{1/n}$$

where n is 2 for A1, 3 for A1 + A2, 4 for A1 + A3 and 6 for A1 + A2 + A3. Other assumptions lead to similar results; a finite starship lifetime (produced by technological advances) linked to a linearly increasing number of planets producing starships would produce a relation similar to A1. If we examine just two of these:

Case 1 . . . $r \propto t^{1/2}$ Case 2 . . . $r \propto t^{1/3}$

and compare them with Hart's assumption (Table 43.2) it can be seen that for Case 1 the maximum time difference between the first and second (ourselves) space-travelling species is 1000 time units, compared to Hart's estimate of 0.5 time units (1 time unit = 2×10^6 yr). For Case 2 several tens of civilizations all 5000 time units old (equal to the age of the Galaxy) could exist independently of ourselves and of each other. Thus the temporal explanation is sufficient to explain Fact A when coupled with reasonable assumptions about future population growth.

Table 43.2.

t (years)	r_{Hart}	Light-years $r_{Case\ 1}$	$r_{Case\ 2}$	t (time units)
200	20	20	20	10^{-4}
2×10^3	200	60	40	10^{-3}
2×10^4	2×10^3	200	100	10^{-2}
2×10^5	2×10^4	600	200	10^{-1}
2×10^6	—	2×10^3	400	I
2×10^7	—	6×10^3	10^3	10
2×10^8	—	2×10^4	2×10^3	100
2×10^9	—	6×10^4	4×10^3	10^3
2×10^{10}	—	—	10^4	10^4

Perhaps They Have Come

It may seem unnecessary to examine this possibility since the three explanations above provide sufficient causes for Fact A. On the other hand it may be that there is a nearby (within 100 ly) advanced civilization, in which case the temporal explanation would not be adequate to explain Fact A and it would be necessary to have recourse to a physical explanation to account for why we do not observe extraterrestrials on Earth now. Since the physical explanation discussed in this paper only prevents extraterrestrials from colonizing our planet not from exploring it we might expect them to follow one of two courses, either to return to Earth at intervals to observe the effects of evolution, or to leave an unmanned probe in the Solar System which would report to the civilization when our society reached a sufficiently advanced state, the Sentinel Hypothesis (Clarke, 1972). Either of these explanations would overcome Hart's objections.

Conclusions

The arguments that intelligent life has evolved on many planets in our Galaxy are attractive in that they predict that (1) our descendants are almost certain to encounter other advanced cultures in the Galaxy; and that (2) it may be possible to communicate with other civilizations by radio within the near future. If as Hart suggests there are no other advanced civilizations in the Galaxy then this in itself would be highly significant. For this reason it is of primary importance that the possible explanations for Fact A are examined thoroughly and their assumptions questioned in detail. Of the three main groups of explanations, the physical explanation proposed seems to be the weakest since if we accept the third and most effective possibility we must also explain why our amino acids are all of the L-type. On the evidence which we have at present it seems

that the second possibility is the most probable. Thus the physical explanation proposed is probably inadequate on its own to explain Fact A except in the special case of the last section.

The sociological and temporal explanations, though, are sufficient to account for Fact A. The static population requirement provides the necessary fundamental change in societies at the same stage as ours, whilst the failure of a society to achieve it would provide the conditions for the Self-Destruction Hypothesis to operate. There is no need to assume nuclear warfare here; overpopulation is an equally effective cause of a population crash and it is applicable to all societies whether or not they have developed weapons of mass destruction. Of the three explanations the third, the temporal explanation, is the most effective in accounting for Fact A. Although different assumptions lead to different rates of expansion of explored space, all reasonable combinations of assumptions lead to rates which are much smaller than that calculated by Hart.

I conclude that, with our present evidence, Fact A cannot be regarded as an indication of our priority in the Galaxy.

[Note added by Laurence Cox in 1980]

I think that the physical and sociological explanations I advanced have not lost any of their force over the intervening years, although what I proposed in the section on temporal explanations has been overtaken by subsequent work, particularly that by Eric Jones, published in *Icarus* (1976) inter alia. The essence of this work was to demonstrate that given reasonable assumptions about population growth and emigration rates, the radius of the sphere of explored space did expand at a constant rate, i.e. Hart's case rather than mine, although the constant of proportionality was several orders of magnitude smaller than Hart's assumed value of the ship speed. I would now agree with Eric Jones' analysis in the main part, although we still disagree about the colonizing time of a civilization and whether this can be equated to its lifetime.

44

Michael H. Hart
Habitable Zones about Main Sequence Stars

In an earlier paper (Hart, 1978), hereafter referred to as Paper I, the author has discussed the construction of a computer simulation which follows the evolution of the atmosphere of the Earth over the course of geologic time. That model attempts to explicitly take into account all the major processes which have affected the bulk composition of the atmosphere and the mean surface temperature of the Earth.[1]

The results of those calculations indicated that had the Earth been situated only 5% closer to the Sun a runaway greenhouse effect would have occurred about 4 by (billion years) ago. On the other hand, if the Earth had been situated only 1% further from the Sun, runaway glaciation would have occurred about 2 by ago, when free O_2 appeared in the Earth's atmosphere and all but trace amounts of reducing gases were eliminated (thereby sharply reducing the greenhouse effect).

Those results indicated that the *continuously* habitable zone about the Sun (i.e., the region within which a planet might enjoy moderate surface temperatures continuously throughout the 3 or 4 by needed for advanced life forms as we know them to evolve) is surprisingly narrow. It is natural, therefore, to inquire how wide the CHZ (continuously habitable zone) is about other main sequence stars. It is to that question that this paper is addressed.

In the course of this project a long series of computer runs were made, each one following the evolution of the atmosphere of an Earthlike planet orbiting about a main sequence star. The only inputs varied from run to run were: (a) the mass, and therefore the luminosity, $L(t)$, of the central star; and (b) the radius of the planet's orbit. It was assumed that life would spontaneously arise and continue to evolve on any such planet provided that the surface temperature was moderate enough for there to be liquid water at some location on the planet's surface, and that evolution to the stage of photosynthetic organism would only require $\sim 8 \times 10^8$ years.

Not surprisingly, the results show that the CHZ (continuously habitable zones) about smaller, less luminous stars are closer in than the CHZ about the Sun. More importantly, those results (which are presented in Section 3 of this paper) indicate that the CHZ about small stars are relatively narrower than the CHZ about the Sun. That is, if r_{outer} and r_{inner} represent the outer and inner radius of the CHZ, the ratio r_{outer}/r_{inner} is smaller for less massive stars.

Even about the Sun, the ratio r_{outer}/r_{inner} is only about 1.05. For a star of 0.83—a typical K1 star—we find that $r_{outer}/r_{inner} = 1.0$. In other words, there is *no* continuously habitable zone about most K or M stars.

This perhaps surprising conclusion depends, of course, not only on the model used for the Earth's atmosphere, but also on the form of the mass-luminosity law for main-sequence stars, and on the evolution rates of main-sequence stars. The derivation of that input data is described in Section 2 of this paper. Such data, however, no matter how carefully derived, may be in error. The sensitivity of our results to the input data on luminosities is discussed in Section 4. The sensitivity of our results to some of the possible errors in the underlying computer simulation (the one described in Paper I) is discussed in Section 5.

Mass-Luminosity Relationships

The luminosity of a main-sequence star (of a given age) depends on its mass, and can be approximated—for stars whose mass is not too different from the Sun's—by $L = K(M/M_s)^p$. The appropriate value to be used for p depends, however, on the age of the stars involved. Let us

---

[1]The possibility exists, of course, that there are other processes, not taken into account in the model, which had a major effect on the Earth's surface temperature or atmospheric composition. Similarly, even the processes included in the model might have been approximated inadequately.

Reprinted from *Icarus*, 37 (1979), 351. Copyright © 1979 by Academic Press, Inc.

denote by m the appropriate value of p for the zero-age main sequence, and let us denote by n the value appropriate for a set of main-sequence stars of age 4.5 by. Since low mass stars evolve more slowly than heavier stars, it is plain that $n > m$.

A star's rate of evolution, of course, depends on its initial composition. Since those few stars whose masses are well known are of varying compositions, it is impractical to estimate m and n from astronomical observations. A more practical way of estimating m is to use the theoretical stellar models of Demarque and Larson (1964). They constructed models of zero-age main-sequence stars having various masses, but all having the composition $X = 0.67$; $Y = 0.30$; and $Z = 0.03$. The best fit to their results, for stars in the range $0.80 < M/M_s < 1.03$, is obtained by using $m = 5.139$.

On the other hand, the best fit to the zero-age models of Iben (1967)—who used the composition $X = 0.708$; $Y = 0.272$; $Z = 0.020$—is obtained by using $m = 5.021$. Copeland (1970) used $X = 0.70$; $Y = 0.27$; and $Z = 0.03$, and the best fit to his results is $m = 4.902$. Maeder (1976) used the same composition as Copeland, but his results are best matched by using $m = 5.015$. All these figures are close to each other, which indicates that minor differences in composition (or in the computations of stellar structure) do not greatly affect the value of m. We should therefore not be too far off by using their average, $m = 5.02$. (Of course, it is logically possible that the theories of all those authors are off systematically.)

If we estimate n from the evolutionary models of Demarque and Larson (1964) we get $n = 6.037$. If, instead, the models of Maeder (1976) are employed, the result is $n = 5.850$. The average of those two figures is $n = 5.94$.

Table 44.1, which formed the basis of our main computer runs, was constructed by using the values $m = 5.02$; $n = 5.94$ and by assuming that $L(0.0)/L(4.5$ by$) = 0.75$ for the Sun (Ulrich, 1975).

It is plain that if m and n are each increased by the same amount, the figures in column 4 of Table 44.1 will not be altered. But changing $(n - m)$ will affect that ratio, and will therefore affect the relative width (i.e., r_{outer}/r_{inner}) of the CHZ about a main sequence star.

Results for $(n - m) = 0.92$

The principal series of computer runs which were made employed as $L(t)$ linear interpolations from the data given in columns 2 and 3 of Table 44.1. The results of those runs are exhibited in Table 44.2.[2] It can be seen that the CHZ about a G8 star is very narrow, and that there is no CHZ about main sequence stars later than K0.

The computer simulations indicate that on planets which are *much* closer to the central star than r_{inner} it is always too hot for oceans to condense. On planets having

Table 44.1.

MASS–LUMINOSITY RELATION FOR MAIN-SEQUENCE STARS

M/M_s	L_{ZAMS}/L_s	$L_{4.5\,by}/L_s$	$L_{4.5\,by}/L_{ZAMS}$	SpT
1.20	1.873	2.953	1.577	F7
1.15	1.513	2.294	1.516	F8
1.10	1.210	1.761	1.456	F9
1.05	0.958	1.336	1.395	G0
1.00	0.750	1.000	1.333	G2
0.95	0.580	0.737	1.271	G5
0.90	0.442	0.535	1.210	G8
0.85	0.332	0.381	1.148	K0
0.80	0.245	0.266	1.086	K2

an orbital radius only slightly less than the critical distance, oceans will exist in the early stages of the planet's history; but the buildup of atmospheric gases, combined with the increase in luminosity of the central star, causes a runaway greenhouse to occur after about 10^9 years. It is therefore the stellar luminosity after about 10^9 years which is crucial in determining the inner boundary of the CHZ. Since r_{inner} is roughly proportional to $[L(1.0$ by$)]^{1/2}$, and since $L(1.0$ by$) \propto M^{[m+2/9(n-m)]} = M^{5.22}$, we would expect that $r_{inner} \propto M^{2.61}$. Indeed, the formula

$$r_{inner} = 0.958(M/M_s)^{2.61} \text{ AU} \qquad (1)$$

turns out to be a close approximation to the results listed in column 3 of Table 44.2.

In our computer simulations, Earthlike planets which are situated *considerably* further from their Sun than r_{outer} undergo runaway glaciation as soon as most of the reducing gases are eliminated from their atmospheres, which generally occurs about 2.5 by after the planet is formed. But, for planets situated just beyond r_{outer}, runaway glaciation does not occur until about $t = 3.5$ by. It is therefore the stellar luminosity after about 3.5 by which determines the outer boundary of the CHZ. Since

$$L(3.5 \text{ by}) \propto M^{[m+7/9(n-m)]} = M^{5.74},$$

we would expect that $r_{outer} \propto M^{2.87}$. Indeed, the formula

$$r_{outer} = 1.004(M/M_s)^{2.87} \text{ AU} \qquad (2)$$

is an excellent fit to the computer results given in column 4 of Table 44.2.

For planets situated about stars heavier than the Sun

---

[2]Since the input data fed into the computer simulation is not reliable to three decimal places, it is plain that the results of this table cannot have three-figure accuracy either. The extra decimal places in this table were retained only to enable the reader to clearly see the trend of the results, since the trend itself is physically meaningful. Similar comments apply to the figures in Tables 44.3–44.5.

Table 44.2.

HABITABLE ZONES ABOUT MAIN-SEQUENCE STARS LIGHTER THAN THE SUN[a]

Stellar mass (M/M_S)	Approximate spectral type	Continuously habitable zone		
		r_{inner} (AU)	r_{outer} (AU)	width (AU)
1.00	G2	0.958	1.004	0.046
0.95	G5	0.837	0.867	0.030
0.90	G8	0.728	0.743	0.015
0.85	K0	0.628	0.629	0.001
0.835	K1	0.598	0.598	—

[a] Calculated using $(n - m) = 0.92$ (= best estimate).

Table 44.3.

HABITABLE ZONES ABOUT MAIN-SEQUENCE STARS HEAVIER THAN THE SUN[a]

Stellar mass (M/M_S)	Approximate spectral type	Continuously habitable zone		
		r_{inner} (AU)	r_{outer} (AU)	width (AU)
1.00	G2	0.958	1.004	0.046
1.05	G0	1.086	1.150	0.064
1.10	F9	1.240	1.310	0.069
1.15	F8	1.420	1.481	0.061
1.20	F7	1.616	1.668	0.054

[a] Calculated using $(n - m) = 0.92$ (= best estimate).

(see Table 44.3) runaway glaciation occurs somewhat earlier, and the relationship weakens to $r_{outer} \propto M^{2.78}$.

It might be expected that the CHZ is wider about stars which are slightly more massive than the Sun. Such is indeed the case (see Table 44.3), and the CHZ is quite wide about G0 and F8 stars. But, by the time they are 4 by old, stars of mass $\geq 1.10\,M_s$ are emitting considerable amounts of uv radiation—so much, in fact, that on planets orbiting about them at a moderate distance the uv flux is so great as to severely inhibit the spread of life to dry land. (See discussion in Paper I and in Berkner and Marshall, 1964). For such stars (1) underestimates r_{inner}, and a better fit to the inner boundary of the CHZ is given by

$$r_{uv} = 0.927(M/M_s)^{3.05} \text{ AU.} \qquad (3)$$

In any event, main-sequence stars heavier than about $1.2\,M_s$ evolve so quickly that planets which are near enough to avoid runaway glaciation at $t = 3.5$ by invariably become far too hot by $t = 0.4$ by. If intelligent life takes ~4.5 by to develop, we cannot expect it to develop about such stars. (The formulas used to construct Table 44.1 cannot be used for such stars, since such formulas only approximate main-sequence evolution, and do not approximate evolution into the red-giant stage).

Sensitivity of Results to Value Adopted for $(n - m)$

From the foregoing, it is obvious that the width of the CHZ about a star depends on how rapidly that star evolves. The comparative rates of evolution of a group of stars depends, in turn, on the value of the quantity $(n - m)$. Since, even for a specified initial composition, we do not know $(n - m)$ exactly, it is necessary to see how sensitive our results are to the value of $(n - m)$.

First, let us suppose that $(n - m) = 0.5$, rather than 0.92. Using, for simplicity, the values $m = 5.0, n = 5.5$, we can construct a table (not shown) corresponding to

Table 44.1. A set of computer runs were made using luminosities interpolated from such a table. The results of those runs are shown in Table 44.4. We see that even using this very low estimate of $(n - m)$ the computed CHZ is very narrow about stars with $M \leq 0.85M_s$ (roughly K0 stars), and there is no CHZ at all about stars having $m \leq 0.715M_s$ (K5 stars).

Alternatively, we can suppose that $(n - m) = 1.25$. Using $m = 5.0$, and $n = 6.25$, we can construct still another table (not shown) corresponding to Table 44.1. A set of computer runs were made using luminosities interpolated from that table too, and their results are presented in Table 44.5. Those results indicate that if $(n - m)$ is as high as 1.25, then the CHZ is very narrow even about G8 stars, and does not exist at all for K stars.

We conclude that the results are not extremely sensitive to the exact value used for $(n - m)$. For any plausible value of $(n - m)$ there is no CHZ about M stars and the CHZ about K stars is either very narrow or nonexistent. The CHZ is probably quite narrow about late G stars also, with the exact width depending on the details of the assumptions made.

Sensitivity of Results to Assumptions in Computer Simulation

All of the foregoing results have been obtained by making use of the computer simulation described in Paper I. There are, of course, approximations and uncertainties in any such computer simulation. It is therefore reasonable to ask to what degree defects in that computer program will affect the results of this analysis.

That program, for example, computes the inner radius of the CHZ about the Sun to be 0.958 AU. Suppose that, by not using a sufficiently sophisticated method of calculating the greenhouse effect, the program has overestimated that distance, and that a more accurate figure would be 0.93 AU. [That was the lowest value suggested by

Table 44.4.

HABITABLE ZONES ABOUT MAIN-SEQUENCE STARS[a]

Stellar mass (M/M_s)	Approximate spectral type	Continuously habitable zone		
		r_{inner} (AU)	r_{outer} (AU)	width (AU)
1.20	F7	1.543	1.630	0.087
1.15	F8	1.370	1.454	0.084
1.10	F9	1.221	1.292	0.071
1.05	G0	1.083	1.143	0.060
1.00	G2	0.958	1.004	0.046
0.95	G5	0.840	0.874	0.034
0.90	G8	0.732	0.755	0.023
0.85	K0	0.634	0.649	0.015
0.80	K2	0.542	0.551	0.009
0.75	K4	0.460	0.463	0.003
0.715	K5	0.407	0.407	—

[a] Calculated using $(n - m) = 0.5$ (= low estimate).

Table 44.5.

HABITABLE ZONES ABOUT MAIN-SEQUENCE STARS[a]

Stellar mass (M/M_s)	Approximate spectral type	Continuously habitable zone		
		r_{inner} (AU)	r_{outer} (AU)	width (AU)
1.20	F7	1.668	1.692	0.024
1.15	F8	1.456	1.499	0.043
1.10	F9	1.262	1.321	0.059
1.05	G0	1.089	1.155	0.066
1.00	G2	0.958	1.004	0.046
0.95	G5	0.836	0.861	0.025
0.90	G8	0.726	0.732	0.006
0.871	K0	0.664	0.664	—

[a] Calculated using $(n - m) = 1.25$ (= high estimate).

Rasool and de Bergh [1970]. How should this affect our estimates of r_{inner} about other main sequence stars? Since there is some theoretical justification for the value of the exponent in (1), we might reasonably replace that equation by

$$r_{inner} = 0.93(M/M_s)^{2.61} \text{ AU}. \tag{1a}$$

If we then combine (1a) and (2), we find that $r_{inner} = r_{outer}$ for $M = 0.745 M_s$. That would mean that there is some CHZ (albeit a narrow one) about main sequence stars as late as K4.

If r_{inner} about the Sun should really be 0.912 AU (which would mean that the computer simulation of Paper I has underestimated the width of the CHZ, by a factor of 2) then we might plausibly replace (1) by

$$r_{inner} = 0.912(M/M_s)^{2.61} \text{ AU}. \tag{1b}$$

The combination of (1b) and (2) gives $M_{crit} = 0.691 M_s$, which is about a K6 star. Since even larger errors in the computation of r_{inner} cannot be definitely excluded, the possibility exists that there are CHZ about all main sequence K stars, although much narrower than those about G stars.

Defects in the algorithms for estimating r_{outer} are much less likely to affect the conclusions of this paper. In the first place, the calculations for runaway glaciation are somewhat simpler than those for the runaway greenhouse effect. In the second place, the range of possible values of r_{outer} about the sun is severely limited. Most climate models of the Earth indicate that even today a reduction of more than 4% in the solar constant (possibly much less) would cause runaway glaciation (Budyko, 1969; Sellers, 1969; Schneider and Gal-Chen, 1975).[3] It follows that r_{outer} is probably no greater than 1.020. Replacing (2) by

$$r_{outer} = 1.020(M/M_s)^{2.87} \text{ AU} \tag{2b}$$

gives—when combined with (1)—$M_{crit} = 0.786 M_s$. That would allow a narrow CHZ about K2 stars, but none about stars later than K3. This would be only a slight modification of the conclusions stated in Section 3.

Another important assumption is that the planet in question has the same mass and radius as the Earth. Since many of the processes involved depend on the size of the planet (for example, a larger planet might well have a thicker atmosphere and therefore a larger greenhouse effect) it would be useful to know how sensitive the results are to the planet size assumed. Research now in progress by the author indicates that the numbers in Tables 44.2–44.5 are surprisingly sensitive to planet size, although the trends illustrated in columns 3–5 still hold for planets somewhat larger or smaller than the Earth.

Summary and Discussion

Our calculations indicate that the *continuously* habitable zone is fairly narrow, even about G2 stars, and is even narrower about later G stars. There is probably no CHZ about most K and M stars.

These results are fairly insensitive to the initial composition assumed for the central stars, unless it is so extreme as to make $(n - m)$ significantly less than 0.5. That might possibly be the case for very metal-rich stars, having $Z > 0.04$ (see Tinsley, 1976).

The results of the analysis do not seem to be very sensitive to the minor details of the computer simulation. The results do, however, depend on the overall picture

---

[3] Of course, those climate models may not be correct. Currently available climate models still have feedback problems associated with them which cause their results to be uncertain.

which that computer simulation gives of the evolution of our atmosphere, to wit:

(a) The early atmosphere of the earth contained large quantities of reducing gases such as H_2, CO, or CH_4.

(b) Oxygen released by photosynthesis converted that atmosphere into an oxidizing one.

(c) The crossover occurred about 2 billion years ago.

Such a sequence of events seems quite plausible.

Of course, the habitable zone is wider about GO and late F stars. But stars earlier than F7 become red giants in less than 4 by, so it is unlikely that an advanced civilization could originate about such stars. It appears, therefore, that there are probably fewer planets in our galaxy suitable for the evolution of advanced civilizations than has previously been thought.

Acknowledgment

Much of this research was conducted while the author held an NRC Resident Research Associateship at the Laboratory for Planetary Atmospheres in NASA's Goddard Space Flight Center.

45

John A. Ball
The Zoo Hypothesis

The most interesting scientific problem of our age involves the question of the existence of extraterrestrial intelligent life. Arguments summarized below make it likely that intelligence exists on many planets throughout our galaxy and that most of these civilizations are much older than our own. This problem has been the subject of considerable work both theoretical and experimental (see Oparin and Fesenkov, 1960; Cameron, 1963a; Shklovskii and Sagan, 1966; Sagan, 1973a; and other references therein) and our understanding of the subject has certainly progressed rapidly in the last decade or so. However, this problem has proved to be extremely difficult, in part because it involves understanding what a civilization much older than ours might be like. It is difficult enough to predict our own development for a few decades hence, but we need to know about other civilizations that may be older than ours not by decades but by eons.

Among currently popular ideas about extraterrestrial intelligence, the idea that "they" are trying to talk to us has many adherents (see, e.g., Drake, Reading 24). This idea seems to me to be unlikely to be correct and the zoo hypothesis is in fact the antithesis of this idea.

Starting Premises

Three working hypotheses or starting premises are used in most discussions of the problem of extraterrestrial life. These premises are stated below with a discussion of their origin and references to the literature. Although this discussion is brief, these premises are in fact crucial and if any of them proves to be incorrect, then the zoo hypothesis falls.

A. Whenever the conditions are such that life can exist and evolve, it will. Life is to be understood as a chemical reaction that occurs whenever the necessary reactants are present under the appropriate conditions for a sufficient

time. This statement represents a considerable extrapolation of our present knowledge. In fact the opposite hypothesis, that life is statistically unlikely even in ideal conditions, has been expressed (e.g., by Townes, 1971). Discovery of primitive life on Mars or Venus would probably settle this question. Our current understanding of biochemistry seems to support premise A (Shklovskii and Sagan, 1966, Chapter 14; and Calvin, 1961).

B. There are many places where life can exist. Planets are probably quite common in the universe. As many as 20% of all stars may have planets and as many as 10% of these planets may have surfaces on which life can form. (However Oparin and Fesenkov, 1960, think that only one star in 10^5 or 10^6 has a planet with a surface suitable for life. See also von Hoerner, 1961). This statement also represents more than we know at present; no star other than our sun is definitely known to have planets comparable to the earth. Objects that may be planets have been detected around a few other stars (see Shklovskii and Sagan, 1966, Chapter 11; Huang, 1959b; and van de Kamp, 1969b), however these objects are much more massive than the earth. Planets comparable to the earth around almost any other star would go undetected with present techniques. The opposite hypothesis, that the solar system is unique, was believed by Jeans (1929, Chapter XVI), but is now discredited (see, e.g., Levin, 1964, for a summary of current thinking).

C. We are unaware of "them."

Who is Out There?

It is statistically unlikely that there exists anywhere in our whole galaxy any other civilization whose level of development is at all comparable to ours. We would expect to find either primitive life forms, perhaps comparable to those on the earth a few million years ago, or very advanced life forms, perhaps comparable to what will be on earth a few million years hence(!)

There are three general categories of possibilities defin-

Reprinted from *Icarus*, 19 (1973), 347. Copyright © 1973 by Academic Press, Inc.

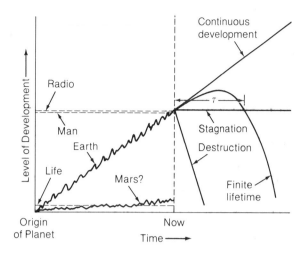

Figure 45.1. This is a sketch of the top level of development, defined in terms of complexity, versatility, and ability to control the environment, either of the organism itself or of the civilization to which it belongs. The various possible extrapolations for our future are discussed in the text.

ing the technological evolution of a civilization:

(1) Destruction (from within or without).

(2) Technological stagnation.

(3) Quasi-continuous technological progress.

Also there are many other mixed possibilities, such as partial destruction and rebuilding, and the surprisingly popular finite-lifetime idea. These possibilities are sketched diagramatically in Figure 45.1 with specific reference to our own extrapolated future. It is likely that some fraction of all civilizations follow each of these possibilities. However, analogy with civilizations on earth indicates that most of those civilizations that are behind in technological development would eventually be engulfed and destroyed, tamed, or perhaps assimilated. So, generally speaking, we need consider only the most technologically advanced civilizations because they will be, in some sense, in control of the universe.

Technological progress may be defined as increasing ability to control one's environment. Already at our level of technology we affect almost everything on earth from elephants to viruses. But we do not always exert the power we possess. Occasionally we set aside wilderness areas, wildlife sanctuaries, or zoos in which other species (or other civilizations) are allowed to develop naturally, i.e., interacting very little with man. The perfect zoo (or wilderness area or sanctuary) would be one in which the fauna inside do not interact with, and are unaware of, their zookeepers.

The Zoo Hypothesis

Premise C above now seems to me to be extremely significant. I believe that the only way we can understand the apparent non-interaction between "them" and us is to hypothesize that they are deliberately avoiding interaction and that they have set aside the area in which we live as a zoo.

The zoo hypothesis predicts that we shall never find them because they do not want to be found and they have the technological ability to insure this. Thus this hypothesis is falsifiable, but not, in principle, confirmable by future observations.

Conclusions

The zoo hypothesis as given here is probably flawed and incomplete. I hope that it can provide some sort of inspiration for further work. Among other hypotheses that one might consider, the laboratory hypothesis is one of the more morbid and grotesque. We may be in an artificial laboratory situation. However, this hypothesis is outside the purview of science because it leads nowhere, it immediately calls into question the premises on which it is based, and it makes no predictions. Or one might suppose that extraterrestrial civilizations have not yet found us or that they know we are here but they are uninterested in us. These latter two hypotheses are probably incompatible with the high level of technological sophistication they undoubtedly possess.

The zoo hypothesis seems to me to be pessimistic and psychologically unpleasant. It would be more pleasant to believe that they want to talk with us, or that they would want to talk with us if they knew that we are here. However the history of science contains numerous examples of psychologically unpleasant hypotheses that turned out to be correct.

Acknowledgments

Although the ideas in this paper are not in the mainstream of current scientific thought about the problem, they are also not new. Science-fiction authors, in particular have toyed with similar notions for many years. And at least a few previous writers have suggested such ideas as a serious possibility.

I thank Sebastian von Hoerner and Mrs. Lyle G. Boyd for pointing out relevant background material and for stimulating discussions. I am grateful to Prof. A. E. Lilley for his encouragement.

46

Michael Papagiannis
Are We Alone, or Could They be in the Asteroid Belt?

The observations that life has a natural tendency to expand into all available space, that advanced technological civilizations should be able to engage with relative ease in interstellar travelling, and that once this threshold is crossed the complete colonization of the entire Galaxy will be accomplished in a very short interval relative to the age of the Galaxy, lead us to the following dilemma: either the entire Galaxy is teeming with intelligent life and hence our solar system must have been colonized hundreds of millions of years ago, or there are no other inhabitants in our solar system and hence most probably neither anywhere else in the Galaxy. Before accepting, however, the bleak verdict that we are all alone in the Galaxy, we must search carefully throughout the solar system for any signs of other technological civilizations. The most logical place to look for them seems to be the asteroid belt because of the many advantages it offers to a galactic society living in space colonies.

A Change of Heart

The euphoric optimism of the sixties and the early seventies that communication with extraterrestrial civilizations seemed quite possible (Sagan, 1973a), is being slowly replaced in the last couple of years (Hart, Reading 42; Jones, 1976; Shklovskii, 1977) by a pessimistic acceptance that we might be the only technological civilization in the entire Galaxy. This change of heart has been happening as a result of the following observations:

(1) Life seems to possess a natural tendency to expand like a gas to occupy all available space. This is evident, e.g., when algae rapidly take over an unattended swimming pool, and certainly has been the characteristic of man who after conquering the entire planet is now ready for new ventures in outer space.

(2) Interstellar travelling seems easily attainable,

especially in the $0.01-0.1\ c$ range, by advanced technological societies. The building of permanent colonies in space, as envisioned by O'Neill (Reading 56), will make it possible for people living all their lives in these colonies to disengage emotionally from the mother planet. Such colonies would have the emotional strength and coherence to undertake trips to the nearby stars that would last for several generations. Nuclear fusion, with a $0.007\ mc^2$ yield, can become a most attractive energy source for such trips even at an efficiency ϵ as low as $10-20$ per cent. As seen from the relation

$$\tfrac{1}{2}\ MV^2 = 0.007\ \epsilon mc^2 \qquad (1)$$

a spaceship of mass M will be able to reach speeds $V = 1-3 \times 10^{-2}c$ with a fuel load m not larger than M.

(3) In the last 100 yr or so, the velocities of long, non-stop voyages (trains to spaceships) have increased from $\sim 10^3$ to $\sim 10^6$ cm/s. It seems reasonable, therefore, to anticipate an additional increase by a factor of 300–1000 in the next 100–200 yr, especially with the use of nuclear fusion. One can be optimistic also about self-sufficient space colonies, which according to the computations of O'Neill could be a reality even before the turn of the century. It appears, therefore, that with steady technological progress and without the need of any new major discoveries, we should be able to undertake stellar missions in a few centuries. This is an interval in cosmic terms as brief as a few minutes in the life of a man, which means that our civilization is extremely close to this critical moment.

(4) Once the threshold of interstellar travelling is crossed the entire Galaxy will be colonized in only a few million years, which is a very short period relative to the 10–15 billion year age of our Galaxy. Even by assuming an expansion rate of one light year per century (say 500 yr for a 10 lt yr trip to a suitable nearby star, and 500 yr for the building of the new colony before it can undertake further stellar missions), we see that the entire Galaxy can be

Reprinted with permission from the *Quarterly Journal of the Royal Astronomical Society,* 19 (1978), 277.

conquered in less than 10 M yr. It is also clear that as new colonies join the colonization wave, there will be so many interstellar travellers that there will be no reasonable place in the Galaxy that will remain unoccupied.

(5) The many attractive features of our solar system (a single, well-behaved, long-lasting, hot star surrounded by a multitude of diverse planets, moons and asteroids) could have not been missed by the colonizers, and therefore our solar system could not have been bypassed as the colonization wave swept through the Galaxy.

(6) The likelihood is that the extraterrestrial colonizers of our solar system, especially after their long interstellar voyages, will have become accustomed to space living. As a result, not only will they not need a habitable planet to settle on, but most probably they would prefer to continue living in space colonies. The orbits of their choice would obviously be these that provide the most efficient access to material and energy resources.

From the above it follows that if hundreds of millions of intelligent civilizations did evolve in our Galaxy over the past several billion years, as suggested by the integration of the Drake–Sagan probability formula over the entire history of the Galaxy (Freeman & Lampton, 1975), then it seems inevitable that some of these galactic civilizations would have achieved interstellar travelling and the whole Galaxy, including our solar system, would have been teeming with advanced technological societies. Conversely, if we are the only technological inhabitants of our solar system, then most likely we are also the only ones of the entire Galaxy. This deduction implies that the values commonly used for one or more of the probability factors of the Drake formula (Shklovskii & Sagan, 1966; Kreifeldt, 1971; Sagan, 1973a; Oliver, Chapter 35) must have been grossly over-estimated (Papagiannis, 1978b).

Could They Be Around?

We have reached, therefore, the stage where the acid test for our dilemma seems to be whether or not our solar system is inhabited by an advanced extraterrestrial society. Of course there are in the literature several reports of UFO sightings and even stories of dramatic encounters with extraterrestrials. There are also several popular books, such as those of von Däniken (1969), in which the intervention of extraterrestrials in this planet is envisioned on countless occasions. Still, however, there is no convincing proof to any of these stories or suggestions; in accordance with the Shklovskii principle that 'all events should be considered natural unless proven otherwise' the scientific community remains unconvinced about visits to Earth by extraterrestrials. As a result, and in accordance with the presently available evidence, we tend to believe that we are the only advanced civilization inhabiting our solar system.

Absence of evidence, however, should not be taken as evidence of absence. Before we resign therefore, to a pessimistic acquiescence that we might be the only technological inhabitants of our Galaxy, we have the responsibility to search exhaustively in our solar system for other advanced societies. The supposition that we are alone in the solar system is based essentially on the assumption that if others were here they would have already made contact with us, or at least we would have become aware of their existence. Neither of these assumptions, however, is necessarily true, though it is possible that some of the thousands of UFO sightings might deserve some further consideration, as suggested by Hynek (1972).

The most intriguing question in the whole problem is the following: if our solar system is indeed inhabited by extraterrestrials, where are they most likely to be found? In earlier days people had tried to identify one of the other planets of our solar system, most frequently Mars (Lowell, 1908), as the abode of an extraterrestrial society. From the above discussion, however, it follows that the colonizers of our solar system are likely to continue to live in space colonies, probably at reasonably close distance to the Sun so as to have a sufficient supply of solar energy for their needs, and most likely near celestial bodies of weak gravity from which they would obtain all the natural materials needed for the continuous prosperity of their civilization.

The Asteroid Belt Choice

Within this framework, it seems that the asteroid belt would be an ideal place for the extraterrestrials to set up their space colonies. Not only would they have an easy access to all natural resources by mining the asteroids, but they would also be close enough to the Sun to have ample solar energy for their needs. Some years ago this suggestion would have sounded unreasonable, mainly because we used to think that the asteroid belt must be full of debris which can be very hazardous for any spaceships permanently stationed in their vicinity. The *Pioneer 10* and *11* missions, however, (Kinard et al., 1974) have found that the density of meteoroids in the asteroid belt hardly differs from any other place in the solar system, and therefore the colonization of the asteroid belt seems quite feasible. There are, of course, also the Kirkwood gaps in the asteroid belt which are practically free of any asteroids. A spaceship could remain there almost indefinitely by simply compensating against the tidal effects of Jupiter with its own propulsion system.

One can even consider the possibility that the large fragmentation of the components of the asteroid belt might be the result of mining projects by the extraterrestrial colonies. It is even conceivable that they have tried to keep the region clean of free floating debris for their safety. The identification of space colonies 1–10 km in size hidden in the asteroid belt would not be an easy task for a terrestrial observer because from a long distance these colonies would be practically indistinguishable from the thousands of natural asteroids. They would also follow the same orbits around the Sun as the asteroids, which would be much more numerous and therefore it would be almost like searching for a needle in a haystack. Still with careful observations in the radio domain we might be able to detect some leakage of radio noise, infrared observations might reveal a higher effective temperature than that which is justified by their distance from the Sun and finally, properly planned space missions to the asteroid belt might do some successful eavesdropping and might even return some direct photographic evidence. The search project, therefore, though admittedly quite difficult, is still within the capabilities of our present technology and in view of the far-reaching consequences of either positive or negative results, should be given a serious consideration.

Why Are They Silent?

As to why they have not yet made contact with us, one can think of several answers, including the zoo hypothesis of Ball (Reading 45). The simplest explanation, however, and hence maybe the most probable one, might be that of confusion and indecision. Our hypothetical neighbours were probably acquainted for millions of years with a lethargic Earth inhabited by life forms not worth any effort of communication. Suddenly, in the last 50 yr or so, which probably is a very short interval for a well-settled galactic society, they have been confronted with an exponentially mushrooming technological society (aeroplanes, radio-communications, nuclear bombs, spacecraft) which undoubtedly must be causing them some serious concern. It is possible, however, that faced with such a sudden technological explosion, a serene cosmic civilization would be perplexed and undecided as to how to handle the situation. They might be debating on whether to crush us or to help us, and therefore they might be simply postponing their decision, waiting to see what we are going to do with ourselves. Meanwhile, the asteroid belt provides a natural hide-out where they can remain inconspicuous for a long time until we decide to search for them.

Conclusions

In conclusion, though the idea that the asteroid belt might be harbouring a number of extraterrestrial colonies sounds like science fiction, the arguments presented above suggest that *if* there are any extraterrestrial colonies in our solar system then the asteroid belt seems to be the most logical place to look for them. Before accepting, therefore, the bleak verdict that we are all alone in the Galaxy, we have an obligation carefully to investigate this possibility, remote as it might seem. ●

47

David G. Stephenson
Extraterrestrial Cultures within the Solar System?

Introduction

Recently Papagiannis (Reading 46) writing in this journal proposed that, if cultures with the capability of interstellar travel exist, then the most probable location for their colonies within the solar system would be amongst the asteroid belt. This paper examines the economies of such colonies in an effort to clarify the possiblities of finding traces of interstellar voyagers within the solar system.

Fermi's Paradox

Papagiannis lists a series of observations as the currently accepted parameters influencing the development of cultures capable of interstellar flight, and their subsequent spread through the Galaxy. At first sight these observations would seem to indicate that interstellar activity is a probable continuation of the development of technically-able cultures paralleling our own. Thus either Hart's (Reading 42) thesis holds, that we are the only culture that has arisen which may develop the ability to cross interstellar space, or that, as Stephenson (1977) proposes, any previous or current visitors to our neighbourhood have taken precautions to ensure we are not aware of them. Otherwise this planetary system should abound in ultra-solar visitors and residents. If the concept of mankind as a unique being is seen to run counter to the modern philosophy and experience of science Fermi's paradox becomes apparent . . . 'Where is everybody?'

Implications of the Current Observations

If the implications of the observations listed by Papagiannis are examined it is found that they are inconsistent. Observation 1 is that life expands to fill all available spaces

Reprinted with permission from the *Quarterly Journal of the Royal Astronomical Society,* 20 (1979), 422.

on Earth, but against this is observation 2 that the interstellar voyagers must be able to 'have the emotional strength and coherence to undertake voyages to nearby stars'. The velocities available for interstellar travel being only a small fraction of the velocity of light the interstellar vessel would probably have to be an ark in which at least one generation of the inhabitants live during the voyage. Only a culture of intelligent individuals that has overcome the natural drive implied by the first observation would be able to do this without a population crisis during the voyage. If this control is available to the colonists for the voyage it would still be there at the destination and therefore there would be no pressure to expand into the planetary system. To interstellar voyagers the structure of a stable population and economy would be as familiar, and more secure, as the human culture based on continuous expansion is to us. Thus the constraints placed on a culture if it is to journey between the stars may also limit its ability to colonize planetary systems. Whenever the expansion of an intelligent species is being considered it must be borne in mind that it is less than 200 years since the hygienic and medical technology was developed to give Man the luxury of restricting his reproductive capacity without the risk of famine or epidemic devastating his numbers. The great colonizing drives of human history have been for the most part the result of a continually changing series of population and resource imbalances as the human population was matched against the destructive consequences of its environment. To use this history as a model for the activities of a culture that must have a fine control of its reproductive capacity would seem to be unwise.

As Papagiannis states, interstellar travel demands the use of a prime mover that is the equivalent of a reactor fusing light atoms with an efficiency of at least 10 per cent. As was pointed out by Stephenson (1977), such a prime mover could be used to dismantle and reconstruct many of the bodies of a planetary system such as ours. The kinetic energy of one kilogram of mass travelling at 0.128 of the velocity of light could ionize 45,000 tons of water ice in deep space. With energies such as these, interstellar cultures could fabricate all their needs from crude matter, that would only have a value with respect to its atomic constituents. During the voyage the prime source of power for the colony would have to supply the life-sustaining light and heat for the voyagers since an interstellar ark would have to be completely self-contained during most of its journey when it is far from any stellar primary. If the colony is adapted to an independent life in interstellar space, the question must be asked as to why such a colony would wish to render itself dependent on a stellar primary, no matter how stable. To an arriving interstellar vessel the environment within the solar system would be different from that to which it is adapted, and would seem alien at best and hostile at worst. Indeed, interstellar craft might be constructed of low-temperature materials that would be seriously disrupted by the solar radiation within the boundaries of the solar system. To a culture that has a stable population the reduction of the genetic uncertainties generated by cosmic radiation would be of vital importance. With its proton storms, X-rays and solar wind the inner regions of the solar system would appear distinctly poisonous.

To assume that a culture that is able to sustain itself in interstellar space for long periods would voluntarily huddle like some primitive planet-borne tribe around a stellar campfire, is like proposing that twentieth-century man could not live outside the tropics unless there was a local source of geothermal heat. Travel from one volcano to another would require that a man carry a chunk of hot lava of sufficient size to prevent injurious exposure during the journey. Thus it can be seen that Papagiannis' points 5 and 6 may be inconsistent. An interstellar culture, with the whole Galaxy to journey in, would find our solar system of little long-term interest. This appearance might be modified by two factors, however: firstly, the presence of a planet teeming with life and a nascent intelligence might give the system a special value, Stephenson (1977). Secondly, the concentration of mass that this system represents might be attractive to a vessel wishing to refurbish, or possibly replicate itself. This latter point will be considered later in this paper.

The first sub-human hominids appeared on Earth only the order of 10^6 years ago, the first true men 10^5 years ago and civilized man is less than 10^4 years old. Unless one assumes that leaving the surface of a mother planet eliminates all social and genetic evolutionary pressures, then the 10^7 years needed to fill the Galaxy (Papagiannis, observation No. 4), implies that the final inheritors of the Galaxy will be vastly different from the first colonists. In all probability the colonists will adapt to the environment in which they spend most time, interstellar space, and may not be able to enter a planetary system such as ours. To impose a time span on intelligent beings longer than the total span of humanity on this planet, and then to require them to cease any social or genetic development is surely an unsound procedure. Today we are already selecting those of us who will function and later live in outer space with great care so that they will best adapt to the alien environment. Is this the first unconscious step towards the ultimate evolution of a possible *homo astris*?

The Economics of an Interstellar Colony

Unlike the current state of human economics in which there is a complex series of interactions between our distinctly limited energy resources and seemingly limitless demands, an interstellar culture would be able to expend the prodigious energies needed to synthesize all their needs from raw cosmic matter. The value of a particular accumulation of matter to the interstellar colony would depend on the atomic abundances within it. The probable primary purposes of various elements within the fusion-based economy are listed in Table 47.1.

From the table it will be seen that the demands for elements corresponds closely to the relative abundances of those elements in the Universe at large. An interstellar colony would have resource demands that are the opposite of our planetary surface economy. Rather than seeking ores in which fractioning of elements has been high, a colony would prefer cosmic matter in as undifferentiated state as possible. In particular, the colony would have only small demands for the heavy atoms that are abundant in the inner solar system.

Although Papagiannis refers to the Sun as a common star, a glance at a list of the nearest stars reveals that by far the most common stellar type is the red dwarf M stars, and G stars like our own Sun are quite unusual. It is tempting to add that there might be even smaller, cooler bodies, that we as yet are unable to locate, present in even greater numbers than the red dwarfs. Little is known about the regions close to these low-temperature stellar bodies, but probably the lack of light and heat would make them more attractive to an interstellar vessel than our system. If stellar radiation is a primary cause of elemental fractioning in planetary systems then the matter around a red dwarf would be closer to the raw state, and therefore again the red dwarf would be a more probable destination. Finally, some of the red dwarfs, such as Barnard's Star, are Population II stars and therefore may have a preponderance of the light elements used as fuel by an interstellar colony.

A Colony in this Solar System?

If an interstellar vessel entered this solar system its first need would be to replenish the light atoms that fuel its prime mover. This would be accentuated by the energy demands during the deceleration into the solar system.

Table 47.1.

USE OF ELEMENTS IN AN INTERSTELLAR CULTURE

Atomic weight	Examples	Purpose	Use[a]	Cycle losses	Demand	Cosmic abundance (percent)	Availability in solar system
Light	H,He,Li,Be,B	Fuel	Single Pass	Consumption	Heavy	+95	Outer planets, Earth's crust Ices on outer moons
Medium	O,C,N,Cl	Life	Recycling	Flushing[b] Dynamic losses	Medium	+1	Outer planets Carbonaceous chondrites (Asteroids?) Earth, Venus crust
Heavy	Fe,Ni,Co, Cu,Ag,Au	Structure	Static	Flushing Erosion Wear	Light (repair and growth)	Trace	Planetary cores Rocky asteroids Inner planets

[a] These are only the primary purposes of the atoms listed, secondary uses such as light atomic ices being used for structure and liquids (water) in the life cycle are probable but should not make significant extra demands for these atoms.

[b] Flushing refers to the process, probably continuous, by which life and structure atoms which have become radio-active are removed to prevent genetic damage in the bio-mass of the colony.

The light elements are most abundant in the region of the gas giant planets and their satellites. To a culture adapted to long periods in interstellar space the most familiar region of the solar system is its outer reaches, where Neptune and Triton are probably rich in light atoms. Having replenished its supplies the colony would desire to return to the environment it knows best, interstellar space. The elements outside the fuel requirements would probably be made available during the collection of fuel as by-products, and under conditions of replenishment there would be no need for a colony to penetrate the solar system.

The only condition that might encourage an interstellar colony to penetrate deeper into the solar system would be a demand for life and structural materials while the colony is expanding or replicating itself. The highest concentration of life atoms in the solar system outside the giant planets themselves is in the Earth's crust or in the carbonaceous chrondrite-like bodies of the solar system. It appears that a considerable proportion of the asteroids exhibit the low albedos of carbonaceous chondrites and forage parties from extraterrestrial colonies might be found in the asteroid belt if a new colony were being constructed. Until the asteroid belt had been denuded of carbonaceous matter there would be no drive to penetrate into the inner solar system, which is rich only in heavy structural elements for which there is least need in an interstellar colony.

The above arguments indicate that the expectation of discovering an extraterrestrial presence in the asteroid belt depends crucially on the model chosen to represent the constructive cycle within an interstellar culture.

The Replication of Extraterrestrial Colonies

As yet mankind has had no direct experience of being part of an interstellar culture, therefore those attempting to estimate the expansion of such cultures have had to assume models based on the activities of human and lower organisms on Earth. So far most authors have followed a model that assumes that interstellar colonies will have a life cycle paralleling that of a virus (see Hart (Reading 42), Cox (Reading 43), Molton (1978), Jones (1978)). In this model the organism, be it a virus or an interstellar vessel, travels through an alien medium with most of the information stored within it in a dormant state and being irrelevant to the surrounding conditions. Upon reaching a suitable host, be it cell or planetary system, the information within the organism is activated, now that conditions render it relevant, to spread through the host and convert it to the creation of new vessels to carry the information to new hosts. The discrepancies between the results of the use of this model come from differing estimates of the time of travel, the reproduction rate and number of organisms released before the host is depleted. As the part of the replication cycle that has no direct replicative result the mobile form of the organism is kept as simple as possible as the information needed to construct it must be minimized to increase the efficiency of the in-host replicative process. A human equivalent of this model is the nineteenth-century colonization of North America by Europeans. During the voyage across the Atlantic, which was a completely alien environment to most of the passengers, the immigrants represented a store of genetic and mechanical information that was of no value in the conduct of the voyage, but upon reaching the New World were rapidly deployed in modifying the existing conditions to parallel the conditions at their departure point.

There is an alternative model that could be applied to interstellar vessels. In this case the vessel parallels a higher organism that is continually in a state of activity with most of its inherent information relevant and in use. This allows the organism to repair any damage to itself, to grow slowly and to replicate as conditions permit, and above all to adapt itself to the conditions in which it is found. This is an essentially independent mode of existence, without the need of a host, be it cell or star, during some stage of its cycle. Replication uses the same information as is normally operative for growth and repair. The information is simply duplicated and the material of the organism is split between the daughters. Such an organism represents a more efficient use of information than a virus and may be more suited to the primordial environment of interstellar space. A human analogy would be a nomadic tribe that has adapted to an environment that provides its normal needs as long as the tribe is mobile. Unlike the previous model this tribe is continually using its knowledge to survive and develop in the environment in which it resides. Only on rare occasions would special knowledge that is usually dormant be called for.

The process of refining the accepted models for the behaviour of an advanced interstellar culture must continue as the search for alien intelligence advances. Of the two models presented here Fermi's paradox seems to question the virus model of interstellar expansion. Indeed, the question of numerical interstellar expansion may have to be examined carefully, since a cultural unit that has balanced its population budget may not wish to replicate. Once launched into interstellar space, colonies of intelligent beings may continue to journey forth coming close to stars at rare intervals for replenishment in the shortest possible time. There may be thousands of interstellar vessels voyaging between the stars, each the result of a mother culture centred on some far-distant star. Having been constructed, the vessels themselves may not give rise to extra craft.

Conclusion

The structure of an advanced interstellar culture must be far different from any that has existed on Earth. Unless great care is taken simplistic models of such cultures will exhibit paradoxes such as Fermi's. If it is hoped to discover the traces of an advanced culture in this solar system it would seem logical to examine those regions that are most like the zone in which an interstellar vessel would spend most time, i.e., the search should concentrate on the outermost bodies of the solar system as being the closest to the interstellar space. It is an intriguing speculation that the unusual orbit and planetary parameters of Pluto might, in part, be due to the activities of visiting interstellar craft that have reduced the body from a conventional outer planet to a cosmic slag tip of unwanted elements.

Acknowledgments

The author wishes to thank the members of the Atmospheric Dynamic Group of the Institute of Space and Atmospheric Studies of the University of Saskatchewan for their criticism and assistance in preparing this paper.

48

Sebastian von Hoerner
Where Is Everybody?

How frequently may intelligence and technology have developed on the planets of other far-away stars, and what are our chances of detecting any radio signals from these alien and mostly much older races? I seriously think that trying to establish contact with other beings in the universe is our next great task, and that success would mean the largest step in the evolution of mankind since our development of speech about a million years ago. And meanwhile, as a fringe benefit, meditating about life in space may give us a helpful distance and a better perspective for looking at our own terrestrial affairs and problems.

On a small scale, we have already started SETI, the *S*earch for *E*xtra*t*errestrial *I*ntelligence; and very large future SETI projects are being prepared and could be undertaken. But what chances do we have? Previous estimates have mostly given quite optimistic results, but during the last years some rather pessimistic arguments have turned up, which are quite puzzling and hard to beat. In the following, I will emphasize this pessimistic puzzle, hoping that someone else will come up with a convincing optimistic solution.

Assuming Us to be Typical

First, I will give a short summary of the previous estimates for SETI. The only thing we can go by, the only case of life we know, is our own life here on Earth. We should not assume to be anything special. Our Sun is just an average star, one out of $200 \cdot 10^9$ which make up our stellar system, the Galaxy. And from our theories of star formation (von Hoerner, 1975b) we think that planets are quite common, formed together with their central star from a contracting cloud of gas and dust. Sun and Earth are about $5 \cdot 10^9$ years old.

We know that our Sun is not an old star, and that most

stars are about $5 \cdot 10^9$ years older because star formation was about 30 times more productive at the beginning than it is now. The very first stars will have had no solid Earth-like planets, because the original big-bang of the universe provided only hydrogen and helium, while all the heavy elements needed for solid planets are produced by stellar evolution and supernova explosions. But still we expect a very large number of habitable planets similar to Earth, most of them being about $4 \cdot 10^9$ years older than Earth.

Regarding life, we should assume similar developments under similar conditions. Which means that life and intelligence have developed wherever a planet similar enough to our Earth was going in the right orbit at the proper distance around a star, similar to our Sun and old enough. Many scientists have estimated how frequently it happens that all conditions for life are fulfilled (Cameron, 1963a; Sagan, 1973a; Bracewell, 1974; Dole, 1970). Within a factor of ten, say, the results mostly state that about 0.5% of all stars should have a planet where life similar to our own could be expected. Our Galaxy then would have about 10^9 of such habitable planets, and our nearest neighbors would be about 20 light years away. If our estimate of the percentage is wrong by a factor of ten, then the number of planets is wrong by a factor of ten, but the distance is only wrong by a factor of two ($10^{1/3} = 2.15$). This then would be the distance we had to travel in future spaceships, in search for habitable planets and for higher life.

If we want to communicate by some technical means, however, for example by radio waves, then our partner must be in a comparable state of mind, in order to use similar means and to understand each other. We should not assume that our present state of mind is the final goal of all evolution; it will be surpassed by completely different interests and activities which we cannot guess at our present state. Unfortunately, we have nothing to go by for estimating the duration of a "comparable technical state," because we are mere beginners. Using a wild guess of a 100,000 years, say, this would yield about 10,000 such

Reprinted from *Naturwissenschaften*, 65 (1978), 553. Copyright © 1978 by *Die Naturwissenschaften*.

civilizations in our Galaxy, and the nearest partners would be almost 1000 light years away from us. If our guess is wrong by a factor of ten, then the distance is only wrong by a factor of two.

What do these distances practically mean? Regarding radio contact, we could assume, for example, the same technical effort on both sides. If we as well as our partner would use equipment like our existing telescope at Arecibo in Puerto Rico, with its present best transmitter and receiver, we could already talk to a partner 40 light years away (provided we had agreed, before that, on the exact frequency and a narrow bandwidth, and knowing already at which star to point the telescope and when). And even a distance of 1000 light years could be bridged if we design a transmitter and receiver for this very purpose, which might take about three years and a million dollars.

Of course, guessing the right frequency channel, being lucky with the right time and finding a partner is a lot more difficult and time-consuming. Either we must assume that strong radio beacons are aimed at us, or we must build a large array of many big telescopes.

The distances also mean a very long waiting time for answers. No signal can go faster than light, and if our partner is 1000 light years away, then the answer to any question will take 2000 years. Not individuals, only whole civilizations can talk to each other. A two-way communication, with questions and answers, is only possible if the general time-scale of development is longer than the waiting time, so that there still is interest in the answer to a question asked so long ago. Our own hectic development is a lot faster than that, which would leave only a one-way communication (being given a wealth of general information without having asked for it). But this should never be underestimated: our whole western culture has been strongly influenced by the ancient Greeks from whom we got nothing but a few books and pieces of art in a one-way communication. Actually, all the traditions and cultural values we inherit from our ancestors are one-way communications. And our interstellar partners will be tremendously far ahead of us since we are mere beginners, and most stars are older than our Sun.

Regarding space travel, nothing can move faster than light (except in science fiction). From basic physical considerations one can show (von Hoerner, Reading 37) that space travel will probably be limited to $1/10$ the speed of light, $v \leqslant c/10$. Relativistic time effects then play no role. With our presently known technology and a good financial effort (Dyson, 1968) we could already achieve $v = c/100$ or maybe $c/30$. Thus, even a trip to the nearest habitable planets would take at least 200 years, with $v = c/10$ and a distance of 20 lightyears, much longer than the lifetime of a crew. Signals then look a lot more promising than travel, and radio waves between 1 cm and 30 cm wavelength seem most economical because there the background noise is smallest.

These promising estimates have led to several SETI searches for radio signals from outer space, beginning with Frank Drake's "Project Ozma" in 1960 at Green Bank (Cameron, 1963a). Meanwhile, about a dozen of searches have been done, with better equipment and new techniques, but all with existing equipment and no extra cost except the telescope time. Some hundred nearby solar-type stars have been occasionally searched, some star clusters and even a few other galaxies. So far, without any success.

The opposite SETI approach has been taken since 1970 by a group at Ames Research Center of NASA (and at JPL at Pasadena), called "Project Cyclops" (Oliver and Billingham, 1972): to find out what kind of equipment we should develop and build in order to have a really good chance of success within, say, 30 years of a dedicated search. The answer is a growing array of up to 1500 telescopes, 100 meter diameter each, with special receivers and correlators (doing normal radio astronomy about $1/10$ of the time). One would start with one such telescope and gradually build up more, searching all the time, and would stop building up whenever success is achieved. The total cost of all 1500 telescopes and receivers would be about 10^{10} dollars. This sounds ridiculously expensive; on the other side, this is just three months of what we actually did pay for the war in Vietnam.

If We Were Typical, We Should Not Exist

Second, I will describe a puzzling line of thought. The basic assumption of our previous estimates, that we on Earth are about average and nothing special, seems to lead to a serious contradiction. So far, we have considered and generalized only our previous and present activities. But what about our future ones? What will we probably be doing in the next few hundred years? And to what consequences does it lead if we generalize that again, assuming a similar development for other civilizations in space?

We have good reasons to believe that our space exploration will continue and expand. There is our great general curiosity, the drive to explore and to use our near and far surroundings. In addition, our planet Earth is getting unpleasantly crowded, and we are using up many of our limited resources at an alarming speed. Within the next few generations we must develop complete recycling systems for many metals and rare elements, and we must develop new energy sources. The next logical step then is to set up space colonies for mining (O'Neill, Reading 57; 1977), in large shelters or underground, on Moon and Mars and even on the many thousands of asteroids (little planets, big chunks of minerals and metals, orbiting around our Sun between Mars and Jupiter). These colonies will become self-supporting and multiplying, they will grow with their own babies and grandchildren. In

September I was at the 28th International Astronautical Congress at Prague, where many such plans were presented and discussed. A wealth of general problems, and even a lot of solid engineering tasks, have already been worked out in an amazing detail. We actually could start all this activity right now, if we wanted it and if we would be willing to spend the money. Sure, mining the asteroids is at present not economical, but that will change when our resources on Earth are getting used up.

Another future aspect must be mentioned. At present we live with a frightening arms race which has completely gotten out of hand, and which cannot continue for too long. Either we learn to be more reasonable, or we will blow each other to pieces. All the big nations spend $1/10$ of their gross national product on the development and manufacture of more and more powerful weapons, which was a worldwide total of $334 \cdot 10^9$ dollars per year in 1976. And nobody wants to be left behind: in 1976, the underdeveloped countries received $13 \cdot 10^9$ dollars for economical help, but bought weapons for $18 \cdot 10^9$ dollars, from the industrialized countries (Sampson, 1977). And the total destructive power of all nuclear bombs was, in 1972, about 40,000 megatons of TNT, which is the same as 10 tons of dynamite for each living person on Earth (von Hoerner, 1975a). Just try to imagine a solid round ball of dynamite with 2 meter diameter, one for each person.

Our chances for blowing up are really quite large. But in case we don't, if we somehow manage to stop the arms race, then a very large amount of money, a large work force of labor and engineering, must be redirected to other activities, and space engineering may come in quite handy and naturally. But especially if we do *not* stop the arms race, we should very soon build large space colonies: for the survival of the human race and culture in case of severe catastrophes on Earth.

Let us assume that after a hundred years or so, if we survive, we have many large self-supporting colonies in space, with people living there for generations, in shelters or underground; maintaining large-scale industries of mining, manufacturing, engineering, and exploration. After a few generations the cultural and emotional ties to the home planet will become less important. It seems to be the next logical step to send out large colonies (in mobile homes) on interstellar trips lasting many generations, in order to explore and colonize other planetary systems. If we do not destroy ourselves, we will probably do all this. Life, in whatever form we know it, has a strong tendency to expand and to fill out every possible niche, into all its corners. Plants, animals and man have settled in hot and dry deserts as well as close to the cold ice-covered poles.

Space travel becomes easy (even for our beginner's technology) if we drop the prejudice that it must be finished within an individual's lifetime. After we have colonized the nearby planetary systems, these settlements will grow, and after a while some of them will send out

their own colonies in mobile homes to the next systems, and so on (just as hundreds of islands spread out over thousands of miles of the Pacific Ocean have been colonized long ago by Polynesians in their small primitive boats). We will have started a continuous wave of growing colonization, spreading out with about $1/100$ the speed of light. And in this way we can colonize our whole Galaxy, from one end to the other, in less than ten million years (Jones, 1976). This is only a very short time in astronomical terms. Even in biological terms it is not too long. The higher mammals are about 20 million years old; and 400 million years ago the first plants and animals moved out of the oceans and started conquering the continents of our Earth which took about 200 million years. Furthermore, ten million years would be needed for our present technology, and it might be reasonable to expect that a further-developed future technology could colonize the whole Galaxy within one million years, going with $1/10$ the speed of light.

Now let us generalize again; let us assume that we on Earth are about average and that similar developments are to be expected on similar planets, and that such planets are provided by 0.5% of all stars which means there are 10^9 such planets in our Galaxy (these are the basic assumptions for our previous and future SETI projects).

Assuming similar developments on these planets then leads to the following conclusion (Hart, Reading 42; Papagiannis, 1977): All what we just have described as our own probable future, large-scale exploration and colonization, all this should have happened long ago. It could have been started by any one out of 10^9 planetary civilizations $4 \cdot 10^9$ years ago, and it would have been finished only $1–10 \cdot 10^6$ years after it started, colonizing the whole Galaxy with everyone of its habitable planets. The whole Galaxy should be teeming with life, so obvious that there is no question about it. And, first of all: we humans should be the descendants of long-ago settlers from somewhere else. Which we certainly are not.

Furthermore it seems clear, from all we know, that all life on Earth had a common origin, and has developed here without any outside interference. All humans, animals and plants use the same basic organic chemistry, the same amino acids and even the same complicated genetic code. And the long chain of development from simple life forms to highly organized ones seems well enough described by mutations and natural selection. As to the origin of life (Sagan, 1977), we know from lab experiments that many large organic molecules, even amino acids and nucleotide bases and sugars (the most important building blocks of living matter), are formed abundantly from water and the gases of the original Earth's atmosphere, if energy is provided by ultraviolet radiation of the Sun or by electric discharges of lightning. We still do not know how the first self-reproducing organisms formed, or the genetic code, but the further development seems more or less clear. Also the development of human intelligence seems to follow quite naturally, since it provides such a wonderfully large and manyfold niche.

Not only our Earth, our whole planetary system appears uninhabited by any outside settlers. Otherwise, their large-scale mining industry or active radio communication would have been very obvious to us, the moment we invented optical and radio telescopes, since all this would be so nearby. There just are no extraterrestrials here.

We then are forced to the sad conclusion that our basic assumption was wrong, the assumption that we are not unique but are typical for the general development, and that life, intelligence and technology have developed in a similar way at a very large number of similar planets. But if all that is wrong, how can we ever expect to detect radio signals from other beings in space? If life is not abundant, and if we are not typical, then there is no chance of success for any SETI project.

We have shown that if our own development and attitude were typical, then the Earth should have been colonized long ago, there should be outside settlers here instead of us homegrown humans. With other words: "If we were typical, we should not exist." Which is such a revolting contradiction that we just cannot be typical. But then SETI has no chance.

The Large-Number Problem

How could we explain the absence of extraterrestrials in our system, or in general the absence of any obvious evidence of higher technology in our Galaxy? Because if even our beginner's technology is able to travel large distances and to send out strong radio signals, then the technology on old planets, after billions of years of further development, should surely be able to do all kinds of "astro-engineering" (Dyson, 1966), plainly visible and obvious, none of which we observe.

Where is everybody? This question has puzzled many scientists. Maybe the origin of self-reproducing life is so improbable that it almost never occurs. Maybe intelligence is so dangerous that it always ends up in self-destruction. Or, if this and other crises are to be mastered, any surviving civilization must be so highly regimented and stabilized that complete stagnation excludes any further technical progress (von Hoerner, 1975a). Maybe not a single one of 10^9 old civilizations felt the desire to colonize the Galaxy, or, first attempts of space colonization and astro-engineering have always met disastrous failures and have since been given up. Another nice thought is the "Zoo Hypothesis": that we live in a region of the Galaxy which has been declared a zoo or wildlife area, and a perfect zoo is one in which the animals don't see their waiters and spectators (Bova, 1963; Ball, Read-

ing 45). Maybe the answer is "change of interest", meaning that science and technology are only a very short-lived phase of any long development to be surpassed by completely different future activities, which we cannot guess and which do not produce any obvious evidence (von Hoerner, 1975a). Well, most of these explanations would not give SETI much chance, and, anyway, none of them sounds very convincing and plausible.

The main difficulty is what I like to call the "Large-Number Problem". We should expect 10^9 habitable planets, with higher life $4 \cdot 10^9$ years old, and both are very large numbers. In order to explain the absence of extraterrestrial evidence, either habitable planets must occur only once in 10^9 cases, or life originates only once on 10^9 planets within $4 \cdot 10^9$ years; or intelligence destroys itself or stagnates without any exception in 10^9 cases, or poachers in wildlife areas are less frequent than one in 10^9 during $4 \cdot 10^9$ years, or not one in 10^9 governments granted funds for space mining, and so on. Any one of these explanations would sound much more acceptable if we had to consider only a small number of cases, but not so for 10^9 of them.

One way of avoiding the large-number problem would be to assume that during these billions of years, with lots of interstellar communications between the old civilizations, all these many original cultures have completely merged into one single galactic superculture which now acts as a whole (ant hill or bee hive), or at least with only a small number of subcultures. This would make several explanations possible, and it might give us some chances for SETI. But it works only if the galactic travel time of communication, 30,000 years from one half to the other for radio signals, were always much shorter than the timescale of individual planetary development, which does not sound plausible. It also leaves an unpleasant afterthought: what about all these newcomers, like us, about one per 100 years in the Galaxy? Those of the old establishment would have to investigate them, early enough and carefully, for finding out whether they can be educated into useful future members, or whether they are to be treated as dangerous bacteria. Hard luck for us, probably.

Conclusions

The absence of extraterrestrials in our solar system seems to show that there are none anywhere else in the Galaxy either (at least no "technical" ones). Even if we have no explanation for our amazing uniqueness, it still would follow that our searches for signals cannot have success.

If we want to continue our searches, and especially if we want to invest large sums, we should try to find a scenario which favors interstellar communication but prevents interstellar travel and colonization. Some suggestions have been made, but none so far looks plausible. In the absence of such a plausible scenario, it might be suggested to continue our searches with moderate means, in case the argument given above is wrong or incomplete, but to postpone any great expenditure until we can justify it again. This is the negative aspect I wanted to emphasize. However, different conclusions are also possible and should be discussed.

Even if one agrees that the absence of extraterrestrials seems to speak against SETI, one still could be in favor of spending larger sums for future SETI projects. First, on the general ground that one should never give up something so important before really having tried with all possible means; and especially so because our beginner's ignorance may have used assumptions and drawn conclusions which are all completely wrong. Second, not only would success be of tremendous importance for the human development, also the lack of success (after a dedicated effort) may have great impact, because it would be "quite a responsibility to know that we are the torch bearers of the flame of cosmic consciousness in our entire Galaxy" (Papagiannis, 1977). Third, in a discussion it was mentioned that Columbus started out for an impossible goal under wrong assumptions, but he still discovered America. Fourth, Phil Morrison suggested in a letter that interstellar colonization in its initial state may start as an organized process where the distance covered goes in proportion with time, but later on it would proceed (because of cultural and informational fragmentation) only as a diffusion or random walk, where the distance goes only with the square root of time, in which case there was still not enough time to cover the whole Galaxy and to reach us.

Finally, it seems to me that the Large-Number argument could be applied in the opposite direction, too; there are no extraterrestrials in our solar system because the colonizers have by chance overlooked, or by some reason neglected, at least one in 10^9 of habitable planets and useful mining sites. In this case there are no great odds against our own uniqueness, because our question, "Where is Everybody," *can* have been asked only at one of these few neglected places. Regarding SETI we then should not expect any strong contacting signals pointed at us, but we still may consider things like omnidirectional navigational beacons, or eavesdropping on someone else's local broadcast.

Any further suggestions?

49

William Markowitz
The Physics and Metaphysics of Unidentified Flying Objects

The possibility that life exists on other planets within the solar system and other stellar systems is a question of profound interest. Conceivably, intelligent life may exist on some of these planets, and in some cases the inhabitants may be more advanced technically than we are. We assume, for purposes of discussion, that such technically advanced beings exist.

In recent years a large number of flying objects of uncertain origin have been reported. Some persons believe that these unidentified flying objects (UFO's) are controlled by extraterrestrial beings who are passengers in the spacecraft, or who may be controlling them by radio from the moon or from another planet. Others doubt this.

In a recent letter to *Science,* J. Allen Hynek urges the scientific investigation of a residue of puzzling UFO cases by physical and social scientists (1966b). He says there are a number of misconceptions concerning UFO reports; two of the misconceptions that he cites (with his comments) are as follows:

> *. . . UFO's are never reported by scientifically trained people.* This is unequivocally false. Some of the very best, most coherent reports have come from scientifically trained people. It is true that scientists are reluctant to make a public report. They also usually request anonymity which is always granted.
> *. . . UFO's are never seen at close range and are always reported vaguely.* When we speak of the body of puzzling reports, we exclude all those which fit the above description. I have in my files several hundred reports which are fine brain teasers and could easily be made the subject of profitable discussion among physical and social scientists alike.

This letter is surprising because Hynek, who has been a consultant to the U.S. Air Force for nearly 20 years, had written in the *Encyclopaedia Britannica* that there were no reports of UFO's by trained observers (1967). He wrote:

> U.S. air force investigators long recognized that most originators of UFO reports are sincere, interested in the wel-

fare and security of their country and honestly puzzled by the sightings they report. Their frequent readiness to ascribe a UFO to extraterrestrial sources, their emotional attachment to this explanation and their reluctance to take into account the failure of continuous and extensive surveillance by trained observers to produce such sightings is surprising. It appears unreasonable that spacecraft should announce themselves to casual observers while craftily avoiding detection by trained observers.

I have been interested in the flight of spacecraft from the standpoint of celestial mechanics and physics for a number of years and have published a method of interstellar navigation (1963).[1] With the publication of the appeal by Hynek I decided to make a new study of the dynamics of flight and compare this with published reports and with the reports which Hynek had.

Aristotle wrote on natural phenomena under the heading "Physics" and continued with another section called "Metaphysics" or "beyond physics." I use a similar approach here. First, I consider the physics of UFO's when the laws of physics are obeyed. After that I consider the case when the laws of physics are not obeyed. The specific question to be studied is whether UFO's are under extraterrestrial control.

Laws of Physics

The laws of physics to which I refer are those taught in any accredited college. They are the laws on which our automotive, space, and nuclear energy technologies, for example, are based. They include the elemental laws of celestial mechanics and physics, including special relativity. Some of these laws are as follows:

1. Every action must have an equal and opposite reaction.

---

[1]The method is based on the fact that the observed time of minimum of an eclipsing binary depends upon the position of an observer in space relative to the sun.

2. Every particle in the universe attracts every other particle with a force proportional to the product of the masses and inversely as the square of the distance.

3. Energy, mass, and momentum are conserved.

4. No material body can have a speed as great as c, the speed of light.[2]

5. The maximum energy which can be obtained from a body at rest is $E = mc^2$.

If anyone wishes to reject these laws I have no quarrel. Let us see, however, what the consequences are if these laws are accepted.

Possible and Impossible

Some people claim that nothing is impossible. This is not so. The laws of mathematics and physics, if accepted, do provide limitations on what can be done. However, one must be careful to state the assumptions under which he says that something is possible or is not possible.

If we accept the properties of real numbers, Euclidean geometry, and the laws of physics, then the following statements hold:

1. It is impossible to find two integers a and b such that $a^2/b^2 = 2$.

2. It is impossible to construct a regular polygon of seven sides using only a straight edge and compass. This was proved to be impossible by an 18-year-old schoolboy named Carl.[3] Incidentally, he showed that it was possible to construct a regular polygon of 17 sides, which no one had previously imagined possible.

3. It is impossible to construct a 2000-kilogram automobile which can be driven from a standing start to the top of a mountain 1600 meters (1 mile) high through the chemical combustion of 0.5 kilogram of gasoline.[4]

4. It is impossible for a man to lift himself by his bootstraps and remain in the air.

5. It is impossible to construct a perpetual motion machine. This principle has been accepted by one agency of the U.S. Government—namely, the U.S. Patent Office—which states (U.S. Government, 1966):

> The views of the Office are in accord with those of the scientists who have investigated the subject, and are to the effect that mechanical perpetual motion is a physical impossibility. These views can be rebutted only by the exhibition of a working model. . . . [In] no instance has the requirement of the Patent Office for a working model ever been complied with. . . .Alleged inventions of perpetual motion machines are refused patents.

Flight Principles: Speed, Energy, Thrust

The principles of celestial mechanics which govern the flight of bodies under the action of gravitation were enunciated by Newton in 1687. They are still valid today for speeds that are small in comparison to the speed of light. For high speeds we must use the modifications of Einstein—that is, the equations of relativity.

The dynamics of rocket flight have been studied intensively during the past 40 years. The equations for space flight by chemical rockets, ions, nuclear engines, and photons (pressure of light), and the effects of relativity, had been derived by 1952 (Carter, 1957). Many of the equations are now contained in textbooks. Here I give equations without derivation.

Table 49.1 gives the speeds, relative energies, and times of flight for a number of hypothetical missions. The term E_{kg} is the kinetic energy per kilogram of rest mass. The round-trip time for people on the earth is t, and for people in the spacecraft, τ,

To achieve the required speed, which can be done in steps, acceleration, a, is required. For a gravity field,

$$a = (\text{thrust} - \text{weight})/\text{mass}. \qquad (1)$$

The weight term is negligible when the craft is in space, but it is important at launch. All power generated is wasted until thrust exceeds weight.

Apart from propeller and balloon action, a spacecraft can generate thrust only by expelling mass. This mass may consist of material particles, whose speed is less than that of light, or equivalent photons, which move with the speed of light. The thrust is

$$F = \dot{m}v_e \qquad (2)$$

where $\dot{m}$ is the mass expelled per second and v_e is the exhaust speed relative to the rocket. The initial acceleration is small for a chemical rocket or a nuclear-powered spacecraft which expels a propellant. The acceleration increases as fuel or propellant is expelled and mass is reduced.

Let v be the speed of the rocket relative to the rest frame (the earth, effectively), $S = v/v_e$, and let R be the ratio of the initial to the final mass. In the absence of gravity and for $v_e < c$ the following equation holds:

---

[2]Electrons given an energy of 1×10^9 ev in an electron synchrotron should have a speed of $62c$, according to Newtonian mechanics. However, from measurements of the synchrotron diameter and the frequency of the alternating field it is readily determined that the speed is nearly $1c$. The speed, according to relativity theory, is $0.999\,999\,87c$ ($c = 3 \times 10^8$ meters per second).

[3]His last name was Gauss.

[4]The energy required is 3.1×10^7 joules; combustion of 0.5 kilogram of gasoline produces energy of 2.4×10^7 joules (1 joule = 0.74 foot-pound or 9.5×10^{-4} BTU).

Table 49.1.

APPROXIMATE SPEEDS, RELATIVE KINETIC ENERGIES, AND FLIGHT TIMES
FOR VARIOUS HYPOTHETICAL FLIGHT MISSIONS

No.	Speed	E_{kg} (joules/kg)	Mission
1	8 km/sec	3×10^7	Orbit, near earth; period, 90 minutes
2	13 km/sec	8×10^7	To moon and return; t, 1 week
3	20 km/sec	2×10^8	To nearby planet and return; t, 3 years
4	100 km/sec	5×10^9	To α Centauri and return; t, 25,000 years
5	$0.5c$	1×10^{16}	To α Centauri and return; t, 17 years; τ, 15 years
6	$(1 - 10^{-11})c$	2×10^{22}	To Andromeda Galaxy and return; distance, 2×10^6 light years; t, 4×10^6 years; τ, 18 years

$$R = e^s. \qquad (3)$$

The speed v can exceed v_e, but R becomes excessively large, for practical purposes, if S approaches 2. Multistaging is used to obtain values as large as 5.

To get an idea of what is required for space exploration, let us consider the Apollo spacecraft.[5] This is designed to take three men to the moon, land two, and return all three to the earth in about 1 week. Its characteristics are as follows: height, 110 meters (364 feet), mass on launching pad, 3×10^6 kilograms (6.5×10^6 pound-mass); initial thrust, 3.3×10^7 newtons (7.5×10^6 pounds); initial acceleration, $0.15g$; acceleration at first-stage burnout, $4g$; first-stage fuel consumption, 14,000 kilograms per second for 150 seconds; exhaust speed, 2.5 kilometers per second; mass of reentry package on return to earth, 5400 kilograms.

Thus, we require about 550 kilograms on the launching pad for every kilogram which is to travel to the moon and return. This mass ratio would be enormously greater for any similar mission to a planet, even to a nearby planet such as Mars or Venus. A single Saturn V vehicle, large as it is, cannot accomplish such a mission.

Manned exploration of the planets will be very difficult with chemical rockets alone. Studies under way envisage the use of ion propulsion and nuclear engines after the spacecraft has been removed from the earth by chemical rockets.

The value of v_e obtained with chemical rockets is small, about $8 \times 10^{-6}c$. In theory, nuclear reactions might be used to obtain high speeds. The products of fission of U^{235} have speeds of about $0.03c$. If we could form helium from the fusion of hydrogen, the speed of the helium would be $0.12c$. A practical problem would remain: the products formed would fly off in all directions.

In practice, nuclear engines operate by heating a

---

[5]Information about the Apollo spacecraft was kindly supplied by NASA.

propellant—hydrogen, for example—and expelling it. When the propellant is gone, the engine is dead. There is a gain over chemical heating, and the gain makes this type of engine potentially useful for planetary exploration.

If matter and antimatter could be stored in a rocket and then brought together, gamma-ray photons, traveling with speed c, would be produced in all directions. If the radiation could be aligned and the process were 100-percent efficient, then the following equation would hold:

$$R = (1 + v/c)^{1/2}/(1 - v/c)^{3/2} \qquad (4)$$

A round trip to another star would require two accelerations and two decelerations. The overall mass ratio would be $Q = R^4$. For $v = 0.5c$, $Q = 9$; for $v = 0.9c$, $Q = 361$. If a voyage of exploration were made to three stars and back, the mass ratio would be R^8.

The thrust that would be obtained if the radiation from the annihilation of matter could be aligned is $F = \dot{m}c$, where $\dot{m}$ is the annihilation rate. The power is $P = \dot{m}c^2$. The ratio of power to thrust is $P/F = c$, and 3×10^8 watts must be generated for each newton of thrust (1.33×10^9 watts per pound).

To lift a spacecraft of mass 5000 kilograms (weight, 49,000 newtons) with an acceleration of $1g$ from the earth would require a power of about 3×10^{13} watts. (This is about 30 times the electrical generating capacity of the entire world.) If the 3×10^{13} watts were radiated from a surface of 10 square meters (108 square feet), then the surface temperature, according to the Stefan-Boltzmann law, would be about 85,000°C. Reflectors would be required to send the radiation backward, and if these absorbed even 1 percent of the radiation they would vaporize. This fundamental difficulty in interstellar flight was pointed out in 1952 by Shepherd (in Carter, 1957, p. 408).

The possible use of interstellar matter for fuel was investigated by Pierce and found not feasible (1959).

We have assumed above that we could control mass-to-energy conversions involving fission, fusion, and annihilation with equipment having negligible mass. Even if this could be done, the basic problem of aligning the motions of the particles or the radiation would remain. At the speeds involved, the particles or the radiation would interact with the atoms of the enclosure; they would not bounce back as the combustion products in a chemical rocket do.

Comparison between Theory and Reports

Published reports of unidentified flying objects usually describe objects seen in flight at a distance. Such sightings can give only angular diameters and angular speeds—not masses, linear dimensions, or linear speeds. Similarly, radar measurements cannot give masses or linear dimensions. For this reason, sightings of distant flying objects are useless for comparison with the laws of physics. I do not take issue with reports of sightings and will not try to explain them away. I agree that unidentified objects exist. The question remains, however, of whether objects seen were under extraterrestrial control.

If an extraterrestrial spacecraft is to land nondestructively and then lift off it must be able to develop a thrust slightly less than its weight on landing, and twice its weight for an acceleration of $1g$ on lift-off. This requirement forms a critical test for comparing UFO reports with physical theory.

The published reports generally describe objects about 5 to 100 meters in diameter, which land and lift off without the use of launching pads and gantries. No similarity to the giant undertaking of a launching from Cape Kennedy has ever been reported. If nuclear energy is used to generate thrust, then searing of the ground from temperatures of $85,000°$ C should result, and nuclear decay products equivalent in quantity to those produced by the detonation of an atomic bomb should be detected. This has not happened. Hence, the published reports of landing and lift-offs of UFO's are not reports of spacecraft controlled by extraterrestrial beings, if the laws of physics are valid.

Unpublished Reports

On 20 December 1966 I wrote to Hynek, asking him for reports in his files of landings and lift-offs. He wrote that he had no reliable reports concerning landings and lift-offs. Also, he told me in a telephone conversation that he had no records of cases wherein a reliable witness visited an extraterrestrial craft or talked with an occupant. Hynek's letter to me (private communication, 1967) states:

To sum up my answer to your request then: the cases I mentioned in *Science* do not deal with landings or takeoffs with the possible exception of the Socorro case which is useless from a quantitative standpoint. To obtain the information you require would be a several months' job in going through 10,000 or more cases. If the Air Force had accepted my recommendation a long time ago to have all this material in machine readable form, we could in a matter of a moment or two query the tapes and bring forth all this information for you in tabular form.

The latter part of the letter is puzzling. I was not interested in the Air Force files—I was specifically interested in the cases in Hynek's files.

To check further, however, I telephoned Major Hector Quintanilla, head of Project Blue Book of the U.S. Air Force. He told me that he did not know of a single case in the U.S. Air Force files of a confirmed report of a landing and lift-off. His comments on the New Mexico case, which he released to the press, are as follows:

Conclusion: The investigators at Wright-Patterson have not been able to identify or determine what type of vehicle or object Mr. Lonnie Zamora observed on 24 April 1964 at Socorro, New Mexico. The object or vehicle displayed flight characteristics well within the State-of-the-Art and the sighting cannot be attributed to atmospheric or astronomical phenomena. In this respect, I can categorically state that the vehicle or object observed by Mr. Lonnie Zamora was not an inter-planetary space vehicle visiting the planet Earth. This case is still open and the investigation is still in progress.

Metaphysics

Let us now consider the possibility that the laws of physics are not valid. One idea frequently suggested is that extraterrestrial beings have discovered gravity shields. This, however, would not solve the problem of propulsion because inertia would remain; reaction would still be needed to obtain acceleration.

If we could cancel gravity on the earth, an object would lift very slowly, (i) because of the buoyancy of the air and (ii) because the object would begin traveling in a straight line whereas the earth would continue to revolve around the sun. In the story *The First Men in the Moon*, written by H. G. Wells about 1900, a shield was used to cancel the attraction of the earth but not that of the moon. The initial acceleration would be $3.5 \times 10^{-6}g$, which is not that reported for UFO's.

We can reconcile UFO reports with extraterrestrial control by assigning various magic properties to extraterrestrial beings. These include "teleportation" (the instantaneous movement of material bodies between planets and stars), the creation of "force-fields" to drive space ships, and propulsion without reaction. The last of these would permit a man to lift himself by his bootstraps. Anyone who wishes is free to accept such magic properties, but I cannot.

Semi-Magic

On another level are semi-magic hypotheses, which are proposed by scientists. These are based in part on sound scientific laws but include magic properties not explicitly stated. The general theme is that, through the use of nuclear energy and the time-dilation effect of relativity, everything is possible. Little attention, however, is paid to the practical difficulties of converting matter into energy and of utilizing it in a spacecraft without burning up the occupants. Physical processes are carried out with practically 100-percent efficiency, and complicated equipment never breaks down. Thus, we have been given theories to the effect that travel between galaxies is feasible, that a colony of extraterrestrial beings may be living on the back side of the moon, and that we may use planets of other stars to take care of overpopulation.

Intergalactic travel is fascinating. From row 6 of Table 49.1 we see that travel to the nearest external galaxy requires at least 4 million years between the sending forth of a spacecraft and its return. The speed given in Table 49.1 is $v_G = 0.999\ 999\ 999\ 99c$. Propulsion would be achieved, it has been suggested, by drawing in interstellar hydrogen over an area of thousands of square kilometers and converting this to helium.

At speed v_G an interstellar particle of dust of diameter 2.5×10^{-6} meter (0.0001 inch) would meet the spacecraft with a kinetic energy of 3×10^8 joules. (The kinetic energy of a 2000-kilogram automobile whose speed is 100 kilometers per hour is 8×10^5 joules.) The energy of a proton which meets the spacecraft would be 2×10^{14} ev. Survival of the spacecraft and its occupants is unlikely.

Interstellar Communication

A question now under discussion is whether it is better to try to contact extraterrestrial beings by interstellar flight or by interstellar radio communication. The former seems impossible but the latter may be possible.

Whether we shall ever receive a message from intelligent beings on a planet outside our solar system depends upon the distance of the nearest civilization which is signaling. If it is within 100 light years, we may pick up signals. If the distance is greater than 1000 light years, the signal-to-noise ratio may be too small. Only one experiment for receiving signals has been made, Project Ozma (Drake, Reading 25). Although no contact was made, the experiment was valuable because information, even if negative, was obtained.

Considerable thought has been given to methods of exchanging information by radio. Ingenious methods of transmitting pictorial information have been proposed (Oliver, in Cameron, 1963a). The exchange of information will not be rapid, even if achieved, because of the large distances between stars. Hundreds of years might be needed for one exchange.

Much statistical work has been done on the probabilities of finding intelligent life. The mathematics is irreproachable, but we do not know whether the assumptions are valid. We do not know, for example, whether, given a suitable planet, living beings capable of transmitting radio signals are bound to evolve (Simpson, Reading 39). Since we cannot compute with certainty where extraterrestrial life exists, we do not know if we shall ever communicate with planets of other stars.

Lack of Definite Evidence

If extraterrestrial beings are not bound by the laws of physics and do visit us, then we should expect to see them or their spacecraft. The facts are as follows:

1. No extraterrestrial spacecraft or passenger thereof has ever been presented to Congress, to any state legislature, or to any recognized scientific society in the United States.

2. No reliable report exists of anyone having talked with an extraterrestrial visitor.

3. No accident has ever occurred at landing or lift-off which has left an extraterrestrial spacecraft on the ground, despite thousands of alleged landings.

Believers in the possibility of interstellar travel have great difficulty in trying to explain why the visitors make no attempt to communicate with us after a voyage which supposedly has lasted tens, hundreds, or thousands of years. We would expect the visitors to contact us, take close-up photographs, and study us in detail before starting the long voyage home.

Hynek explains the lack of contact by asking, Why should extraterrestrial visitors try to communicate with us? He states that we would observe, but wouldn't try to communicate with, a new species of kangaroo in Australia (Hynek, 1966a). This is not a convincing explanation. Intelligent, human beings are not in the class of kangaroos. A more appropriate case is that of Columbus. When he landed in the New World he did communicate with the natives.

Hard-Data Cases

Calls for investigation of UFO's have been made by Hynek, director of the Dearborn Observatory of Northwestern University, and two associates, W. T. Powers and Jacques Vallee, based on "hard-data" cases. These cases are defined as reports of responsible witnesses from which sightings traceable to balloons, satellites, and meteors have been excluded. None, however, of the close-range cases in Hynek's files has been published in any scientific journal (Hynek, 1953). The information which has been published contains inconsistencies.

An article by Hynek published recently in the *Saturday Evening Post* (1966a) includes four pictures of flying saucers. Three had captions, but there is no reference to or comment on these photographs in the text. The one showing purported objects in Sicily in 1954 was included in a book by Menzel and Boyd, who described it as a fake (1963, p. 205 and plate VI).

In a letter to *Science* of 7 April 1967 (Powers, 1967), Powers mentions "our best five or six hundred reports," and says, "In 1954, over 200 reports over the whole world concerned landings of objects, many with occupants." Powers seems unaware that Hynek had already informed me that he had no reliable reports of landings.

Jacques Vallee is the author or co-author of books on flying saucers (1965, 1966). These do not report the cases completely. Some examples follow.

1. In describing the Chiles-Whitted case of 25 July 1948, *Challenge to Science* (Vallee, 1966) mentions that two pilots in a DC-3 reported seeing a metallic, cigar-shaped object about 30 meters long with two rows of portholes, shining with supernatural brilliance; also, that a jet of flame from the object rocked the airplane (pp. 117, 119, 185). The book fails to mention that Hynek has identified the object as an undoubted meteor in his report of 30 April 1949 to the Air Force (Menzel and Boyd, 1963, p. 108). This omission is curious because Hynek wrote a foreword to *Challenge to Science*.

2. Vallee describes the sighting of a geometric formation by Clyde Tombaugh on 20 August 1949, leaving the impression that the discoverer of Pluto had observed a flying saucer (1965, p. 96; 1966, p. 90). Vallee does not mention Tombaugh's statement that he regarded this as being a natural optical phenomenon, not an extraterrestrial spacecraft (Menzel and Boyd, 1963, p. 266).

3. Vallee implies that an intra-Mercurial planet (a planet whose orbit is within that of Mercury) was known to exist and had been lost by astronomers (1965, p. 35; 1966, p. 115). He does not mention that Simon Newcomb (1895) had found that the hypothesis of its existence is not tenable.

The question at issue in the above cases is not what interpretation is correct or whether the authors accept the reports made. It is whether complete information has been given.

Investigations

The intense public interest in UFO's is due to the possibility that they carry extraterrestrial beings. Were it not for this fact there would be no demands for special inquiries, to be conducted by Congress or scientific panels. Unidentified flying objects have been the subject of countless articles in newspapers, magazines, and flying-saucer reviews, because of their sensational nature. The discussion of UFO's in scientific journals, however, has been almost nil. This is not because scientists are reluctant to study the phenomenon. It is because no reports of hard-data cases, detailed and documented, have been published in scientific journals. Such reports would have provided the basic material needed for study and discussion by scientists.

The search for extraterrestrial life is one of the most interesting problems of our times. Various methods of search have been proposed, including the manned and unmanned exploration of Mars. The wisdom of spending vast sums of money on such projects has been questioned (Simpson, Reading 39), but at least the projects are sound; they are based on accepted principles of physics and engineering. This is not the case, however, for investigations of UFO's, because the extraterrestrial control of reported UFO's is contrary to accepted principles of physics.

Unidentified flying objects have been investigated by the U.S. Air Force and its scientific consultants for nearly 20 years, and not a single extraterrestrial spacecraft or occupant has been produced. As early as 1953 a panel headed by H. P. Robertson reported that UFO's are not a threat to the security of the United States. No hard-data cases which would justify the holding of additional investigations have been made public. This is not to say that airplane pilots, for example, have not seen strange phenomena. However, these cases could be studied through publication of reports in scientific journals.

From the material published it appears doubtful that any Air-Force-sponsored investigation will change the following conclusions:

1. UFO's are not under extraterrestrial control.

2. The laws of physics do not need revision to accommodate UFO sightings.

3. UFO's are not a threat to the security of the United States.

It is suggested therefore that, to save money and manpower, the U.S. Air Force should cease investigating UFO's. (Major Quintanilla raised no objections when I mentioned this suggestion.) Further studies should be left to any interested scientist or amateur. In particular, on-the-spot investigations by the Air Force should be terminated.[6] This would free the Air Force from the charge of imposing secrecy. Surveillance of the skies for defense would continue.

In regard to secrecy, the charge that the U.S. Air Force is withholding information that UFO's are extraterrestrial is absurd. The prestige of announcing the existence of extraterrestrial beings would be so great that no scientist,

---

[6]Hynek, in the *Saturday Evening Post* (1966a), describes the near hysteria which accompanies the chase of UFO's.

journalist, politician, or government—whether of the United States, England, France, the U.S.S.R., or China—would hesitate for a moment to release the news. It could not be kept a secret.

Summary and Discussion

In response to the request made in *Science* (Hynek, 1966b), I have investigated UFO's and report as follows:

1. The control of reported UFO's by extraterrestrial beings is contrary to the laws of physics.

2. The data published do not justify the holding of investigations.

The U.S. Air Force has been able to resist pressures to declare that UFO's are under extraterrestrial control, but not pressures for the repetition of investigations. However, if the U.S. Patent Office can take a position on the feasibility of constructing perpetual motion machines, then the Air Force should be able to take a position on closing out its investigations of UFO's.

We have been reminded (Hynek, 1966b) that 21st-century science will look back on us. This is true. We, ourselves, look back on eras when many people believed in the existence of centaurs, mermaids, and fire-breathing dragons. I am afraid that 21st-century science will contemplate with wonder the fact that, in an age of science such as ours, the U.S. Air Force was required to sponsor repeated studies of UFO's.

I have no quarrel with anyone who wishes to believe that UFO's are under extraterrestrial control. As for me, I shall not believe that we have ever been visited by any extraterrestrial visitor—either from the moon, from a planet of our solar system, or from any other stellar system—until I am shown such a visitor.

50

David G. Stephenson
Extraterrestrial Intelligence

In recent years there has been an upsurge of serious interest in the problem of searching for extraterrestrial intelligences (ETI). Some authors, for example, Sturrock in the 1978 December issue of [the *Quarterly Journal of the Royal Astronomical Society*], propose that some of the unexplained reports of unidentified flying object (UFO) sightings may be explicable as the result of penetration missions by extraterrestrial intelligences into the human sphere of influence. Before such suggestions can be taken seriously one outstanding inconsistency would have to be explained.

As yet we have no sure model for the distribution of intelligent cultures throughout this region of the galaxy, but the distances between the stars is sufficiently large to demand that an interstellar culture must have a technical capability, and experience of using that capability, far in advance of our own. This capability would probably include an efficient reactor, which would fuse light elements, as a prime mover; advanced computational systems with electromagnetic sensors both active (radar) and passive (telescopes—radio, and optical) for navigational purposes; advanced life-sustaining systems to preserve the biostructure of the inhabitants of a vessel as it traverses the space between the stars; and not least, an advanced cultural structure that permitted the coherent effort that constructed and launched the interstellar vessel in the first place. If unidentified flying objects are to be considered as manifestations of extraterrestrial visitors it must now be assumed that upon reaching our local environment this highly advanced culture must abandon many of its advanced abilities. No modern reconnaissance aircraft would penetrate an alien air space on Earth without a full complement of decoys and electronic counter-measures designed to mask and confuse any sensory systems directed to detecting its presence. These counter measures may completely hide the presence of the penetrating aircraft by mimicking innocuous natural objects, or by con-

fusing the observer with a series of spurious ghost images, 'angels'. A model of extraterrestrial visitors that denies them similar but far more advanced capabilities is surely deficient. Given these capabilities, our current means of detecting extraterrestrial visitors would be totally incapable of distinguishing their presence from naturally occurring phenomena. For example, if an extraterrestrial culture had observed our planet's weather system and what, to them, would be our primitive radar systems, it would not be a difficult task to direct their computers to model our climate structures to predict those times when naturally occurring weather patterns would produce large numbers of mirages and ghost radar images. At this time a penetration probe could pass through our air space and be indistinguishable from naturally occurring 'angels'. If we ascribe an advanced technology to an interstellar culture then we must accept that that technology would give it the power to observe us while denying us the chance of recognizing them or the results of their activities.

If we are an object of interest that is being covertly observed, there must be some unknown factor in our culture that makes us interesting. Therefore there remains a faint possibility of an accidental recognition of an extraterrestrial visitor. Presumably the observers have their own reasons for not making themselves apparent to us, so that they would take action to prevent such an accidental observation from resulting in a significant distortion of our ecumenical world culture. The best way of masking the effect of an impulse that could cause a significant effect on a cybernetic system such as our culture is to bury that impulse in extraneous noise. Since the Second World War, large areas of this planet have begun radiating intimate details of their population's social structures on a day-to-day basis by means of television. These VHF signals are in a simple format and readily penetrate the Earth's ionosphere and thus would be accessible to an observing station in inter-planetary space. If after a visitor's penetration mission the television signals from surrounding localities revealed a large number of dangerously significant 'unidentified flying objects' observations

Reprinted with permission from the *Quarterly Journal of the Royal Astronomical Society*, 20 (1979), 481.

we may presume that our observers would take action to dampen the effect of this event. In this context, social noise is in the form of ridicule and confusion, and would be induced into the population of the areas surrounding the significant observations. The generation of a large number of spurious lights in the sky, radar echoes, and other abnormal effects to create inconsistent and conflicting information would not be a great problem to a culture that can traverse the distances between the stars. In an extreme case the deliberate construction of physical representations in the form of a human psychological archetype, such as 'little green men' would not be impossible. These could then be planted in such as way as to be transiently observed by fringe members of the human culture to stimulate doubt and ridicule, and effectively mask any possibility of us recognizing the true nature of an occasional observation of extraterrestrial craft.

Of all the methods of detecting the presence of extraterrestrial cultures the examination of unidentified flying objects as their manifestation would seem to be the least likely to be fruitful. Only a culture far in advance of our own could successfully transport some of its members to our solar system. Given that this is the case let us be realistic and accept that such a culture could choose to remain hidden or to reveal itself as it wishes, and until human beings approach a similar advanced state of cultural and technical development we will have no means of trapping possible visitors into revealing their presence unambiguously. This does not imply that radio or optical searches of the heavens for intelligence-generated signals may be unsuccessful, for these could be generated by cultures only slightly more advanced than our own, who have not yet developed any thoroughgoing interstellar capability, and whose culture may be sufficiently close to our own state of development to be comprehensible to us.

51

David W. Schwartzman
The Absence of Extraterrestrials on Earth and the Prospects for CETI

There is a possibility that the oldest and most advanced civilizations on distant stars have in fact reached the level of permanent intercommunication and have formed, as it were, a club of communicating intellects of which we have only just qualified for membership and are probably now having our credentials examined. In view of the present chaotic political and economic situation of the world, it is not by any means certain that we would be accepted.

Bernal (1967)

The reason they do not intervene in the affairs of Earth is due to the fact that cosmic humanitarianism is not alien to them.

Fialkowski (1977)

The prevailing belief within the scientific community is in the abundance of extraterrestrial life (ETL), with perhaps millions of probable civilizations in our own Galaxy. However, a recent paper by Hart (Reading 42) has called into question this view by a simple argument: if other advanced civilizations exist they should have colonized Earth—since they are not here they do not exist, assuming colonization is a probable strategy of extraterrestrials. Given this assumption of noncontact the case is persuasively argued. Physical, sociological, and temporal objections to the conclusion are dealt with in a convincing way. For example, explanations which link the lack of colonization to difficulties in fast interstellar space flight or lack of motivation are unconvincing, given technological extrapolation of even present physics and the high probability that at least one civilization will be interested in galactic exploration (see also Kuiper and Morris, Reading

Reprinted from *Icarus*, 32 (1977), 473. Copyright © 1977 by Academic Press, Inc.

34). Jones (1976) supports this argument with more detailed calculations of probable colonization rates.

Cox (Reading 43) has challenged Hart's argument by a closer look at the limits to colonization. He argues that the time allowed in the history of the Galaxy may be too short for colonization to occur for a *handful* of civilizations (the physical and sociological arguments are trivial in comparison). However, even his assumptions [very conservative, in my opinion, for technologies close to "magic" (Sagan, Reading 29)] lead to colonization in a few million years for $N > 50,000$, where N is the number of advanced civilizations in the Galaxy. Further, he points out that colonization is an extreme assumption; things are much easier if extraterrestrials simply explore and leave Bracewell probes (Bracewell, 1974). I believe this is highly probable if $N > 1$ even if colonization never occurs. Hart (private communication, 1976) suggests an alternative argument against $N > 1$; the first advanced civilization in the Galaxy would have preceded the second by some millions of years, thus allowing for complete colonization of the Galaxy by the first. He argues that since we are not products of this colonization, we are therefore alone. I differ with this interpretation by assuming that colonization is not a probable extraterrestrial strategy, but that surveillance and eventual contact are. This of course is close to Ball's (Reading 45) zoo hypothesis, which Hart dismisses as untestable. The search for Bracewell probes in our solar system is a near recognition of the zoo hypothesis and is part of the Soviet program for CETI (Academy of Sciences of the U.S.S.R., 1975).

Hart's argument is in sharp contrast to most other estimates of N, e.g., those of Shklovskii and Sagan (1966). Further, they argue that "efficient interstellar spaceflight" is likely to be developed by a civilization substantially in advance of our own. Given their preferred values of N (10^6), lifetime L (10^7 yr), and one contact yr^{-1} civilization^{-1}, this results in an average contact frequency per planet of 10^{-5} yr^{-1} for intelligent pretechnical planetary communities. Sagan notes that L and the colonization rate might be increased significantly by the "feedback effect" (von Hoerner, 1961) of mutual communication via simple electromagnetic radiative contact.

However, Hart's and Jones' arguments are quite convincing given Hart's simple assumption that "they" are not here. Hart calls the rejection of this assumption the "UFO hypothesis" [i.e., UFOs are of extraterrestrial intelligent (ETI) origin]. He dismisses this by saying that "since very few astronomers believe the UFO hypothesis it seems unnecessary to discuss my own reasons for rejecting it." Hart is of course correct in maintaining that very few astronomers believe UFOs are extraterrestrial craft. Sturrock (1977) found that 1356 members of the American Astronomical Society estimated the probability of UFOs being "alien" (i.e., extraterrestrial) as 3% compared to a probability of 78% for some conventional explanation. However, 23% responded that the UFO problem "certainly deserves" scientific study, while 30% responded that it "probably" does. Note that although Sagan, too, dismisses the "UFO hypothesis" (Sagan, 1972), he does suggest, on the basis of previously mentioned contact frequencies, that extraterrestrial contact may have occurred within historic times (Shklovskii and Sagan, 1966).

I believe that a case can be made for the ETI explanation of at least some UFOs (see, e.g., Friedman, 1975). The definition of UFOs used in this paper is that of Hynek (1972): those objects (lights) that remain "unidentified after close scrutiny of all available evidence by persons who are technically capable of making a common sense identification." Of course, surveillance by ETI may be taking place quite independently of the UFO phenomenon (e.g., by an automatic station in the solar system). However, in my opinion, the plausibility that at least some UFOs are of ETI origin is greatly increased by Hart's arguments. I will not review the case for at least examining the UFO phenomenon in a serious scientific study. The references cited above along with others of the "invisible college" (Vallee, 1975) have presented this argument well. Rather, the purpose of this communication is to show that present surveillance of Earth by ETI is the best reconciliation of Hart's *arguments* with those that uphold the probable high value of N. I hope to encourage the dialogue started at the Stanford Workshop held in 1974 (Carlson and Sturrock, 1975).

To summarize, I assume the following: (1) $N \gg 1$; the Galactic Club (Bracewell, 1974), i.e., a "vast network of intelligent civilizations in productive mutual contact" (Shklovskii and Sagan, 1966), exists. This means all civilizations in the Galaxy are much more advanced than we (the step function effect in the emergence of a civilization capable of communication in the life of the Galaxy), equivalent to Shklovskii and Sagan's principle of mediocrity (see Morrison, 1974). (2) UFOs are of ETI origin, craft of the Galactic Club (following Sagan's arguments

for the likelihood of direct contact). These assumptions provide a reason behind the obvious lack of "official contact" by ETI. We *are* unique, not in any geocentric chauvinist sense, but in the sense of happening to be on the verge of becoming a member of the Galactic Club.

The motivation behind the present apparent surveillance by ETI is speculative, but some reasons can be suggested [see Friedman (1975) for a more complete list]. The Galactic Club has a science of comparative noogenics (Kamshilov, 1973), encompassing exobiology and the laws of the interaction of civilizations and "nature." We are *at present* a unique or at least a very rare object of this science, with perhaps 1000 other similar cases in the last 10,000 years (a rate of 10^{-1} yr^{-1}). Entrance into the Galactic Club may take only a few hundred years after a civilization is communicative. Perhaps the observed behavior of UFOs is intended to condition us to the acceptance of this fact [Vallee's (1975) "cultural manipulation"]. This argument is close to the zoo hypothesis (Ball, Reading 45). Ball assumes that "we are unaware of them." Many on Earth are aware of them, even though admittedly most awareness is of a very primitive, often quasi-religious character. "We" as a world scientific community or planetary entity are not.

The main argument that Sagan has offered against the ETI explanation for UFOs is his Santa Claus analogy, with every UFO craft representing a separate interstellar visit (Sagan, 1972, 1973b). Friedman (1973) has pointed out the absurdity of this argument by simply distinguishing between the local surveillance craft ("UFOs") and the presumed primary contact vehicles. Sagan does raise this possibility but dismisses it by saying it only changes the frequency by 10 to 100 times at most. How he computed this is not explained. Suppose only one visit has occurred in the last 10,000 years. Is it really so improbable that surveillance has continued since then from a base in our solar system? Sagan maintains that "looking for UFOs remains an unprofitable investment of terrestrial intelligence—if we are truly interested in the quest for extraterrestrial intelligence" (1973b). He advocates the search via radiotelescopes. Yet, he seems pessimistic about the likelihood of detecting ETI. He argues persuasively that it is "a great conceit, the idea of the present Earth establishing radio contact and becoming a member of the galactic federation" (1973b), since advanced civilizations would be unlikely to "make their presence known to emerging civilizations via antique communication modes" (Reading 29).

Where, then, do we stand on a strategy for CETI? In the light of the previous discussion a program for CETI should include the following: (1) Radio (and other parts of the EM spectrum) search for Type II or Type III civilizations (Kardashev, Reading 28) among nearer galaxies (Sagan, Reading 29), and a search for Bracewell probes (Academy of Sciences of the U.S.S.R., 1975). (2) A much more serious study of the UFO phenomenon to produce "harder" data (e.g., spectra from glowing UFOs), a systematic search using radar networks, infrared sensors from space, etc. (see Baker, 1968; McCampbell, 1973). Studies of possible historical ETI contacts should also be continued. Shklovskii and Sagan (1966) are open to this approach. Temple's (1976) study of the Dogon mythology is an example of serious work of this kind. (3) Perhaps the only method likely to produce positive results: the ongoing effort via political and social practice toward the creation of a unified planetary civilization by elimination of obsolete political–economic formations, i.e., passing the entrance requirements of the Galactic Club. (Bernal, 1967).

Acknowledgments

John Carlson and Michael Hart made helpful suggestions on a first draft of this paper. Thanks are also given for the helpful suggestions of a reviewer.

52

Walter Sullivan
Astronomer Fears Hostile Attack; Would Keep Life on Earth a Secret

Sir Martin Ryle, Nobel laureate in physics and Britain's Astronomer Royal, is trying to persuade the radio astronomers of the world to refrain from making known the existence of intelligent life on this planet, lest the earth be invaded by hostile beings.

He has addressed an appeal to the International Astronomical Union, urging that no attempts to communicate with other civilizations be undertaken, at least until there is international agreement on such a step. Copies of the appeal have been sent to Dr. Margaret Burbidge, President of the American Astronomical Society, and others.

So far as is known to radio astronomers here, no attempts to send signals to other worlds are under way or planned to date, and none of a serious nature have been undertaken. However, they point out that normal transmissions from the earth, dating from the development of high-powered radars and other transmitters, have by now reached out at least 20 light years in all directions. One light year is the distance traveled by light in a year.

Runs Counter to View

Sir Martin's concern, as expressed in his appeal to the Astronomical Union and in his correspondence with American colleagues, is that another civilization might see the earth as a tempting place for colonization or for extraction of mineral resources.

His suggestions in this regard run counter to the widely held view that travel across the vast distances separating stars and their planetary systems would be hard to justify for any conceivable purposes. Travel times in each direction would probably run to centuries.

Proponents of the search for signals from other worlds argue, on the other hand, that to learn that such civilizations exist and how they may have overcome the prob-

lems currently besetting this world could help mankind to survive.

Radio astronomers in the United States and the Soviet Union are actively seeking signals of intelligent origin. The emphasis is on listening, with virtually no discussion of transmission, at least until signals from space indicate in what direction and how to communicate.

It is significant, in this regard, that the name of the American effort, under the National Aeronautics and Space Administration, has evolved from CETI to SETI. The former acronym stood for "communication with extraterrestrial intelligence," whereas the title now refers to the "search" for such intelligence.

Sir Martin, who shared the 1974 Nobel Prize for his innovative development of antenna systems, became concerned last July when he saw a report on BBC television indicating that powerful signals were being transmitted at a 21-centimeter wavelength with the giant dish, 1,000 feet in diameter, at Arecibo, Puerto Rico. The alleged purpose was to attract the attention of civilizations far out in space.

The 21-centimeter wavelength (equivalent to a frequency of 1420 megahertz) is the "landmark" frequency of radio astronomy, being emitted by free-drifting clouds of hydrogen throughout the universe. It and its neighbouring wavelengths have long been regarded as the most logical radio-frequency rendezvous for technological societies trying to make contact.

Sir Martin wrote to Dr. Frank Drake at Cornell University, who directs the National Astronomy and Ionosphere Center, which operates the Arecibo observatory, who replied that no signals were being sent at 21 centimeters. That wavelength, by agreement, is protected to allow unimpaired observations.

In 1974, as part of the ceremonies dedicating the resurfaced Arecibo dish, an ingenious pictorial message was transmitted for three minutes at two other frequencies. The power of the transmission, Dr. Drake said in his reply, was "trivial" compared to the radar pulses from Arecibo and Goldstone, Calif., used in investigating and mapping nearby planets.

From the *New York Times*, November 4, 1976, p. 46. Copyright © 1976 by The New York Times Company. Reprinted by permission.

267

Other Transmission Sources

Some military radars also transmit powerful signals and the total energy sent into space by television stations probably exceeds all other sources. These transmissions, however, are spread out both in frequency and in time.

Sir Martin then proposed a global commitment to refrain from any attempt at signalling. He noted that when pulsars were first detected by his colleagues at Cambridge University they were suspected to be of artificial origin, implying that the next time it might be the real thing.

His appeal for action by the International Astronomical Union, the umbrella organization of world astronomers, was apparently transmitted to the union—but without endorsement—by Sir Bernard Lovell, the British pioneer in radio astronomy. However, no action was taken on it when that organization recently held its General Assembly in Grenoble, France.

Meanwhile, Dr. Drake's colleague at Cornell, Dr. Thomas Gold, has proposed that widely separated civilizations may be using clouds of gas in certain regions of the Milky Way Galaxy as masers to achieve enormous amplification of their signals. Such clouds appear to be amplifying emissions of natural origin to great intensity.

53

Should Mankind Hide?

An extraordinary warning has come from Sir Martin Ryle, Britain's Astronomer Royal and Nobel laureate in physics. Sir Martin is worried that some of his fellow scientists may give away the fact that the human race exists to alien intelligences elsewhere in the cosmos.

Such a historic security leak, Sir Martin fears, might result in the invasion and conquest of Earth by hostile beings who might view Earth and its inhabitants as a source of food, slaves, minerals and the like. Put another way, Sir Martin has raised the question: Should mankind hide in this obscure corner of a tiny solar system attached to a minor star?

Some observers, cognizant of the daily volume of electronic transmissions on this planet, may wonder whether mankind *can* hide, whether signals from radio, television or other stations may not already have been picked up.

Too many uncomfortable precedents from Earth's own history argue against a cavalier dismissal of Sir Martin's warning. Why should potential visitors from Alpha Centauri, or some more distant star, be expected to be more merciful to earthlings than Europeans were to American Indians only a few centuries ago? Or, if the visitors were of another physical form, why should they not see the same virtues in domesticating human beings that men realized long ago when they domesticated cattle, horses, dogs and cats? Or impressed other human beings into slavery?

But to live is to accept dangers. On balance, the chances of gain from communication with alien intelligence greatly exceed the chances of harm. Men have already returned from landings on the moon; instruments sent by human beings have already reported from Mercury, Venus, Mars and Jupiter. There is no evidence that humanity has close, intelligent neighbors.

Given the immensity of space, the nearest intelligent beings are not likely to exist closer than hundreds or thousands of light years away, at the very least. Should they detect the existence of Earth's civilization, the odds would seem to be high that they know far more than mankind about utilizing nature's bounty and thus have no need for such crude techniques of appropriation as were employed by those who followed Columbus to the New World and other human conquerors of new territories. The universe seems too rich to require an advanced race to look hungrily on Earth's eager patrimony.

Sir Martin appears to pay too little heed to the possibility of gain from such communication. The vastness of the Universe might contain beings who long ago found the cure for cancer, solved the problems of taming thermonuclear energy, and routinely practice genetic engineering for the benefit of their species. In short, information transmitted from alien intelligences could improve human existence. Despite Sir Martin's eminence there is no reason to assume that alien intelligence among the stars must be hostile or predatory.

VI

Bread on the Waters

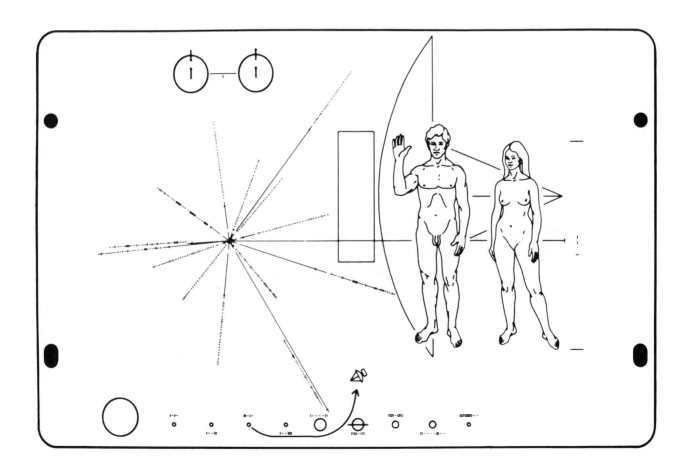

Humans love to communicate, and this collection of articles stands as a monument to this desire, sharpened to direct itself outward to the stars. When Carl Sagan and Frank Drake persuaded the National Aeronautics and Space Administration to add message plaques to the Pioneer 10 and Pioneer 11 spacecraft, engraved with human figures and accompanying diagrams, drawn by Linda Salzman Sagan, they probably exceeded their modest expectations of awakening their fellow creatures to the possibilities of interstellar messages.

As Sagan later wrote, "The greater significance of the Pioneer 10 plaque is not as a message to out there; it is as a message to back here." Traveling at a mere 40,000 kilometers per hour, the spacecraft has only recently passed beyond Uranus's distance from the sun; a radio message needs only a few hours to surpass the range that the Pioneers took ten years to cover. But as a message to "back here," the Pioneer plaques do just fine, provoking an instant reaction in anyone who sees them. As the *New York Daily News* summed them up in inimitable fashion, "Nudes and Map Tell about Earth to Other Worlds." The details can be found in the article by Sagan, Sagan, and Drake.

Five years after the Pioneer spacecraft set out, the more advanced Voyager 1 and Voyager 2 spacecraft were launched toward Jupiter, Saturn, and (in the case of Voyager 2) Uranus and Neptune. Instead of message plaques, the two Voyager spacecraft carry message *records,* gold-anodized discs that can be "played" to give not only sound but also pictures. Again Sagan, Sagan, and Drake, along with Ann Druyan, Timothy Ferris, and Jon Lomberg, were intimately involved in selecting the messages and pictures incorporated in the two records. Anyone who wishes to follow these messages in detail should read *Murmurs of Earth,* by the authors mentioned above. It is also reassuring to note that the suggestion made by Lewis Thomas, the author of *Lives of a Cell,* that the music of Bach should dominate any such message was followed to the extent that Bach does begin the music section of the Voyager records.

Like the Pioneer plaques, the Voyager records basically provide messages to ourselves. Real messages, those with a chance of covering interstellar distances in human lifetimes, consist of photons, probably radio photons. Certainly this is true of human messages toward the stars; the radio signals sent from the Arecibo Observatory in Puerto Rico on November 16, 1974, provide a reasonable way to signal our presence. No doubt from prudence, this message was directed toward a group of stars so distant that 50,000 years would be required to receive a return message. Nonetheless, the same technique would work (even better!) for planets that might exist around stars no more than a few dozen light years away, for which a single generation of human life would suffice to receive an answer—if one is sent.

Included in this section is a now-famous article by Gerard O'Neill on "The Colonization of Space." Published in 1974, O'Neill's work drew a tremendous response among the technologically oriented members of society, who saw it as a ray of hope in an increasingly gloomy world. In effect, O'Neill suggests we begin to colonize the region around Earth as the first step toward moving from Type I to Type II status among civilizations. In a relatively few generations, O'Neill concludes, we could have far more humans living in space habitats than upon our overcrowded, ever more polluted planet.

O'Neill's suggestions have the merit of leaving no one cold: People have an immediate response to the thought that most of us should live in space. Those who love the idea tend to see themselves as part of the accelerated evolution of humanity; those who hate it think of themselves more as the guardians of human tradition. The speed with which O'Neill's proposals become reality—if they ever do so—depends on our local development; the possibility that other civilizations may have long ago moved through this stage has already drawn our consideration. All that remains is to find our neighbors and ask how it went.

54

Carl Sagan, Linda Salzman Sagan and Frank Drake
A Message from Earth

Pioneer 10 is the first spacecraft that will leave the solar system. Scheduled for a launch no earlier than 27 February 1972, its 630- to 790-day-long flight will take it within two planetary radii of Jupiter, where, in a momentum exchange with the largest planet in the solar system, the spacecraft will be accelerated out of the solar system with a residual velocity at infinity of 11.5 km/sec. The spacecraft is designed to examine interplanetary space between the earth and Jupiter, perform preliminary reconnaissance in the asteroid belt, and make the first close-up observations of Jupiter and its particles and fields environment.

It seemed to us appropriate that this spacecraft, the first man-made object to leave the solar system, should carry some indication of the locale, epoch, and nature of its builders. We do not know the likelihood of the Galaxy being filled with advanced technological societies capable of and interested in intercepting such a spacecraft. It is clear, however, that such interception is a very long term proposition. With a residual interstellar velocity of 11.5 km/sec, the characteristic time for Pioneer 10 to travel 1 parsec (pc)—slightly less than the distance to the nearest star—is some 80,000 years. From the simplest collision physics, it follows that the mean time for such a spacecraft to come within 30 astronomical units (1 A.U. $= 1.5 \times 10^{13}$ cm) of a star is much longer than the age of the Galaxy. Consequently there is a negligible chance that Pioneer 10 will penetrate the planetary system of a technologically advanced society. But it appears possible that some civilizations technologically much more advanced than ours have the means of detecting an object such as Pioneer 10 in interstellar space, distinguishing it from other objects of comparable size but not of artificial origin, and then intercepting it and acquiring the spacecraft.

But if the intercepting civilization is not within the immediate solar neighborhood, the epoch of such an interception can only be in the very distant future. Accord-

ingly, we cannot see any conceivable danger in indicating our position in the Galaxy, even in the eventuality, which we consider highly unlikely, that such advanced societies would be hostile. In addition we have already sent much more rapidly moving indications of our presence and locale: the artificial radio-frequency emission which we use for our own purposes on Earth.

Erosional processes in the interstellar environment are largely unknown, but are very likely less efficient than erosion within the solar system, where a characteristic erosion rate, due mainly to micrometeoritic pitting, is of the order of 1 Å/year. Thus a plate etched to a depth $\sim 10^{-2}$ cm should survive recognizably at least to a distance ~ 10 pc, and most probably to $>> 100$ pc. Accordingly, Pioneer 10 and any etched metal message aboard it are likely to survive for much longer periods than any of the works of man on Earth.

With the support of the Pioneer Project Office at NASA's Ames Research Center in Mountain View, California, and of NASA Headquarters in Washington, D.C., it was agreed to prepare a message on a 6- by 9-inch surface of 6061 T6 gold-anodized aluminum plate, 50/1000 inch thick. The mean depth of engraving is 15/1000 inch. The plate is mounted in an exterior but largely protected position on the antenna support struts, behind the ARC plasma experimental package, on the Pioneer 10 spacecraft.

The question of the contents of such a message is not an easy one. The message finally agreed upon (Figure 54.1) is in our view an adequate but hardly ideal solution to the problem. A time interval of only 3 weeks existed between the formulation of the idea of including a message on Pioneer 10, achieving NASA concurrence, devising the message, and delivering the draft message for engraving. We believe that any such message will be constrained, to a greater or lesser degree, by the limitations of human perceptual and logical processes. The message inadvertently contains anthropocentric content. Nevertheless we feel that an advanced civilization would be able to decipher it.

Reprinted from *Science*, 175 (1972), 881. Copyright 1972 by the American Association for the Advancement of Science.

274

HYPERFINE TRANSITION OF
NEUTRAL HYDROGEN

SILHOUETTE OF
SPACECRAFT

BINARY EQUIVALENT
OF DECIMAL 8

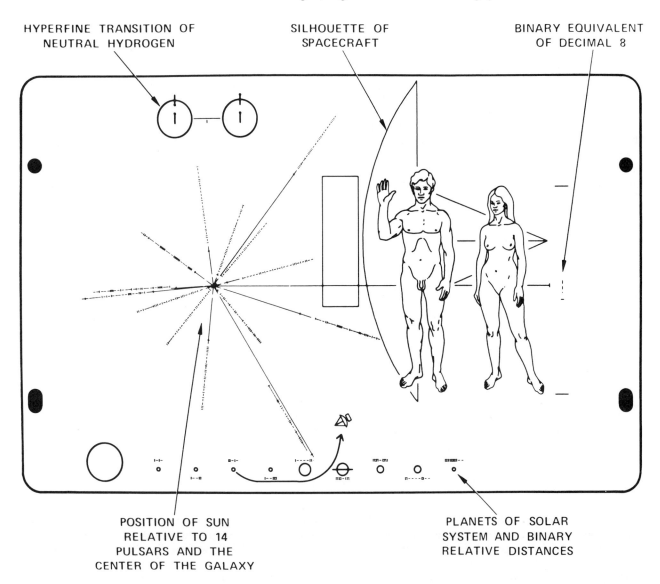

POSITION OF SUN
RELATIVE TO 14
PULSARS AND THE
CENTER OF THE GALAXY

PLANETS OF SOLAR
SYSTEM AND BINARY
RELATIVE DISTANCES

Figure 54.1. The engraved aluminum plate carried aboard Pioneer 10. It contains information on the position, epoch, and nature of the spacecraft.

At top left is a schematic representation of the hyperfine transition of neutral atomic hydrogen. A transition from antiparallel nuclear and electronic spins to parallel nuclear and electronic spins is shown above the binary digit 1. So far the message does not specify whether this is a unit of length (21 cm) or a unit of time $[(1420 \text{ MHz})^{-1}]$. This fundamental transition of the most abundant atom in the Galaxy should be readily recognizable to the physicists of other civilizations. As a cross-check, we have indicated the binary equivalent of the decimal number 8 along the right-hand margin, between two tote marks corresponding to the height of the human beings shown. The Pioneer 10 spacecraft is displayed behind the human beings and to the

same scale. A society that intercepts the spacecraft will of course be able to measure its dimensions and determine that 8 by 21 cm corresponds to the characteristic dimensions of the spacecraft.

With this first unit of space or time specified we now consider the radial pattern at left center. This is in fact a polar coordinate representation of the positions of some objects about some origin, with this interpretation being a probable, but not certain, initial hypothesis to scientists elsewhere. The two most likely origins in an astronomical interpretation would be the home star of the launch civilization and the center of the Galaxy. There are 15 lines emanating from the origin, corresponding to 15 objects.

Fourteen of these objects have a long binary number attached, corresponding to a 10-digit number in decimal notation. The large number of digits is the key that these numbers indicate time intervals, not distances or some other quantity. A civilization at our level of technology (as evinced from the Pioneer 10 spacecraft itself) will not know the distances to galactic objects useful for direction-finding to ten significant figures; and, even if we did, the proper motion of such objects within the Galaxy would render this degree of precision pointless. There are no other conceivable quantities that we might know to ten significant figures for relatively distant cosmic objects. The numbers attached to the 14 objects are therefore most plausibly time intervals. From the unit of time, the indicated time intervals are all ~ 0.1 second. For what objects might a civilization at our level of advance know time periods ~ 0.1 second to ten significant figures? Pulsars are the obvious answer. Since pulsars are running down at largely known rates they can be used as galactic clocks for time intervals of hundreds of millions of years. The radial pattern therefore must indicate the positions (obtained by us from the observed dispersion measures) and periods at the launch epoch of 14 pulsars, plus one additional object which is the most distant.

The problem thus reduces to searching the astronomical records to find a locale and epoch within the galaxy at which 14 pulsars were in evidence with the denoted periods and relative coordinates. Because the message is so overspecified, and because the pulsar periods are given to such precision, we believe that this is not an extremely difficult computer task, even with time intervals $\gg 10^6$ years between launch and recovery. The pulsars utilized, with their periods in seconds and in units of the hydrogen hyperfine transition, are indicated in Table 54.1. The hyperfine period of $(1.420405752 \times 10^9 \ \mathrm{sec}^{-1})^{-1}$, a fraction of a nanosecond, is just small enough so that all the known digits of the pulsar periods can just be written to the left of the decimal point. Accordingly decimals and fractions are entirely avoided with no loss of accuracy and without many noninformative digits. The presence of several consecutive terminal zeros (Table 54.1), particularly in pulsars 1240 and 1727, imply that, for these two pulsars, we have given a precision greater than we now have. The problem of which end of a number is the most significant digit is expressed automatically in this formulation, since all binary numbers start with a 1 but end in a 1 or a 0. The binary notation, in addition to being the simplest, is selected in order to produce a message that can suffer considerable erosion and still be readable. In principle, the reader only need determine that there were two varieties of symbols present, and the spacings alone will lead to a correct reconstruction of the number.

Those radial lines for which the earth-pulsar distance is not accurately known are shown with breaks. All three spatial coordinates of the pulsars are indicated. The (r, θ)

Table 54.1.

THE 14 SELECTED PULSARS

Pulsar	Period (1970/1971 epoch) (second)	Period (units of H hyperfine transition)
0328	$7.145186424 \times 10^{-1}$	1.014906390×10^9
0525	3.745490800	5.320116676×10^9
0531	$3.312964500 \times 10^{-2}$	4.705753832×10^7
0823	$5.306595990 \times 10^{-1}$	7.537519468×10^8
0833	$8.921874790 \times 10^{-2}$	1.267268227×10^8
0950	$2.530650432 \times 10^{-1}$	3.594550429×10^8
1240	$3.880000000 \times 10^{-1}$	5.511174318×10^8
1451	$2.633767640 \times 10^{-1}$	3.741018705×10^8
1642	$3.876887790 \times 10^{-1}$	5.506753717×10^8
1727	$8.296830000 \times 10^{-1}$	1.178486506×10^9
1929	$2.265170380 \times 10^{-1}$	3.217461037×10^8
1933	$3.587354200 \times 10^{-1}$	5.095498540×10^8
2016	$5.579533900 \times 10^{-1}$	7.925202045×10^8
2217	$5.384673780 \times 10^{-1}$	7.648421610×10^8

coordinates are given in the usual polar projection. The tick marks near the ends of the radial lines give the z coordinate normal to the galactic plane, with the distances measured from the far end of the line. The reconstruction of pulsar periods will indicate that the origin of (r, θ) coordinates is not the center of the Galaxy. Accordingly the long line extending to the right, behind the human beings, and which is not accompanied by a pulsar period, should be identifiable as the distance to the galactic center. Since the tick mark of this line is precisely at its end, this should simultaneously confirm that the ticks denote the galactic z coordinate, and that the longest line represents the distance from the launch planet to the galactic center. The tick marks were intended to be asymmetric about the radial distance lines, in order to give the sign of the galactic latitude or z coordinate. In the execution of the message this convention was inadvertently breached. But the sign of the z coordinate should be easily deducible without this aid. There is an initial ambiguity about whether the (r, θ) presentation is from the North or South Galactic Pole, but this ambiguity would be resolved as soon as even one pulsar was identified.

The 14 pulsars denoted have been chosen to include the shortest period pulsars which give the greatest luminosity; they are, therefore, the pulsars of greatest use in this problem where interception of the message occurs only in the far future. They are also selected to be distributed as evenly as possible in galactic longitude. Included are both pulsars in the vicinity of the Crab Nebula; the second (PSR 0525) has the longest known period. Fourteen pulsars were included to provide redundancy for any position and time solutions, but also to allow for the good possibility that pulsar emission is highly beamed and that not all pulsars are visible at all view angles. We expect that some of the 14 would be observable from all locales. In addition a very advanced civilization might have information on astronomy from other locales in the Galaxy. If the spacecraft is intercepted after only a few tens of millions of years (having traveled several hundred parsecs), all 14 pulsars may still be detectable.

The reconstruction of the epoch in which the message was devised should be performable to high precision: With 14 periods, almost all of which are accurate to nine significant figures in decimal equivalent, a society which has detailed records of past pulsar behavior should be able to reconstruct the epoch of launch to the equivalent of the year 1971. If past records of pulsar "glitches" (discontinuities in the period) are not kept or reconstructable from the physics it should still be possible to reconstruct the epoch to the nearest century or millennium.

Fortuitously, two of the pulsars are very near Earth. If either is correctly identified, it can be used to place the position of our solar system in the galaxy to approximately 20 pc, thereby specifying our location to approximately 1 in 10^3 stars.

To specify our position to greater accuracy, we have included a schematic solar system at the bottom of the diagram. Because of the limited plate dimensions, the solar system was engraved with the planets not in the solar equatorial plane. (If this were an accurate representation of our solar system it would identify it very well indeed!) Relative distances of the planets are indicated in binary notation above or below each planet. The serifs on the binary "ones" are presented to stress that the units are different from those of pulsar length and period. The numbers represent the semimajor axes of the planetary orbits in units of one-tenth the semimajor axis of the orbit of Mercury, or 0.0387 A.U., approximately. There is no way for this unit of length to be deciphered in the message, but the schematic sizes and relative distances—given to three significant figures in decimal equivalent—of the planets in our solar system, as well as the schematic representation of the rings of Saturn seen edge-on, should easily distinguish our solar system from the few thousand nearest stars if they have been surveyed once. Also indicated is a schematic trajectory of the Pioneer 10 spacecraft, passing by Jupiter and leaving the solar system. Its antenna is shown pointing approximately back at Earth. The cross-correlation between this stage of solar system exploration and the instrumentation and electronics of the Pioneer 10 spacecraft itself should specify the level of contemporary human technology with some precision.

The message is completed by a representation at right of a man and woman drawn to scale before a schematic Pioneer 10 spacecraft. The absolute dimensions of the human beings are specified in two ways: by comparison with the Pioneer 10 spacecraft and in units of the wavelength of the hyperfine transition of hydrogen, as described above. It is not clear how much evolutionary or anthropological information can be deduced from such a sketch drawing. Ten fingers and ten toes may provide a clue to man's arboreal ancestry, and the fact that the distance of Mercury from the sun is given as 10 units may be a clue to the development of counting. It seems likely, if the interceptor society has not had previous contact with organisms similar to human beings, that many of the body characteristics shown will prove deeply mysterious. We rejected many alternative representations of human beings for a variety of reasons; for example, we do not show them holding hands lest one rather than two organisms be deduced. With a set of human representations to this degree of detail, it was not possible to avoid some racial stereotypes, but we hope that this man and woman will be considered representative of all of mankind. A raised outstretched right hand has been indicated as a "universal" symbol of good will in many human writings; we doubt any literal universality, but included it for want of a better symbol. It has at least the advantage of displaying an opposable thumb.

Among the large number of alternative message contents considered and rejected were radioactive time markers (rejected because of interference with the Pioneer radiation detectors), star map position indicators (rejected because of stellar proper motions and serious data-handling problems in decoding), and schematic representations of the vascular, neurological, or muscular apparatus of human beings or some indication of the number of cortical neural connections (rejected because of the ambiguity of the envisioned representations). It is nevertheless clear that the message can be improved upon; and we hope that future spacecraft launched beyond the solar system will carry such improved messages.

This message then is a first attempt to specify our position in the Galaxy, our epoch and something of our nature. We do not know if the message will ever be found or decoded; but its inclusion on the Pioneer 10 spacecraft seems to us a hopeful symbol of a vigorous civilization on Earth.

We thank the Pioneer Project Office at Ames Research Center, especially Charles Hall, the Program Manager, and Theodore Webber; and officials at NASA Headquarters, particularly John Naugle, Ishtiaq Rasool, and Henry J. Smith, for supporting a small project involving rather longer time scales than government agencies usually plan for. The initial suggestion to include some message aboard Pioneer 10 was made by Eric Burgess and Richard Hoagland. A redrawing of the initial message for engraving was made by Owen Finstad; the message was engraved by Carl Ray. We are grateful to A. G. W. Cameron for reviewing this message and for suggesting the serifs on the solar system distance indicators; and to J. Berger and J. R. Houck for assistance in computer programming.

55

Lewis Thomas
Notes of a Biology-Watcher: Ceti

Tau Ceti is a relatively nearby star that sufficiently resembles our sun to make its solar system a plausible candidate for the existence of life. We are, it appears, ready to begin getting in touch with Ceti, and with any other interested celestial body in more remote places, out to the edge. CETI is also, by intention, the acronym of the First International Conference on Communication with Extraterrestrial Intelligence, held a few months ago in Soviet Armenia under the joint sponsorship of the National Academy of Sciences of the United States and the Soviet Academy, involving eminent physicists and astronomers from various countries, most of whom are convinced that the odds for the existence of life elsewhere are very high, with a reasonable probability that there are civilizations, one place or another, with technologic mastery matching or exceeding ours.

On this assumption, the conferees thought it likely that radioastronomy would be the generally accepted mode of interstellar communication, on grounds of speed and economy. They made a formal recommendation that we organize an international co-operative program, with new and immense radio telescopes, to probe the reaches of deep space for electromagnetic signals making sense. Eventually, we would plan to send out messages of our own, and receive answers, but at the outset it seems more practical to begin by catching snatches of conversation between others.

So, the highest of all our complex technologies in the hardest of our sciences will soon be engaged, full scale, in what is essentially biologic research—and with some aspects of social science, at that.

The earth has become, just in the last decade, too small a place. We have the feeling of being confined—shut in; it is something like outgrowing a small town in a small county. The views of the dark, pocked surface of Mars, still lifeless to judge from the latest photographs, do not seem to have extended our reach; instead, they bring

closer, too close, another unsatisfactory feature of our local environment. The bright blue noonday sky, cloudless, has lost its old look of immensity. The word is out that the sky is not limitless; it is finite. It is in truth only a kind of local roof, a membrane under which we live, luminous but confusingly refractile when suffused with sunlight; we can sense its concave surface a few miles over our heads. We know that it is tough and thick enough so that when hard objects strike it from the outside they burst into flames. The color photographs of the earth are more amazing than anything outside: we live inside a blue chamber, a bubble of air blown by ourselves. The other sky beyond, absolutely black and appalling, is wide-open country, irresistible for exploration.

Here we go, then. An extraterrestrial embryologist, having a close look at us from time to time, would probably conclude that the morphogenesis of the earth is coming along well, with the beginnings of a nervous system and fair-sized ganglions in the form of cities, and now with specialized dish-shaped sensory organs, miles across, ready to receive stimuli. He may well wonder, however, how we will go about responding. We are evolving into the situation of a Skinner pigeon in a Skinner box, peering about in all directions, trying to make connections, probing.

When the first word comes in from outer space, finally, we will probably be used to the idea. We can already provide a quite good explanation for the origin of life, here or elsewhere. Given a moist planet with methane, formaldehyde, ammonia and some usable minerals, all of which abound, exposed to lightning or ultraviolet irradiation at the right temperature, life might start off almost anywhere. The tricky, unsolved thing is how to get the polymers to arrange in membranes and invent replication. The rest is clear going. If they follow our protocol, it will be anaerobic life at first, then photosynthesis and the first exhalation of oxygen, then respiring life and the great burst of variation, then speciation, and finally some kind of consciousness. It is easy, in the telling.

I suspect that when we have recovered from the first

Reprinted by permission from the *New England Journal of Medicine,* 286 (1972), 306.

easy acceptance of signs of life from elsewhere, and finished nodding at each other, and finished smiling, we will be in for a shock. We have had it easy, relatively speaking, being unique all these years, and it will be hard to deal with the thought that the whole, infinitely huge, spinning, clock-like apparatus around us is itself animate, and can sprout life whenever the conditions are right. We will respond, beyond doubt, by making connections after the fashion of established life, floating out our filaments, extending pili, but we will end up feeling smaller than ever, as small as a single cell, with a quite new sense of continuity. It will take some getting used to.

The immediate problem, however, is a much more practical, down-to-earth matter, and must be giving insomnia to the CETI participants. Let us assume that there is, indeed, sentient life in one or another part of remote space, and that we will be successful in getting in touch with it. What on earth are we going to talk about? If, as seems likely, it is 100 or more light years away, there are going to be some very long pauses. The barest amenities, on which we rely for opening conversations—Hello, are you there?, from us, followed by Yes, hello, from them—will take 200 years at least. By the time we have our party we may have forgotten what we had in mind.

We could begin by gambling on the rightness of our technology, and just send out news of ourselves, like a mimeographed Christmas letter, but we would have to choose our items carefully, with durability of meaning in mind. Whatever information we provide must still make sense to us two centuries later, and must still seem important, or the conversation will be an embarrassment to all concerned. In 200 years it is, as we have found, easy to lose the thread.

Perhaps the safest thing to do at the outset, if technology permits, is to send music. This language may be the best we have for explaining what we are like to others in space, with least ambiguity. I would vote for Bach, all of Bach, streamed out into space, over and over again. We would be bragging, of course, but it is surely excusable for us to put the best possible face on at the beginning of such an acquaintance. We can tell the harder truths later. And, to do ourselves justice, music would give a fairer picture of what we are really like than some of the other things we might be sending, like Time, say, or a history of the UN, or presidential speeches. We could send out our science, of course, but just think of wincing at this end when the polite comments arrive 200 years from now. Whatever we offer as today's items of liveliest interest are bound to be out of date and irrelevant, maybe even ridiculous. I think we should stick to music.

Perhaps, if the technology can be adapted to it, we should send some paintings. Nothing would better describe what this place is like, to an outsider, than the Cezanne demonstrations that an apple is really part fruit, part earth.

What kinds of questions should we ask? The choices will be hard, and everyone will want his special question first. What are your smallest particles? Did you think yourselves unique? Do you have colds? Have you anything quicker than light? Do you always tell the truth? Do you cry? There is no end to the list.

Perhaps we should wait awhile, until we are sure we know what we want to know, before we get down to detailed questions. After all, the main question will be the opener: Hello, are you there? If the reply should turn out to be Yes, Hello, we might want to stop there and think about that, for quite a long time.

56

Jonathan Eberhart
The World on a Record

Describe the world. Not just that multicolored ball in the spacecraft photos, but the *world*—its place in space, its diverse biota, its wide-ranging cultures with their life-styles, arts and technologies—everything, or at least enough to get the idea across. And do it on one long-playing record.

Oh, there's one stipulation: Assume not only that your audience doesn't speak your language, but that it has never even heard of the earth or the rest of the solar system. An audience that lives, say, on a planet orbiting another star, light-years away from anything you would recognize as home.

This fascinating challenge was taken up this year by a group of friends and associates with just such an extraterrestrial audience in mind. Their inducement was the chance to have their message actually delivered, or at least tossed into space like a message in a bottle in a cosmic ocean.

The means of delivery was also the impetus for devising the message: the two Voyager spacecraft, bound for encounters with Jupiter and Saturn and their many moons, with the possibility that one of the probes will press on to Uranus and perhaps even to Neptune. Beyond their planetary objectives, however, the Voyagers will become only the third and fourth manmade vehicles ever to leave the solar system, the third and fourth chances against huge—but not infinite—odds to say "hello" to the members of a truly unearthly civilization. The pathfinders were the Pioneer 10 and 11 spacecraft, which also visited Jupiter (Pioneer 11 will pass close to Saturn in 1979), and which carry small metal plaques portraying man, woman and the position of the earth in space.

The new message is far more elaborate. When Voyager project manager John Casani asked Cornell astronomer Carl Sagan to chair a committee that would consider some sort of message for the Voyager probes, Sagan's first idea was just "a souped-up Pioneer plaque." Indeed, the mere

presence of the spacecraft should be very significant to any civilization advanced enough to detect it. But the committee members, perhaps imagining what their own reactions would be if confronted with a crewless alien probe, opted for a message with a considerably higher "information density."

The chosen medium was essentially a phonograph record, a 12-inch copper disk to be played at $16\frac{2}{3}$ revolutions per minute using a ceramic cartridge and stylus sent along in the spacecraft for the purpose. Pictorial instructions on the lid of the disk's aluminum container show how the cartridge is to be used, with other notations indicating the proper playing speed and the playing time of each side. The group believes that the sealed container and the benign space environment should enable the disk to survive a journey of more than a billion years.

But how do you describe an entire world? The disk is a mixed-media affair, its grooves containing information that a technologically inclined civilization could readily reconstruct into both sounds and pictures. But which sounds, and which pictures? Technical suggestions from Frank Drake, another Cornell astronomer who has spent many years considering interstellar communications, enabled the number of pictures on the disk to be raised from 6 to 116, but that didn't make the choosing any easier.

The picture sequence as adopted begins with a "calibration circle," designed to show extraterrestrial viewers the proper height-to-width ratio of the images that follow. Many people contributed suggestions—the original committee grew considerably during the process—including Toronto artist Jon Lomberg, who prepared a number of diagrams such as DNA structure and continental drift for the project. The sequence flows from a view of the solar system down through the planets to the earth, then up to the most basic levels of biology and up through human beings, other animals and plants. It then moves through the diversity of cultures and the ways in which they live, work and play, finally wending through technology to space flight and ending with sunset—and a violin.

Box 56.1.

PICTURES (in sequence)

calibration circle
solar location map
mathematical definitions
physical unit definitions
solar sys. parameters (2)
the sun
solar spectrum
Mercury
Mars
Jupiter
Earth
Egypt, Red Sea, Sinai
   Pen., Nile (from orbit)
chemical definitions
DNA structure
DNA structure magnified
cells and cell division
anatomy (8)
human sex organs (drawing)
conception diagram
conception photo
fertilized ovum
fetus diagram

fetus
diag. of male and female
birth
nursing mother
father and daughter (Malaysia)
group of children
diagram of family ages
family portrait
continental drift diagram
structure of earth
Heron Island (Australia)
seashore
Snake River, Grand Tetons
sand dunes
Monument Valley
leaf
fallen leaves
sequoia
snowflake
tree with daffodils
flying insect, flowers
vertebrate evolution diag.
seashell (Xancidae)

dolphins
school of fish
tree toad
crocodile
eagle
S. African waterhole
Jane Goodall, chimps
sketch of bushman
bushmen hunters
Guatemalan man
Balinese dancer
Andean girls
Thai craftsman
elephant
Turkish man with beard
   and glasses
old man with dog and
   flowers
mountain climber
Cathy Rigby
Olympic sprinters
schoolroom
children with globe

cotton harvest
grape picker
supermarket
diver with fish
fishing boat, nets
cooking fish
Chinese dinner
licking, eating, drinking
Great Wall of China
African house construction
Amish construction scene
African house
New England house
modern house (Cloudcroft)
house interior with
   artist and fire
Taj Mahal
English city (Oxford)
Boston
UN building (day)
UN building (night)
Sydney Opera House
artisan with drill

factory interior
museum
X-ray of hand
woman with microscope
Pakistan street scene
India rush-hour traffic
modern highway (Ithaca)
Golden Gate Bridge
train
airplane in flight
airport (Toronto)
Antarctic expedition
radio telescope
   (Westerbork)
radio telescope (Arecibo)
book page (Newton's *System
   of the World*)
astronaut in space
Titan Centaur launch
sunset with birds
string quartet
violin with score

GREETINGS IN MANY TONGUES (alphabetically)

Akkadian
Amoy (Min dial.)
Arabic
Aramaic
Armenian
Bengali
Burmese
Cantonese
Czech
Dutch

English
French
German
Greek
Gujarati (India)
Hebrew
Hindi
Hittite
Hungarian

Ila (Zambia)
Indonesian
Italian
Japanese
Kannada (India)
Kechua (Peru)
Korean
Latin
Luganda (Uganda)

Mandarin
Marathi (India)
Nepali
Nguni (SE Africa)
Nyanja (Malawi)
Oriya (India)
Persian
Polish
Portuguese

Punjabi
Rajasthani
Roumanian
Russian
Serbian
Sinhalese (Sri Lanka)
Sotho (Lesotho)
Spanish
Sumerian

Swedish
Telugu (India)
Thai
Turkish
Ukranian
Urdu
Vietnamese
Welsh
Wu (Shanghai dial.)

SOUNDS OF EARTH (in sequence)

whales
planets (audio
   analog of
   orbital velocity)
volcanoes
mud pots
rain

surf
cricket frogs
birds
hyena
elephant
chimpanzee
wild dog

footsteps and
   heartbeats
laughter
fire
tools
dogs (domestic)
herding sheep

blacksmith shop
sawing
tractor
riveter
Morse code
ships
horse and cart

horse and carriage
train whistle
tractor
truck
auto gears
Saturn 5 rocket
   liftoff

kiss
baby
life signs:
   EEG, EKG
pulsar

MUSIC (in sequence)

Bach: Brandenberg Concerto #2, 1st m.
Java: court gamelan— "Kinds of Flowers"
Senegal: percussion
Zaire: "Pygmy girls" Initiation song
Australia: horn and totem song
Mexico: mariachi— "El Cascabel"
Chuck Berry: "Johnny B. Goode"
New Guinea: men's house
Japan: shekuhachi (flute)—
   "Depicting the Cranes in Their Nest"

Bach: Partita #3 for violin
Mozart: "Queen of the Night"
   (from "The Magic Flute")
Georgia (USSR): folk chorus—"Chakrulo"
Peru: pan pipes
Louis Armstrong: "Melancholy Blues"
Azerbaijan: two flutes
Stravinsky: "Rite of Spring" conclusion
Bach: Prelude and Fugue #1 in C Major
Beethoven: Symphony #5, 1st m.

Bulgaria: shepherdess song—
   "Izlel Delyo hajdutin"
Navajo: night chant
English 15th cent.: "The Fairie Round"
Melanesia: pan pipes
Peru: woman's wedding song
China: ch'in (zither)—"Flowing Streams"
India: raga—"Jaat Kahan Ho"
Blind Willie Johnson: "Dark Was the Night"
Beethoven: String Quartet #13, "Cavatina"

One conspicuous lack is that of nude human beings, which might have been informative to an alien species wondering about the creatures that built the spacecraft. A nude man and woman appear on the Pioneer plaques, but the National Aeronautics and Space Administration got a little protest mail about that ("using the taxpayers' money to send smut into space") and vetoed the idea for Voyager. The included representation of human sex organs is a drawing from a biology textbook.

Following the picture sequence is a message from President Carter, a list of the congressional leaders and committee members who enabled Voyager to be funded, and a message from United Nations Secretary General Kurt Waldheim. The list and both messages are also presented visually; it is not until the next section that the record produces its first sound.

It is a greeting, spoken in Sumerian, perhaps the oldest known language in the world. There are greetings in 55 languages on the disk, including the 25 most widely spoken ones and ending, in English, with a little voice saying, "Hello from the children of planet earth."

Next comes an "essay in sound," compiled from many sources and organized primarily by writer Ann Druyan, who also served as "creative director" for the project. Like the pictures, the sounds flow in a progression, moving from the recorded "greeting call" of a whale to the natural sounds of the earth, its animals and signs of human beings. There is fire, then tools, then domestication of animals, manual labor, the coming of machines and on to the roar of a Saturn 5 rocket taking off. Then, almost as a coda, there is a kiss, a baby's voice and the sounds of medical instruments indicating that life continues to be reborn.

The record ends with music, which posed the hardest choices of all. It has been proposed in the past that the intricate music of Bach might be appropriate as a message to distant civilizations, suggesting artistic sophistication amid technological development. The Voyager message-makers, however, wanted to present the diversity of a world—and spent 10 weeks deciding how to do it. Their choice runs from Bach to the blues, an Indian raga and a Navajo chant, often juxtaposed to suggest an underlying unity by aligning similar instruments or moods from different traditions.

About 30 organizations contributed to the overall project, says Sagan, as did many individuals. Timothy Ferris of Rolling Stone produced the record, and Columbia Records manufactured it as a public service. But after all the effort, will anyone ever receive the message?

Certainly not for tens or hundreds of millennia. Michael R. Helton of Jet Propulsion Laboratory ran a computer analysis of where the Voyagers will go after they leave the solar system. Voyager I will pass Pluto's orbit late in 1987 and head toward the constellation Ophiuchus. Voyager 2, assuming that it goes to Uranus but not to Neptune (a reasonable bet, since a close look at Uranus's newly discovered rings would prohibit the Neptune visit), will leave the system in mid-1989 on the way to Capricornus. In about 40,000 years, both craft should pass within 1 to 2 light-years of a fourth-magnitude star (AC + 79 3888)— not exactly grabbing distance. Voyager 2 should pass a similar distance to another star (AC −24 2833-183) 110,000 years later, and about 375,000 years after that, Voyager I will pass perhaps 1.5 light-years from DM + 21 652 in Taurus.

None of those encounters are close ones, and tiny errors could make large differences in the trajectories anyway. But the numbers are all so vast as to make such calculations nearly irrelevant. The record's mentors, however, feel that the possibility of actual reception is only one reason for sending the message. Another is simply to get earthlings thinking about the chance that there are beings even capable of receiving it. And then there is the significance of looking at one's whole home planet through a few representative details.

Try it. Make your own list. Or imagine: If you, as an alien, got this message, what would you think?

57

Gerard K. O'Neill
The Colonization of Space

New ideas are controversial when they challenge orthodoxy, but orthodoxy changes with time, often surprisingly fast. It is orthodox, for example, to believe that Earth is the only practical habitat for Man, and that the human race is close to its ultimate size limits. But I believe we have now reached the point where we can, if we so choose, build new habitats far more comfortable, productive and attractive than is most of Earth.

Although thoughts about migration into space are as old as science fiction, the technical basis for serious calculation did not exist until the late 1960's. In addition, a mental "hangup"—the fixed idea of planets as colony sites—appears to have trapped nearly everyone who has considered the problem, including, curiously enough, almost all science-fiction writers. In recent months I learned that the space pioneer Konstantin Tsiolkowsky, in his dreams of the future, was one of the first to escape that hangup.

By chance, and initially almost as a joke, I began some calculations on the problem in 1969, at first as an exercise for the most ambitious students in an introductory physics course. As sometimes happens in the hard sciences, what began as a joke had to be taken more seriously when the numbers began to come out right. There followed several years of frustrating attempts to get these studies published.

Friends advised that I take my ideas "to the people" in the form of physics lectures at universities. The positive response (especially from students) encouraged me to dig harder for the answers to questions about meteoroid damage, agricultural productivity, materials sources, economics and other topics. The results of that study indicate that

● we *can* colonize space, and do so without robbing or harming anyone and without polluting anything.

● if work is begun soon, nearly all our industrial activity could be moved away from Earth's fragile biosphere within less than a century from now.

● the technical imperatives of this kind of migration of people and industry into space are likely to encourage self-sufficiency, small-scale governmental units, cultural diversity and a high degree of independence.

● the ultimate size limit for the human race on the newly available frontier is at least 20,000 times its present value.

How can colonization take place? It is possible even with existing technology, if done in the most efficient ways. New methods are needed, but none goes beyond the range of present-day knowledge. The challenge is to bring the goal of space colonization into economic feasibility now, and the key is to treat the region beyond Earth not as a void but as a culture medium, rich in matter and energy. To live normally, people need energy, air, water, land and gravity. In space, solar energy is dependable and convenient to use; the Moon and asteroid belt can supply the needed materials, and rotational acceleration can substitute for Earth's gravity.

Space exploration so far, like Antarctic exploration before it, has consisted of short-term scientific expeditions, wholly dependent for survival on supplies brought from home. If, in contrast, we use the matter and energy available in space to colonize and build, we can achieve great productivity of food and material goods. Then, in a time short enough to be useful, the exponential growth of colonies can reach the point at which the colonies can be of great benefit to the entire human race.

To show that we are technically able to begin such a development now, this discussion will be limited to the technology of the 1970's, assuming only those structural materials that already exist. Within a development that may span 100 years, this assumption is unrealistically conservative. We shall look at the individual space communities—their structure and appearance and the activities possible for their inhabitants, their relation to the space around them, sources of food, travel between communities as well as to Earth, the economics of the colonies and plans for their growth. As is usual in physics, it is valuable to consider limiting cases: for this study, the limits are an eventual full-size space community on a scale

Reprinted from *Physics Today*, 27 (1974), 32. Copyright © 1974 American Institute of Physics. Reprinted with permission.

established by the strength of materials, and a first model, for which cost estimates can reasonably be made. The goals of the proposal will be clearer if we first discuss the large community.

A Cylindrical Habitat

The geometry of each space community is fairly closely defined if all of the following conditions are required: normal gravity, normal day and night cycle, natural sunlight, an earthlike appearance, efficient use of solar power and of materials. The most effective geometry satisfying all of these conditions appears to be a pair of cylinders. The economics of efficient use of materials tends to limit their size to about four miles in diameter, and perhaps about 16 miles in length. (See Figure 57.1). In these cylinder pairs, the entire land area is devoted to living space, parkland and forest, with lakes, rivers, grass, trees, animals and birds, an environment like most attractive parts of Earth; agriculture is carried on elsewhere. The circumference is divided into alternating strips of land area ("valleys") and window area ("solars"). The rotation period is two minutes, and the cylinder axes are always pointed toward the Sun.

Because the Moon is a rich source both of titanium and of aluminum, it is likely that these metals will be used extensively in the colonies. For conservatism, though, the calculation of the cylinder structure has been based on the use of steel cables, to form "longerons" (longitudinal members carrying the atmospheric forces on the end caps) and circumferential bands (carrying the atmospheric force and the spin-induced weights of the ground, of the longerons and of themselves. For details of this calculation and the assumptions it includes, see Box 57.1). The steel cables are bunched to form a coarse mesh in the window areas. The bands there subtend a visual angle of 2.3×10^{-4} radians, about equal to the diffraction limit for the sunlight-adapted human eye, and so are nearly invisible. The windows themselves are of glass or plastic, subdivided into small panels.

There is no sharp upper limit on the size of a space-community cylinder; with increasing size, though, a larger fraction of the total mass is in the form of supporting cables. The figure 3200 meters for radius R is somewhat arbitrary. Economy would favor a smaller size; use of high-strength materials, or a strong desire for an even more earthlike environment, would favor a larger. Independent of size, the apparent gravity is earth-normal, and the air composition as well as the atmospheric pressure are those of sea level on Earth. For R equal to 3200 meters, the atmospheric depth is that of an Earth location at 3300 meters above sea level, an altitude where the sky is blue and the climate habitable: At any radius r within the cylinder we have

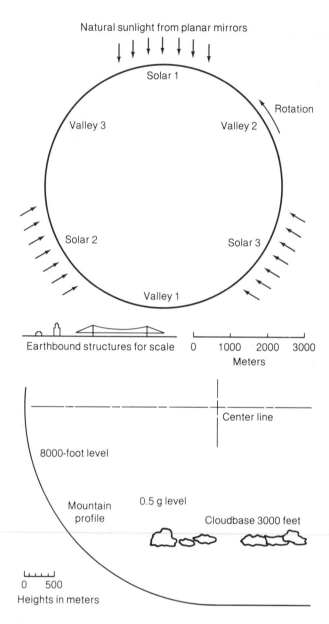

Figure 57.1. Section of a space-community main cylinder (top). The circumference is divided into alternating strips of land area (valleys) and window area (solars). Although the space-community valleys offer new landscaping opportunities and architectural possibilities, it is reassuring to note that certain Earth features can be recreated: the side view of a cylinder end cap (bottom) includes a mountain profile taken from an aerial photograph of a section of the Grand Teton range in Wyoming.

$$p = p_0 e^{-a(R^2 - r^2)}$$

where

$$a \equiv g \, \rho_0 / 2Rp_0$$
$$= (1/2R)(1.2 \times 10^{-4}/\text{meter})$$

Box 57.1. Steel Structure

For the structure, steel cables are assumed to be formed into longerons (average thickness Δr_L) and circular bands (average thickness Δr_B). The value of Δr_L required is

$$\Delta r_L = R p_0 / 2T$$

where R is the cylinder radius, p_0 the atmospheric pressure and T the tension. For land density ρ_L and depth x_L, and bands of density ρ_F, the total equivalent internal pressure p_T is

$$p_T = p_0 + \rho_L x_L g + \rho_F r_B g + \rho_F r_L g$$

To solve for p_T we note that

$$\Delta r_B = p_T R / T$$

so that

$$p_T = (p_0 + g \rho_L x_L + g \rho_F R / T) / (1 - g \rho_F R / T)$$

For an average soil depth of 150 cm, with an average density of 1.5 gm per cc,

$$p_0 = g \rho_L x_L = 1.23 \times 10^5 \text{ newtons/m}^2$$

To arrive at a conservative value for T, we note that half a century ago, the working stress for suspension-bridge cables was 70,000 to 80,000 pounds per square inch (Hool and Kinne, 1943; Steinman, 1929). At that time, D. B. Steinman argued for the use of stresses over 100,000 psi. If we use 1920's steels, hardened to bring the yield point to 90% of the ultimate strength, and work at 75% of the yield point, the working stress can be 152,000 psi. If we take T as 150,000 psi and R as 3200 meters, the averaged surface mass density is 7.5 tons per square meter.

In the window (solar) areas, the longerons can be 0.8-meter cables in stacks of four at 14-meter intervals. The bands can be in the same arrangement, but with a 1.5-meter diameter, and the mesh transparency will then be 84%. Considerably larger values of R would result from the extensive use of titanium in the structure, together with a thinner layer of earth.

The length of a day in each community is controlled by opening and closing the main mirrors that rotate with the cylinders. The length of day then sets the average temperature and seasonal variation within the cylinder. Each cylinder can be thought of as a heat sink equivalent to 3×10^8 tons of water; for complete heat exchange, the warmup rate in full daylight would be about 0.7 deg C per hour. As on Earth, the true warmup rate is higher because the ground more than a few centimeters below the surface does not follow the diurnal variation.

Bird and animal species that are endangered on Earth by agricultural and industrial chemical residues may find havens for growth in the space colonies, where insec-

ticides are unnecessary, agricultural areas are physically separate from living areas, and industry has unlimited energy for recycling.

As we can see in Figure 57.1, it is possible to recreate certain Earth features: the mountain profile is taken from an aerial photograph of a section of the Grand Teton range in Wyoming. The calculated cloud base heights as seen in the figure are typical of summer weather on Earth: For a dry adiabatic lapse rate of 3.1 deg per 300 meters and a dew-point lapse rate of 0.56 deg per 300 meters, 50% relative humidity and a temperature range between zero and 32° C, the cloud base heights range between 1100 and 1400 meters.

Environmental Control

The agricultural areas are separate from the living areas, and each one has the best climate for the particular crop it is to grow. Gravity, atmosphere and insolation are earthlike in most agricultural cylinders, but there is no attempt there to simulate an earthlike appearance. Selected seeds in a sterile, isolated environment initiate growth, so that no insecticides or pesticides are needed. (The evolution time for infectious organisms is long, and resterilization of a contaminated agricultural cylinder by heating would not be difficult.) All food can be fresh, because it is grown only 20 miles from the point of use. The agricultural cylinders can be evenly distributed in seasonal phase, so that at any given time several of them are at the right month for harvesting any desired crop.

Figure 57.2 shows side and end views of a space community as a complete eco-system. The main mirrors are made of aluminum foil and are planar. Moving these mirrors varies the angle at which sunlight hits the valleys (controlling the diurnal cycle), and the Sun appears motionless in the sky, as it does on Earth. The solar power stations, which consist of paraboloidal mirrors, boiler tubes and conventional steam-turbine electric generators, can provide the community with sufficient power, easily up to ten times the power per person now used (10 kw) in highly industrialized regions (Singer, 1970). For such energy-rich conditions (120 kw per person) the power needed for a cylinder housing 100,000 people is 12,000 megawatts: The solar power incident on a cylinder end cap is 36,000 megawatts, adequate if the thermal efficiency is 33%. Extra power plants near the agricultural ring would be needed for higher population density. Waste heat is sent into space by infrared radiators of low directionality.

The communities are protected from cosmic rays by the depth of the atmosphere and by the land and steel supporting structure, the bands and longerons being distributed where visual transparency is unnecessary. Meteoroid damage should not be a serious danger. Most meteoroids are of cometary rather than asteroidal origin and are dust

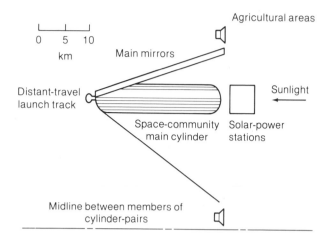

Figure 57.2. Space community as a whole is seen in side (left) and end (right) views. For the end view, 37 of the 72 agricultural cylinders in a ring are shown; the ring does not rotate as a whole. Note the lines of symmetry in both sections of the figure.

conglomerates, possibly bound by frozen gases (NASA, 1969); a typical meteoroid is more like a snowball than like a rock. Spacecraft sensors have collected abundant and consistent data on meteoroids in the range 10^{-6} to 1 gram, and the Apollo lunar seismic network is believed to have 100% detection efficiency for meteoroids (Latham et al., 1973) above 10 kg: Data from these sources are consistent with a single distribution law.

The Prairie Network sky-camera data (McCrosky, 1965) after substantial correction for assumed luminous efficiency, agree with data from the National Aeronautics and Space Administration for 10-gm meteoroids. The spacecraft and seismic data indicate a mean interval of about one-million years for a strike by a heavy (one ton) meteoroid on a space community of cross section 1000 square kilometers. Even such a strike should produce only local damage if the structure is well designed. For 100-gram meteoroids, the mean interval for a strike is about three years. From the combined viewpoints of frequency and of momentum carried, the size range from one to ten grams may need the most care in window design and repair methods. For total breakage of one window panel, Daniel Villani at Princeton has calculated a leakdown time of about 300 years. Meteoroid-damage control is, then, a matter of sensing and of regular minor repair rather than of sudden emergencies.

Axial Rotation and Transport

A key element in the design of the space colony is the coupling of two cylinders by a tension cable and a compression tower to form a system that has zero axial angular momentum and is therefore able to maintain its axis

pointed toward the Sun without the use of thrusters. The force and torque diagram for this arrangement is seen in Figure 57.3. To accelerate the cylinders up to the required rotational speed, static torque is transmitted through the compression framework that joins the two cylinders of a pair. For a spin-up time of three years, a constant 560 000 horsepower is needed; this is 3% of the generator capacity of a cylinder. After spinup, the same motors can provide maintenance power for frictional losses and for attitude control about the spin axis. Each cylinder's angular momentum is 1.5×10^{18} kg^2 rad per sec; the torque needed to precess this angular momentum once each year is 3×10^{11} newton meters, corresponding to a constant force of 1200 tons on a 26-km lever arm.

The phase difference of seasons between the two cylinders permits "seasonal counterpoint," midsummer in one cylinder during midwinter in the other. Travel between the two requires no power and only nine minutes of time. They are only 90 km apart, and engineless vehicles can unlock from the outer surface of one cylinder at a preset time, move in free flight with the tangential velocity (180 meters per sec or 400 miles per hour) and lock on to the other cylinder at zero relative velocity.

Travel between communities can also be carried out with simple engineless vehicles, accelerated in a computed direction by a stationary cable-pulling electric motor and decelerated by an arresting cable at the destination. The "cable-car" vehicles for such free flight need no fuel, no complex maintenance nor a highly trained crew, and should be inexpensive. Vehicle speeds permit travel among a total population larger than that of Earth within flight times of seven hours. (I have here assumed communities spaced at 200-km intervals, so that the maximum dimension of a planar cluster housing 4 billion people is

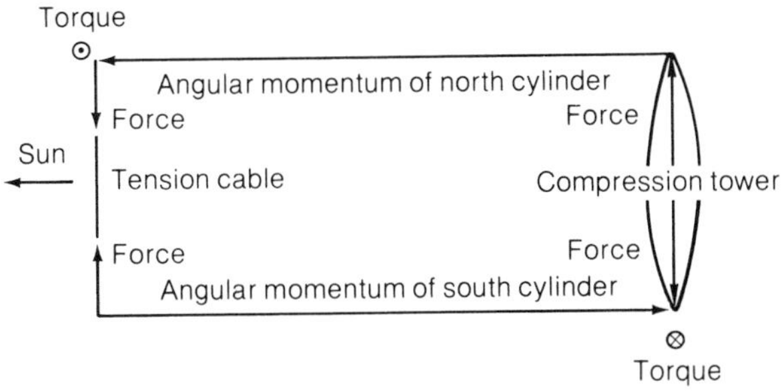
Torque
Angular momentum of north cylinder
Force
Force
Sun
Tension cable
Compression tower
Force
Force
Angular momentum of south cylinder
Torque

Figure 57.3. Force and torque diagram for a cylinder. Nondissipative static forces are used to precess the spin angular momenta, so that cylinder axes always point toward the Sun.

29,000 km. For a vehicle with acceleration 1g and the required travel time of seven hours, the acceleration length is 66 km.) With no need for aerodynamic design, the vehicles can be far more roomy and comfortable than the typical earthbound commercial jet.

Life in the Colonies

The key statements so far have been based on known facts, on calculations that can be checked and on technology whose costs can be estimated realistically. The discussion, however, would be sterile without some speculations—speculations that must, of course, be consistent with the known facts.

With an abundance of food and clean electrical energy, controlled climates and temperate weather, living conditions in the colonies should be much more pleasant than in most places on Earth. For the 20-mile distances of the cylinder interiors, bicycles and lowspeed electric vehicles are adequate. Fuel-burning cars, powered aircraft and combustion heating are not needed; therefore, no smog. For external travel, the simplicity of engineless, pilotless vehicles probably means that individuals and families will be easily able to afford private space vehicles for low-cost travel to far distant communities with diverse cultures and languages. The "recreational vehicles" of the colonial age are therefore likely to be simple spacecraft, consisting of well furnished pressure shells with little complexity beyond an oxygen supply and with much the same arrangement of kitchen facilities and living space as are found today in our travelling homes.

All Earth sports, as well as new ones, are possible in the communities. Skiing, sailing, mountain climbing (with the gravity decreasing linearly as the altitude increases) and soaring are examples. As an enthusiastic glider pilot, I have checked the question of thermal scales: The soaring pilots of the colonial age should find sufficient atmospheric instability to provide them with lift. At high altitudes, man-powered flight—a nearly impossible dream on Earth—becomes easy. A special, slowly rotating agricultural cylinder with water and fish can have gravity 10^{-2} or 10^{-3} times that on Earth for skin diving free of pressure-equalization problems. Noisy or polluting sports, such as auto racing, can easily be carried out in one of the cylinders of the external ring.

The self-sufficiency of space communities probably has a strong effect on government. A community of 200,000 people, eager to preserve its own culture and language, can even choose to remain largely isolated. Free, diverse social experimentation could thrive in such a protected, self-sufficient environment.

If we drop our limitation to present technology, the size of a community could be larger. One foreseeable development is the use of near-frictionless (for example, magnetic) bearings between a rotating cylinder and its supporting structure, which need not be spun. For eight tons per square meter of surface density and a tensile strength of 300,000 psi R would be 16 km, the total area would be 50,000 km^2, and the population would be between five million (low density) and 700 million (the ecological limit, the maximum population that can be supported).

In Table 57.1 we see my estimate of the earliest possible schedule for space colonization, beginning with a model community in the late 1980's. From about the year 2014, I assume a doubling time of six years for the colonies; that is, the workforce of a "parent" colony could build a "daughter" colony within that time. In making these estimates I have calculated that the first model community would require a construction effort of 42 tons per man-year, comparable to the effort for large-scale bridge building on Earth. Full-size communities at high population density require 50 tons per man-year, and up to 5000 tons per man-year for low population density. For comparison, automated mining and shipping in Australia now reaches 200 tons per man-year averaged over a town (MacLeish, 1973).

In the long run, space-colony construction is ideally suited to automation. A colony's structure consists mainly of cables, fittings and window panels of standard modular form in a pattern repeated thousands of times. The assembly takes place in a zero-gravity environment free of the vagaries of weather. By the time that the colonies are evolving to low population density, therefore, I suspect that very few people will be involved in their construction. Most of the workforce will probably be occupied in architecture, landscaping, forestry, zoological planning, botany and other activities that are nonrepetitive and require a sense of art and beauty.

Table 57.1.

POSSIBLE STAGES IN THE DEVELOPMENT OF SPACE COMMUNITIES

Model	Length (km)	Radius (m)	Period (sec)	Population[a]	Earliest estimated date
1	1	100	21	10,000	1988
2	3.2	320	36	$100\text{-}200 \times 10^3$	1996
3	10	1000	63	$0.2\text{-}2 \times 10^6$	2002
4	32	3200	114	$0.2\text{-}20 \times 10^6$	2008

[a] Population figures are for double unit; higher figures are the approximate ecological limits, for conventional agriculture.

Our New Options

It is important to realize the enormous power of the space-colonization technique. If we begin to use it soon enough, and if we employ it wisely, at least five of the most serious problems now facing the world can be solved without recourse to repression: bringing every human being up to a living standard now enjoyed only by the most fortunate; protecting the biosphere from damage caused by transportation and industrial pollution; finding high-quality living space for a world population that is doubling every 35 years; finding clean, practical energy sources; preventing overload of Earth's heat balance.

I hesitate somewhat to claim for space-colonization the ability to solve one other problem, one of the most agonizing of all: the pain and destruction caused by territorial wars. Cynics are sure that humanity will always choose savagery even when territorial pressures are much reduced. Certainly the maniacal wars of conquest have not been basically territorial. Yet I am more hopeful; I believe we have begun to learn a little bit in the past few decades. The history of the past 30 years suggests that warfare in the nuclear age is strongly, although not wholly, motivated by territorial conflicts; battles over limited, nonextendable pieces of land.

From the viewpoint of international arms control, two reasons for hope come to mind. We already have an international treaty banning nuclear weapons from space, and the colonies can obtain all the energy they could ever need from clean solar power, so the temptations presented by nuclear-reactor byproducts need not exist in the space communities.

To illustrate the power of space-colonization in a specific, calculable situation, we trace the evolution of a worst-case example: Suppose the present population-increase rate were to continue on Earth *and* in the space colonies. In that case the total human population would increase 20,000-fold in a little over 500 years. Space-colonization would absorb even so huge a growth, as we shall see from our calculations.

The total volume of material needed in a full-size community is 1.4×10^9 cubic meters, and the material available in the asteroid belt (from which the later communities will be built) is estimated to be 4×10^{17} cubic meters, about one twenty-five hundredth the volume of Earth. For a present world population of 3.9×10^9 people and a growth rate (Population Reference Bureau, 1970) of 1.98% per year (the 1965-71 average), the asteroidal material would last 500 years, corresponding to a 20,000-fold population increase at low population density.

In Figure 57.4, we see the development of this worst-case problem. To hasten the solution of that problem, the initial space community population density is taken as the ecological limit; the maximum number of people that can be supported with food grown within the communities, with conventional agriculture. Richard Bradfield has grown enough to feed 72 people per hectare by the techniques of double planting and multiple cropping, and with the use of cuttings for livestock feed. These results (Bradfield, 1972), as published and also as described to me by Bradfield, were obtained in the Phillipines, which has only a nine-month growing season and less than ideal weather conditions. Calculations based on his figures, but assuming an ideal twelve-month season, indicate that the colonies should be able to support 143 people per hectare with a diet of 3000 calories, 52 grams of usable protein and 4.3 pounds of total food per person per day (Lappe, 1971). Much of the protein would come from poultry and pork. The two main cylinders of Model 1 should then be able to support up to 10,800 people, and the corresponding ecological limit for a full-size community would be 20 million people. At this limit, all the colonists would have a high standard of living, but in apartment-house living conditions, looking out over farmland. For a community limit of 13 million people, the main cylinders could be kept free of agriculture.

By about 2050, then, Figure 56.4 indicates that emigration to the colonies could reverse the rise in Earth's population, and that the acceleration of the solution could be dramatically fast: Within less than 30 years, Earth's population could be reduced from a peak of 16.5 billion people to whatever stable value is desired. I have suggested 1.2 billion as a possible optimum; it corresponds to the year 1910 in Earth history. The reduction in population density in the space communities could be equally rapid, and within another 40 years new construction could thin out the communities to a stable density of 1.43 people per hectare, about one hundredth of the ecological limit. The total land area in the colonies would then be more than three times that of Earth.

We can hope that, in contrast to this worst-case example, some progress toward zero population growth (Club of Rome, 1972) will be made in the next 75 years. Any such progress will hasten the solution, reduce Earth's population peak, and hasten the day when the population

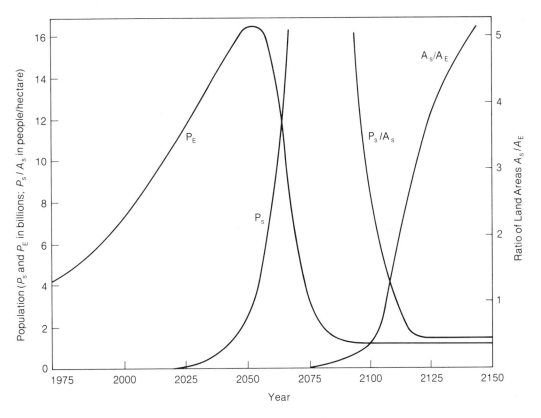

Figure 57.4. Effectiveness of space colonization in solving a hypothetical "worst case" population-growth problem. The case considered assumes no reduction of population growth rate either on Earth or in the space colonies. Here P_E is the population of Earth, P_s that of space, and A_s/A_E the ratio of land area in space (all usable to total land area of Earth.

Both P_E and P_s/A_s reach stable, relatively low values. Changes within wide limits in the assumed input numbers do not affect the reaching of a stable solution, nor do they affect the final stable values of P_E and P_s/A_s. This figure is an example of the power of space-colonization, not a prediction.

densities on Earth as well as in the colonies can be reduced to an optimum value.

Building the First Colony

A responsible proposal to begin the construction of the first colony must be based on a demonstration, in some detail, of one workable plan with realistic cost estimates. I emphasize two points about any such plan: The details presented should be thought of simply as an existence proof of feasibility; and many variations are possible. The optimum design and course of action can only be decided on after study and consultation among experts in a number of fields.

The nominal values for the first model colony are taken as: construction force, 2000 people; population, 10,000; total mass, 500,000 tons. When the design and cost analysis are done in detail for the entire enterprise, the need to fit a budget may force some reduction in size. The

initial estimates have been aimed at holding the cost equal to that of one project we have already carried through: Apollo. The choice of 10,000 as a target population ensures that, even with some reduction, Model 1 will be large enough to obtain economies of scale and to serve as an effective industrial base for the construction of Model 2. A much reduced colonization project would be little more than a renamed space station, perhaps able to maintain itself but incapable of building the larger models that are necessary if the program is ultimately to support itself. It is an essential feature of the colonization project that Earth should no longer have to support it after the first two or three stages.

Ultimately, colonization could take place in the entire sphere, 3×10^{17} km^2 in area, that surrounds the Sun at the distance we have evolved to prefer (the so-called "Dyson sphere"). For the first colony it is probably best to choose a particular point on that sphere, within easy range of both Earth and Moon, not so close as to be eclipsed often, and preferably stable against displacements in all three coordinates. The L_4 and L_5 Lagrange libration points satisfy all these conditions. They have the further advantage of forming only a very shallow effective-potential well (Michael, 1963).

Earth, Moon, Sun and the colony form a restricted four-body gravitational problem, for which the full solution has only been worked out within the past several years (Kolenkiewicz and Carpenter, 1968; Kamel, 1970). The stable motion is a quasielliptical orbit, of large dimensions, about L_5. The maximum excursions in arc and radius are several tenths of the Earth-Moon distance. On the stable orbit there is room for several thousand colonies; a long time will pass before colonization can fill so big an orbit.

Cost Minimization

There are several key problems involved here, each of which appears to yield to an efficient solution in principle: reducing freight-shipment cost from the Earth to L_5, the colony site; minimizing the mass needed from Earth; designing a device for low-cost transfer of materials from the Moon to L_5.

The first problem was considered by Robert Wilson (NASA), Eric Hannah and George Hazelrigg (Princeton) at a meeting held 9 and 10 May at Princeton (A *Proceedings* of this meeting will be published). Their conclusion was that the best method during the 1980's will probably be conventional chemical rockets—specifically, the high-quality engines already being developed for the space shuttle. Among several variations possible, the common feature was reusability, and the cost estimates for shipment varied from $190 to $400 per pound, in 1972 dollars. The cost summary table (Table 57.4) therefore assumes $425 per pound.

To reduce the mass needed from Earth, most of the repetitive structural members (aluminum) and window panels (glass) must be produced at L_5 from lunar material. A further, important saving is made by getting 89% of the mass of needed water from oxygen in the plentiful lunar-surface oxides, bringing only 11% of the water mass as liquid hydrogen from Earth. Of the 500,000-ton total mass (see Table 57.2) for the Model 1 colony, 98% can be obtained from the Moon. The elements most needed are aluminum, titanium, silicon and oxygen. Lunar surface soil is usable for agriculture, with the addition of nitrates and small amounts of trace elements. The remaining 10,000 tons must come from the Earth.

To bring the total cost within practical limits, we must develop a low-cost method for transporting raw materials from the Moon to the construction site. The discussion of transport methods should be taken as an existence proof rather than as a detailed design. There may very well be better methods than those I have considered; however, it is

Table 57.2.

MASSES OF MATERIALS REQUIRED FOR MODEL 1 (METRIC TONS)

	Total mass required	Mass required from Earth
Aluminum (container, structures)	20,000	—
Glass (solars)	10,000	—
Water	50,000[a]	—
Generator plant	1,000[b]	1,000
Initial structures	1,000	1,000
Special fabricated hardware	1,000	1,000
Machines and tools	800	800
Soil, rock and construction materials	420,000[a]	—
Liquid hydrogen	5,400	5,400
2000 people and equipment	200	200
Dehydrated food	600	600
Totals	>500,000	10,000

[a] Includes replenishable reserves to be used to initiate construction of Model 2, and so are higher than the minima required for Model 1.
[b] For 100 MW plant.

enough to show two solutions that appear to be workable. Both use the two great advantages of the lunar environment: an excellent vacuum and a very low escape velocity, about 1.5 miles per sec, less than one quarter of the escape velocity from Earth. To bring a kilogram to L_5 from the Moon takes less than 5% of the energy needed to take a kilogram from Earth.

Both methods assume electric power from a conventional steam-electric power plant that uses solar energy, and both assume that the system runs only during the lunar day, the night being used for scheduled maintenance, crew rest and possibly materials processing. I have also assumed another factor of two lost to system breakdowns. Overall then, each system is assumed to be running only one week in four.

The first method, called "RPL" for rotary pellet launcher, is a symmetric, two-arm propeller-like device, running at constant speed. (See Box 57.2 for description). To transfer 500 tons in six years, about 26 such RPL's would be needed, for a total power of 32 MW. Precise steering is carried out by a linear electromagnetic deflection-plate system after the launching, to hold down the pellet dispersion and permit easy collection.

The alternative method, called "TLA" for transport linear accelerator, uses the technology of dynamic magnetic levitation and the linear synchronous motor. The TLA is a recirculating system of small, passive vehicles (buckets), each having no moving parts but containing superconducting coils. The bucket accelerates a 9-kg payload to escape speed along a magnetic-levitation, linear-synchronous track. Deceleration then releases the payload, the bucket slows to a moderate speed, and is recirculated to receive another payload. Table 57.3 shows some guideline parameters. The mass estimate is 1500 tons, of which about 80% is in power-generation and

Box 57.2. Rotary Pellet Launcher

The rotary launcher is assumed to be a symmetric two-arm propeller-like device, running at constant speed, with launching arms of ten-meter radius.

Mass	10 tons
Rotation rate	2300 rpm
Tip speed (escape velocity)	2400 m/sec
Power	1600 horsepower

The transfer rate per launcher is 3250 tons per year for the transfer of 5-gm pellets, assuming a 25% duty cycle. The strength-to-mass ratio for the launcher is within the range attainable by boron-filament technology: An aluminum matrix containing boron grown on tungsten cores is calculated to have a yield stress of 322,000 psi and average density 4.1, so that

$$\rho/T = 1.85 \times 10^{-6}$$

Here ρ is the density and T the tension in MKS units. For uniform stress, the ratio of arm radii at the base and the tip (r_1 and r_2) is

$$\log r_1/r_2 = (\rho/4T)v^2$$

where v is the escape velocity. For r_1/r_2 less than 50, ρ/T must be less than 2.08×10^{-6}.

Table 57.3.

GUIDELINE PARAMETERS FOR TRANSPORT LINEAR ACCELERATOR

Acceleration	288 meters/sec^2
Average accelerating force	900 lbs
Maximum field	10,000 gauss
Bucket dry mass	5 kg
Payload	9 kg
Repetition rate	1/sec
Transport rate	750 tons/day
Buckets on accelerator	8
Sector length (accelerator)	50 meters
Inductance per meter	0.6 microHy
Peak stored energy per meter	10.4 KJ
Maximum frequency (LSM)[a]	2500 Hz
Instantaneous length driven	2 meters
Direct current in bucket coil	75×10^3 amp-turns
Peak current in LSM[a]	136×10^3 amps
Acceleration power	40 MW
Ohmic losses (feeders)	15 MW

[a] LSM: linear synchronous motor

power-handling equipment. In six years, running 25% of the time, the TLA can transport over 300 times its own weight. (For a short bibliography of early work on the possibilities of electromagnetic launching, before the development of dynamic magnetic levitation, see Clarke, 1950.)

Bot RPL and TLA may have eventual applications as high-throughput, energetically efficient reaction motors, running on solar power and able to use any kind of asteroidal debris as reaction mass. They could propel very large payloads, in the million-ton range or higher, between the asteroid belt and the L$_5$ site.

The Model 1 colony will be too small to carry out a wide variety of manufacturing processes, but it can perform those tasks that are energy intensive, not labor intensive, and that will produce a large return in total tonnage. One example is the production of aluminum by the Hall process. An installed capacity of 40 MW is enough to produce 20,000 tons of aluminum in two years, for the exostructure of Model 1. Another example is the separation of oxygen from the lunar oxides to combine with hydrogen brought from Earth. (With 50,000 tons of water, Model 1 can have lush vegetation as well as substantial streams and small lakes). In contrast, small, low-mass parts are best brought from Earth.

The later colonies, perhaps beginning with Model 3, will use asteroidal material, which is rich in hydrocarbons as well as in metals. We can speculate that, relatively early in the development of the colonies, the economics of freight transport will probably dictate that the "up" shipments from Earth will consist only of people and labor-intensive, miniaturized products such as computers and calculators. The "down" shipping costs may be lower because of the possibility of atmospheric braking. Between colonies, all shipping and travel costs should be very low. For Model 1, the project cost is summarized in Table 57.4. For comparison, the Apollo project cost about 33 billion (1972) dollars.

We can also see in Table 57.4 that the economic payoff from the construction of the first community will come quickly, during construction of the second. That payoff will be in the form of transport costs saved because tools and fabricated structures will be made from lunar material at Community 1 rather than on Earth. The first colony can apparently pay for itself in one or two years, and, by its presence, can keep the annual cost of building Community 2—with its 100,000 to 200,000 people—at about the same level as for Community 1. After that, construction costs for models 3, 4 and so on, should taper off as space-based industry becomes stronger, and as the wide range of chemical elements in the asteroids are used.

We can speculate that the second or third colony may begin to pay back its construction cost in additional ways, for example by the manufacture of high-strength single crystals (Gatos and Witt, 1974) in the zero-gravity, high-vacuum environment that surrounds it, and by the manufacture of titanium products.

To follow the economics as far as Model 3 would be too speculative; its costs to Earth will mainly be those of transporting its one to two million inhabitants to L$_5$. Its

Table 57.4.

ESTIMATED COST OF BUILDING SPACE COLONIES (IN 1972 DOLLARS)

Item	Model 1		Model 2	
	Unit cost	Total (in 10^9)	Unit cost	Total (in 10^9)
Launch vehicles	0.3×10^9	0.9	0.5×10^5	1.5
Transport E → L$_5$	425/lb.	8.5	250/lb	11.0
People E → L$_5$	1000/lb	2.2	500/lb	8.8
Transport E → M	1000/lb	6.6	500/lb	2.2
Equipment for Moon	400/lb	2.4	400/lb	1.8
Equipment for L$_5$	180/lb	1.2	180/lb	2.0
Machines and tools (L$_5$)	625/lb	1.1	625/lb	2.8
Salaries (L$_5$)	50,000/man-year	0.6	(25% on Earth)	2.0
Sararies (Earth)	30,000/man-year	7.2	(30,000/man-year)	2.0
Totals		30.7		34.1[a]
		(5.1×10^9/yr)		(4.3×10^9/yr[a])

[a] The cost saving due to the presence of Model 1 can be divided as follows: production, 25,000 lbs/man-year; workforce, 4000 people; transport costs, $250/lb. The saving over the eight years needed to complete the colony is thus a total of 200×10^9.

earliest possible completion date is estimated at just after the turn of the century (28 years forward in time; going back the same number of years brings us to the era of the V2 rocket, more than ten years before the first artificial satellite). Around the year 2000, a fully reusable chemical rocket system could transport payloads to L$_5$ at a cost of about $100 per pound (again, in 1972 dollars). A prospective colonist could therefore save enough money (one or two years' salary) to emigrate with his family of three. The near certainty of continued advances in propulsion systems suggests that the actual costs will be lower.

By the middle years of the next century, and possibly earlier, production costs at L$_5$ should be lower than on Earth. My reasons for this belief are that:

• the asteroid belt is a rich source of raw materials, already exposed and differentiated.

• transport from the belt to L$_5$ can be done in a way analogous to ocean freight on Earth; that is, in very large units, with low fuel costs and very small crews. In space, it may be most practical to eliminate the freighter hulls entirely. A TLA-type reaction motor can run on free solar power and transport an entire asteroid to L$_5$, perhaps with no crew at all.

• food-raising costs, production costs and shipping costs among the communities should all be lower than on Earth because of ideal growing conditions, proximity of farms to consumers, availability of unlimited solar power and the convenience of zero-gravity and high-vacuum environments for production and transportation.

If we are so prodigal as to run through the entire material of the asteroid belt in the next 500 years, we can even gain another 500 years by using up the moons of the outer planets. Long before then, I hope we will have slowed the growth of the human population. And I feel sure that long before then a modified version of a space community will have travelled to a nearby star.

I am left with the desire to communicate two aspects of this work more completely. On the one hand, I would like to display for review more of the details of calculations and references than is possible here. And on the other hand, I am acutely aware of the need for discussion outside our own group of physics-oriented people. This work should be discussed and debated as widely as possible, by people with a range of technical and artistic talents, and by people who claim no special talent beyond the ability to work hard for a worthwhile goal. I hope I have conveyed at least a little of the sense of excitement that I have enjoyed over the past few years as each serious problem has appeared to yield to a solution, as well as how much more remains to be done and how much need there is for good ideas and hard work.

For private communications leading to references, I thank Donald Gault, Barry Royce, Richard Johnson, George Hazelrigg and John Breakwell. And it is a special pleasure to thank those who encouraged me to continue this work in the years when it was little known, particularly George Pimentel, Freeman Dyson, Brian O'Leary, Roman Smoluchowski, Richard Feynman and John Tukey. I am also grateful to Michael Phillips of the Point Foundation, which supported the first public meeting on this subject.

58

The Staff at the National Astronomy and Ionosphere Center
The Arecibo Message of November 1974

Major improvements have been made to the facilities at the Arecibo Observatory (Lalonde, 1974), which permit the generation of a radio transmission beam of unprecedented strength. Utilizing maximum transmitter power, the effective radiated power in the beam is approximately 2×10^{13} W, a power some 20 times the generating capacity of the totality of terrestrial electrical generating stations. With bandwidths of the order of 1 Hz or less, this signal is detectable by radio telescopes of the order of sensitivity of the Arecibo instrument throughout the Milky Way Galaxy. As the first use of this remarkable new facility, the Arecibo staff thought it was highly appropriate to send a simple and brief signal to the fringes of the Galaxy telling of the existence and nature of human life.

As part of the ceremonies to dedicate the newly upgraded Arecibo 1000-ft radio/radar telescope, at 1700 GMT on November 16, 1974, the telescope was so used to transmit a message for possible reception by other intelligent creatures. The transmission was made at a radio frequency of 2380 Mhz and bandwidth of some 10 Hz. It utilized the new telescope reflector surface supported by the National Science Foundation, and was the first use of the new radar transmitter supported by the National Aeronautics and Space Administration. The effective average radiated power in the direction of transmission was 3×10^{12} W, which was, we believe, the strongest signal yet radiated by our civilization. The transmission was directed at a globular cluster of stars, the Great Cluster in Hercules, Messier 13, a group of some 300,000 stars 25,000 light years distant whose apparent size closely matches the beamwidth of the transmission. A radio telescope in M13 operating at the transmission frequency, and pointed toward the Sun at the time the message arrives at the receiving site will observe a flux density from the message which will exceed the flux density of the Sun itself by a factor of roughly 10^7. Indeed, at that unique time, the Sun will appear to the receptors to be by far the brightest star of the Milky Way.

Reprinted from *Icarus*, 26 (1975), 462. Copyright © 1975 by Academic Press, Inc.

The Message

The message describes some characteristics of life on Earth which members of the staff of the National Astronomy and Ionosphere Center feel would be of most interest and relevance to other civilizations in space. It consists of 1679 consecutive characters, written in a format such that only two different characters are used. The two characters are denoted "0" or "1" in the message (Figure 58.1). In the actual transmission each character was represented by one of two specific radio frequencies, and the message was transmitted by shifting the frequency of the radio transmitter between these two radio frequencies in accordance with the plan of the message. The frequencies transmitted were continuously adjusted to correct for the Doppler effect of the orbital motion and rotation of the Earth so that none of the frequencies as observed outside the solar system varies from a specific fixed value.

The message is decoded by breaking the message up into 73 consecutive groups of 23 characters each, and arranging these groups in sequence one under the other. That 73 and 23 are prime numbers facilitates the discovery by any recipient that the above format is the correct way to interpret the message. Figure 58.2 shows the message so laid out, with the first character sent or received in the upper right-hand corner. The interpretation of the message is as follows.

The message begins with a "lesson" describing the number system to be used. This system is the binary system, believed to be one of the simplest number systems. Written across the top of the message, from right to left, are the numbers 1–10 in binary notation. Each number is marked with a "number label," a single character which marks the start of a number. A problem which must be dealt with carefully is the writing of a number so large that all its digits cannot be fitted into the available space. The solution to this is a particular point of the number lesson, and is shown in the numbers 8, 9, and 10. The number sequence has purposely been written so that

293

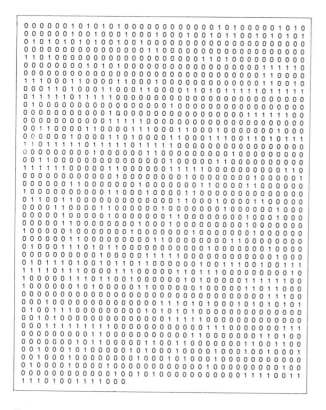

Figure 58.1. The 1679 character message transmitted on November 16, 1974. The characters were sent sequentially, with the transmitter frequency switching between two nearby radio frequencies to perform the transmission of a "0" or a "1."

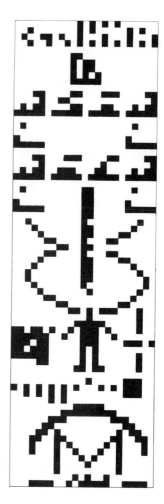

Figure 58.2. The message of Figure 58.1 laid out in a format of 73 lines of 23 characters per line. Zeros are shown as white, and ones as black. The first character sent is in the upper right-hand corner, and the message reads from right to left and down.

there is not enough room to write 8, 9, and 10 on a single line. The means of dealing with this problem is revealed with numbers 8, 9, and 10, where the digits for which there is not room are written "below" the least-significant digits. In the figure these digits appear to the left of the least-significant digits. An important point made here is that the second and successive lines of digits are not written under the number label, but only under characters denoting numerical values. This means that, as we would describe it, the number label always stands by itself in the upper right-hand corner of a number. This then marks a number, the orientation of a written number, and where the number begins. As will be seen, this is necessary to make the interpretation of numbers clearcut and to economize in the use of message characters.

Description of Fundamental Terrestrial Biochemistry

The second key lesson is the next prominent group to be sent, the one just below the numbers. This is recognizable as five numbers: from right to left, the numbers 1, 6, 7, 8,

and 15. This otherwise unlikely sequence of numbers is to be interpreted as the atomic numbers of the elements hydrogen, carbon, nitrogen, oxygen, and phosphorus.

There are 12 groups on lines 12–30 which are similar groups of five numbers. Each of these groups represents the chemical formula of a chemical molecule or radical, with the numbers from right to left in each case giving the number of atoms in the radical or molecule of hydrogen, carbon, nitrogen, oxygen, and phosphorus, respectively.

Since the limitations of the message do not allow us to describe the physical structure of the radicals and molecules, the simple chemical formulas do not in all cases define the precise identity of the radical or molecule. However, these structures are arranged as they are organized in the macromolecule here described and knowledgeable organic chemists anywhere should be able through simple logic to arrive at a unique solution for the molecular structures described here.

The most specific of the structures, which should point the way to the correct pictures of the others, is the structure which appears four times on lines 17–20 and 27–30. This is a structure containing one phosphorus atom and four oxygen atoms, the well-known phosphate group. The outer structures on lines 12–15 and 22–25 give the formula for the sugar molecule deoxyribose. The two sugar molecules on lines 12–15 have between them two structures; the chemical formulas given are, for the left structure, thymine, and for the right, adenine. Similarly, the molecules between the sugar molecules on lines 22–25 are, on the left, guanine, and on the right, cytosine.

This chemical structure is that of deoxyribonucleic acid, DNA, the molecule which contains the genetic information controlling the form, living processes, and inherited behavior of all life on Earth. This structure is actually wound as a double helix, as shown in lines 32–46. The complexity and degree of development of intelligent life on earth can perhaps be described by the number of base pairs in the genetic code, the number of adenine–thymine and guanine–cytosine combinations in the DNA. That there are some 4 billion such pairs in a single human chromosome is shown by the number given in the center of the double helix between lines 27 and 43. Note the use of the number label to establish this as a number and to show where the number begins.

The double helix leads to the head in a crude sketch of a human, establishing the connection between the DNA, the complexity of the helix, and the intelligent creature. To the right of the human is a line extending from the head to the feet of the human, and accompanied by the number 14. This means that the human is 14 units of length in size; the only possible unit of length is the wavelength of the transmission, 12.6 cm, which makes the human 5'9.5" tall. To the left of the human is a number, approximately 4 billion, which denotes the human population, Note that in this number and in the number in the double helix a few insignificant bits have been written as zeros. This is to make these groups look like numbers and not perhaps a drawing of some object.

The Location and Status of Human Civilization

Below the human is a sketch of the solar system—the Sun to the right, then the nine planets with some indication of relative sizes. Planet 3, Earth, is displaced to indicate that there is something special about it. It is displaced toward the human, who is centered on it, and this would suggest that planet 3 is the home of the creatures which sent the message.

Below the solar system, and again centered on planet 3, is an image of a telescope. The concept "telescope" is

described by showing a device which directs rays to a point. The mathematical curve which leads to such a diversion of paths is crudely indicated; accurately, of course, it is a parabola. Note that the telescope is not upside down, but "up" with respect to the symbol for the planet Earth.

At the very bottom of the message the size of the telescope is shown; in this case it is the size of both the largest telescope on Earth and the telescope which sent the message. It is 2430 wavelengths across, or 1004 ft. A number giving a result closer to the correct 1000 ft can be constructed, but infelicitously, it turns out to be uncertain whether the number is 1 or 2 separate numbers. The number actually given in the message can only be interpreted plausibly in one way. Of course, great accuracy is of no importance here. This information tells indirectly, when taken with the strength of our signal, a great deal about the level of our technology. If the message is transmitted repeatedly, a desired impression is made that the message emerges from the telescope.

Message Significance

As the choice of frequency, duration of message, and distance of the target clearly shows, the Arecibo message is very unlikely to produce interstellar discourse in the foreseeable future. Rather it was intended as a concrete demonstration that terrestrial radio astronomy has now reached a level of advance entirely adequate for interstellar radio communication over immense distances. More extensive attempts at the transmission of radio messages from the Earth to extraterrestrial civilizations should be made only after international scientific consultations as recommended by the first Soviet-American conference on communication with extraterrestrial intelligence (Sagan, 1973a).

Message Construction and Transmission

The content of the message was constructed primarily by Frank Drake, Richard Isaacman, Linda May, and James C. G. Walker. Valuable suggestions for improvements were given by a number of people, but particularly by Carl Sagan. The computer programming of the message and of the automatic transmitter control program was made by Bernard Jackson. The high-quality operation of the transmitter owes much to Robert McDonald of Continental Electronics Mfg. Corporation, and Thomas Dickinson of the NAIC. The high-power aberration-correcting line feed was designed and constructed under the supervision of Merle Lalonde.

The message was transmitted on November 16, 1974, at a rate of ten characters a second. It took 169 sec to send.

One minute after completion of the transmission, the message was as far from the Sun as the orbit of Mars. After 35 min it passed the distance of Jupiter, and after 71 min, Saturn. It had already overtaken the two previous potential communications with extraterrestrial intelligent life, the plaques on the Pioneer 10 and 11 spacecraft. Five hours and 20 min after transmission it passed the distance of the outermost planet, Pluto, and thus left the solar system. It became forever the vanguard of what we hope will be a growing body of human intelligence transmitted to space for the benefit of others.

Acknowledgment

The National Astronomy and Ionosphere Center is operated by Cornell University under contract with the National Science Foundation.

Bibliography

Academy of Sciences of the U.S.S.R., Moscow. 1957. *The Origin of Life on the Earth: Reports on the International Symposium, Moscow, August 1957.*

———. 1975. "The Soviet CETI Program." *Icarus,* 26, 377.

Ackeret, J. 1946. *Helv. Phys. Acta,* 19, 103.

———. 1956. *Inter. Avia,* 11, 989.

Adler, A. 1974. "Behold the Stars." *Atlantic Monthly,* 109 (October). (Reading 41 here.)

Alfven, H. 1954. *Origin of the Solar System.* Oxford University Press.

Allen, C. W. 1973. *Astrophysical Quantities.* 3d ed. Athlone Press, University of London.

Allen, R. L., B. Anderson, R. G. Conway, H. P. Palmer, V. C. Reddich, and B. Rowson. 1962. *Monthly Notices Roy. Astron. Soc.,* 124, 477.

Alpatov, V. 1959. "The rocket, the Moon, and life." *Izvest. Akad. Nauk. S.S.S.R., Ser. Bio.,* Sept. 18.

Altenhoff, W., P. G. Mezger, H. Strasse, H. Wendkerr, and G. Westerhout. 1960. *Verentliche Univ.-Sternavarte zu Bohn,* no. 9.

Anders, E., and F. W. Fitch. 1962. "Search for organized elements in carbonaceous chondrites." *Science,* 138, 1392.

Andrus, W. H., and N. J. Gurney, eds. 1973. *Proceedings of the MUFON Symposium, Quincy, Illinois.*

Anfinsen, C. B. 1959. *The Molecular Basis of Evolution.* Wiley.

Arrhenius, S. 1907. "Panspermia." *Scientific American,* 96, 196.

———. 1908. *Worlds in the Making.* Harper.

Avery, L. W., N. W. Broten, J. M. MacLeod, and T. Oka. 1976. *Astrophys. J.,* 205, L173.

Baker, R. M. L., Jr. 1968. Testimony in *Symposium on Unidentified Flying Objects.* Hearings, 90th Cong. July 29. House Committee on Science and Astronautics.

Ball, J. 1973. "The zoo hypothesis." *Icarus,* 19, 347. (Reading 45 here.)

Barath, F. T., A. H. Barrett, J. Copeland, D. E. Jones, and A. E. Liffey. 1963. "Mariner II: Preliminary reports on measurements of Venus. Microwave radiometers." *Science,* 139, 908.

Becquerel, P. 1950. "La suspension de la vie des spores. . . ." *Compt. Rend.,* 231, 1392.

Bell, D. 1976. *Phys. Today,* 29, 46.

Berkner, L. V., and L. C. Marshall. 1964. "The history of the growth of oxygen in the Earth's atmosphere," in Brancazio and Cameron (1964), pp. 102–126.

Bernal, J. D. 1967. *The Origin of Life.* London: Weidenfeld and Nicolson.

Blaauw, A., and M. Schmidt. 1965. *Galactic Structure.* University of Chicago Press.

Black, D. C., and R. Piziali. 1980. Project Orion. NASA-Special Report.

Black, D. C., and G. C. J. Suffolk. 1973a. "Concerning the planetary system of Barnard's Star." *Icarus,* 19, 353. (Reading 32 here.)

———. 1973b. "Some perturbing aspects of the planetary system of Barnard's Star." *Proceedings of the Symposium on the Origin of the Solar System,* Nice.

Bless, R. C., and B. D. Savage. 1972. *Astrophys. J.,* 171, 293.

297

Blum, H. F. 1963. "Negentropy and living systems." *Science,* 139, 398.

Bond, A., and A. R. Martin. 1975. *J. Brit. Interplanetary Soc.,* 28, 147.

Bova, B., 1963. *Amazing Fact Science Fiction,* 37, 113.

Bowen, I. S. 1950. *Publ. Astron. Soc. Pacific,* 62, 91.

Bracewell, R. N. 1960. "Communications from superior galactic communities." *Nature,* 186, 670. (Reading 21 here.)

——— . 1974. *The Galactic Club: Intelligent Life in Outer Space.* San Francisco: W. H. Freeman.

——— . 1978a. "Detecting nonsolar planets by spinning infrared interferometers." *Nature,* 274, 780.

——— . 1978b. "Man's role in the galaxy." *Isaac Asimov's Science Fiction Magazine,* 2, 142.

Bracewell, R. N. and R. MacPhie. 1979. "Searching for nonsolar planets." *Icarus,* 38, 136. (Reading 33 here.)

Bradfield, R. 1972. "Multiple cropping: hope for hungry Asia." *Reader's Digest,* October, p. 217.

Brancazio, P. M., and A. G. W. Cameron, eds. 1964. *The Origin and Evolution of Atmospheres and Oceans.* Wiley.

Breger, I. A., P. Zubovic, J. C. Chandler, and R. S. Clark. 1972. *Nature,* 236, 155.

Budden, K. G., and G. G. Yates. 1951. *J. Atmos. Terr. Phys.,* 2, 272.

Budyko, M. I. 1969. "The effect of solar radiation variations on the climate of the Earth." *Tellus,* 21, 611.

Burbidge, E., G. R. Burbidge, W. A. Fowler, and F. Hoyle. 1957. *Rev. Mod. Phys.,* 29, 547.

Burke, B. F., and K. S. Franklin. 1955. *J. Geophys. Res.,* 60, 213.

Bussard, R. W. 1960. *Astronautica Acta,* 6, 179.

Cairns-Smith, A. G. 1965. *J. Theor. Biol.,* 10, 53.

Calvin, M. 1961. *Chemical Evolution.* Eugene: University of Oregon Press.

——— . 1962. "Communication: from molecules to Mars." *Am. Inst. Biol. Sci. Bull.,* 12 (no. 5), 29.

Cameron, A. G. W. 1962. *Icarus,* 1, 13.

——— , ed. 1963a. *Interstellar Communication.* Benjamin.

——— . 1963b. "Communicating with intelligent life on other worlds." *Sky and Telescope,* 26, 258. (Reading 27 here.)

Carlson, J. V., and P. A. Sturrock. 1975. "Stanford workshop on extraterrestrial civilization: Opening a new scientific dialog." *Origin of Life,* 6, 459.

Carter, L. J., ed. 1957. *Realities of Space Travel: Selected Papers of the British Interplanetary Society.* McGraw-Hill.

Charles, R. H., ed., and W. R. Morfill, trans. 1896. *The Book of the Secrets of Enoch.* Oxford: Clarendon Press.

Clarke, A. C. 1950. *J. Brit. Interplanetary Soc.,* 9, 261.

——— . 1972. "The sentinel," in *The Lost Worlds of 2001.* London: Sidgwick and Jackson.

Club of Rome. 1972. *Limits of Development.* Report by the Systems Dynamics Group, Massachusetts Institute of Technology. Geneva: Club of Rome.

Cocconi, G., and P. Morrison. 1959. "Searching for interstellar communications." *Nature,* 184, 844. (Reading 20 here.)

Committee on the Exploration of Extraterrestrial Space (CETEX). 1959. First report, *ICSU Rev.,* 1, 100; second report, *Nature,* 183, 925.

Conway, R. G., K. I. Kellerman, and R. J. Long. 1963. *Monthly Notices Roy. Astron. Soc.,* 125, 261.

Cooke, A., and N. C. Wickramasinghe. 1977. *Astrophys. Space Sci.*

Copeland, H., J. O. Jensen, and H. E. Jorgensen. 1970. "Homogeneous models for population I and population II compositions." *Astron. and Astrophys.,* 5, 12–34.

Cox, L. J. 1976. "An Explanation for the absence of extraterrestrials on Earth." *Quart. J. Roy. Astron. Soc.,* 17, 201. (Reading 43 here.)

Crick, F. H. C., and L. E. Orgel. 1973. "Directed panspermia." *Icarus,* 14, 341. (Reading 10 here.)

Cronin, J. R., and C. B. Moore. 1976. *Geochim Cosmochim. Acta,* 40, 853.

Davies, R. W., and M. G. Comuntzis. 1959. "The sterilization of space vehicles to prevent extraterrestrial biological contamination." *Proc. Tenth Intern. Astronaut. Congr., London.*

Demarque, P. R., and R. B. Larson. 1964. "The age of galactic cluster NGC 188." *Astrophys. J.,* 140, 544.

Dickerson, R. 1978. "Chemical evolution and the origin of life." *Scientific American,* 239 (no. 3), 70. (Reading 12 here.)

Dieter, N. H. 1972. "Berkeley survey of high-velocity interstellar neutral hydrogen." *Astron. Astrophys. Suppl.* 5, 21.

Dole, S. H. 1970. *Habitable Planets for Man.* 2d. ed. American Elsevier.

Dollfus, A. 1960. "Resultats d'observations indiquant la vie sur la planete Mars." *Proc. First Intern. Space Science Symposium, Nice.*

Drake, F. 1959. "Interstellar communication." *Bull. Phil. Soc. Washington,* 16, 58.

––––––. 1960. "How can we detect radio transmissions from distant planetary systems?" *Sky and Telescope,* 19, 140. (Reading 24 here.)

––––––. 1961. "Project Ozma." *Physics Today,* 14 (no. 4), 40. (Reading 25 here.)

DuBridge, A. 1960. *Introduction to Space.* Columbia University Press.

Dufay, J. 1957. *Nebuleuses galactiques et matière interstellaire.* Translated by A. J. Pomerans as *Galactic Nebulae and Interstellar Matter.* London: Hutchinsons.

Dyson, F. 1960. "Search for artificial stellar sources of infrared radiation." *Science,* 131, 1667. (Reading 22 here.)

––––––. 1966. *Perspectives in Modern Physics: Thoughts on the Search for Extraterrestrial Technology.* Interscience.

––––––. 1968. *Physics Today,* 21, 41.

Eberhart, J. 1977. "The world on a record." *Science News,* 112, 124. (Reading 56 here.)

Ehrensvärd, G. 1962. *Life: Origin and Development.* University of Chicago Press.

Eliade, M. 1959. *Cosmos and History.* Harper.

Elton, C. S. 1958. *The Ecology of Invasions by Animals and Plants.* London: Methuen.

Emerson, A. D., ed. 1975. *Proceedings of the AIAA Symposium on UFOs and the Future, Los Angeles.*

Emmons, G. T. 1911. *Amer. Anthrop.,* N. S., vol. 13. Reprinted in *Primitive Heritage,* ed. M. Mead and N. Calas. Random House.

Everitt, C. W. F. 1977. "Final report on NASA Grant 05-020-019 to perform a gyro test of general relativity in a satellite and develop associated control technology." W. W. Hansen Laboratories of Physics and the Department of Aeronautics and Astronautics, Stanford University, Stanford, Calif.

Faegre, A. 1972. "An intransitive model of the Earth-atmosphere-ocean system." *J. Appl. Meteorol.,* 11, 4–6.

Fialkowski, K. 1977. "A model of imagined reality." *Poland,* no. 2

Florkin, M., ed. 1961. *Some Aspects of the Origin of Life.* London: Pergamon.

Fox, S. W. 1960. *Science,* 132, 200.

––––––. 1963a. *Ann. N.Y. Acad. Sci.,* 108, 467.

––––––. 1963b. *J. Bacteriol.,* 85, 279.

––––––. 1964a. *Nature,* 201, 336.

––––––. 1964b. "Humanoids and protenoids." *Science,* 144, 954. (Reading 40 here.)

––––––. 1978. "The Origin and Nature of Protolife," in *The Nature of Life,* ed. by W. H. Heidcamp. Baltimore: University Park Press.

––––––, ed. 1965. *The Origins of Prebiological Systems.* Academic Press.

––––––, and Dose. 1977. *Molecular Evolution and the Origin of Life.* Marcel Dekker.

Fredrick, L. W., and P. H. Shelus. 1969. "An astrometric study of L726-8." *Bull. A.A.S.,* 1, 23.

Freeman, J., and M. Lampton. 1975. "Interstellar archaeology and the prevalence of intelligence." *Icarus,* 25, 368.

Friedman, S. T. 1973. "Ufology and the search for extraterrestrial intelligent life." In Andrus and Gurney (1973).

––––––. 1975. "A scientific approach to flying saucer behavior." In Emerson (1975).

Gatewood, E. 1976. "On the astrometric detection of neighboring planetary systems." *Icarus,* 27, 1.

Gatland, K. W. 1974. *Spaceflight,* 16, 356.

Gatos, H. C., and A. F. Witt. 1974. "Crystal Growth Studies on Skylab." MIT News Release, May 14.

Gehrz, R. D., and J. A. Hackwell. 1978. "Exploring the infrared universe from Wyoming." *Sky and Telescope,* 55, 466.

Gold, T. 1960. "Cosmic Garbage." *Air Force and Space Digest,* May, p. 65.

Goldman, S. 1953. *Information Theory*. Prentice-Hall.

Greenstein, J., and T. A. Matthews. 1963. *Nature, 197,* 1042.

Gulkis, S., M. Janssen, T. Kuiper, and R. Edelson. 1976. Paper presented at the International Symposium of the IEEE, Univ. of Massachusetts, Amherst.

Haldane, J. B. S. 1929. "The origin of life." *Rationalist Annual*. (Reading 8 here.)

Harada, K., and S. W. Fox. 1964. *Nature, 201,* 335.

Harris, D. E., and J. A. Roberts. 1960. *Publ. Astron. Soc. Pacific, 72,* 237.

Harrison, G. R., J. E. Archer, and J. Camus. 1952. *J. Opt. Soc. Amer., 42,* 706.

Hart, M. H. 1975. "An explanation for the absence of extraterrestrials on Earth." *Quart. J. Roy. Astron. Soc., 16,* 128. (Reading 42 here.)

———. 1978. "The evolution of the atmosphere of the Earth." *Icarus, 33,* 23–39.

———. 1979. "Habitable zones about main sequence stars." *Icarus, 37,* 351. (Reading 44 here.)

Hawrylewicz, E., B. Gowdy, and R. Ehrlich. 1962. "Microorganisms under a simulated Martian environment." *Nature, 193,* 497.

Hedden, R. L. 1976. "A telescope for the infrared astronomical satellite (IRAS)," in Society of Photo-optical Instrumentation Engineers (SPIE) 95, *Modern Utilization of Infrared Technology,* II, 8.

Hess, H. H., et al. 1962. *A Review of Space Research*. Natl. Aca. Sci.–Natl. Res. Council Publ., no. 1079.

Hodges, E. R. 1876. Cory's *Ancient Fragments,* rev. ed. London: Reeves & Turner.

Hool, G. A., and W. S. Kinne. 1943. *Movable and Long Span Steel Bridges*. McGraw-Hill.

Horowitz, N. 1958. "The origin of life." In Hutchings (1958).

———. 1976. "The search for life on Mars." *Scientific American,* 237 (no. 5), 52. (Reading 19 here.)

Hoyle, F. 1950. *The Nature of the Universe*. Harper.

———. 1955. *Frontiers of Astronomy*. Harper.

———. 1961. *Quart. J. Roy. Astron. Soc.,* 1, 28.

Hoyle, F., and N. C. Wickramasinghe. 1976. *Nature,* 264, 45.

———. 1977. "Prebiotic molecules and interstellar grain clumps." *Nature,* 266, 341. (Reading 13 here.)

Huang, S. S. 1957. *Publ. Astron. Soc. Pacific,* 69, 427.

———. 1959a. "Occurrence of life in the universe." *American Scientist,* 47, 397.

———. 1959b. "The problem of life in the universe and the mode of star formation." *Publ. Astron. Soc. Pacific,* 7, 421.

———. 1960. "Life-supporting regions in the vicinity of binary systems." *Publ. Astron. Soc. Pacific,* 72, 106.

———. 1961. *Publ. Astron. Soc. Pacific,* 73, 30.

Hutchings, E., Jr. 1958. *Frontiers of Science*. Basic Books.

Hynek, J. A. 1953. *J. Opt. Soc. Amer.,* 43, 311.

———. 1966a. *Saturday Evening Post,* Dec. 17.

———. 1966b. *Science,* 154, 329.

———. 1967. "Unidentified flying objects." *Encyclopaedia Britannica* (Chicago: Benton), XXII, 696 (in 1964 ed.) or 499 (in 1967 ed.).

———. 1972. *The UFO Experience: A Scientific Inquiry*. Chicago: Regnery.

Iben, I. 1967. "Stellar evolution, VI: Evolution from the main sequence to the red-giant branch for stars of 1 solar mass, 1.25 solar mass, and 1.5 solar mass." *Astrophys. J.,* 147, 624–649.

Jackson, F., and P. Moore. 1962. *Life in the Universe*. Norton.

Javan, A., W. R. Bennett, Jr., and D. R. Herriot. 1961. *Phys. Rev. Letters,* 6, 106.

Jeans, James H. 1929. *Astronomy and Cosmogony*. Cambridge University Press: Dover reprint, 1961.

———. 1942. "Is there life on other worlds?" *Science,* 95, 589. (Reading 16 here.)

Jones, E. M. 1976. "Colonization of the galaxy." *Icarus,* 28, 421–422.

———. 1978. *J. Brit. Interplanetary Soc.,* 31, March, 103.

Jones, H. S. 1949. *Life on Other Worlds*. New American Library.

Kamel, A. A. 1970. "Perturbation theory based on Lie transforms and its application to the stability of motion near Sun-perturbed Earth-Moon triangular libration points." NASA, CR-1622, August.

Kamshilov, M. 1973. "Scientific and technological progress and the evolution of the biosphere." *Social Science,* 4 (no. 14), 53–62.

Kardashev, N. S. 1964. "Transmission of information by extraterrestrial civilizations." *Soviet Astronomy-AJ,* 8, 217. (Reading 28 here.)

Kash, S. W., and R. F. Tooper. 1962. *Astronautics.* 7, 68.

Keilin, D. 1955. "The problem of anabiosis or latent life: History and current concept." *Proc. Roy. Soc. London,* B150, 149.

KenKnight, C. E. 1977. "Methods of detecting extrasolar planets, II: Imaging." *Icarus,* 30, 422.

Keosian, J. 1964. *The Origin of Life.* Reinhold.

Kiess, C. C., S. Karrer, and H. K. Kiess. 1960. "A new interpretation of Martian phenomena." *Publ. Astron. Soc. Pacific,* 72, 256.

Kinard, W. H., R. L. O'Neal, J. M. Alarez, and D. H. Humes. 1974. "Interplanetary and near-Jupiter meteoroid environments: Preliminary results from the meteoroid detection experiment." *Science,* 183, 321–322.

Kolenkiewicz, R., and L. Carpenter. 1968. "Stable periodic orbits about the Sun-perturbed Earth-Moon triangular points." *AEAA Journal,* 6 (no. 7), 1301.

Kornberg, A. 1960. "Biologic synthesis of deoxyribonucleic acid." *Science,* 131, 1503.

Kreifeldt, J. G. 1973. "A formulation for the number of communicative civilizations in the Galaxy." *Icarus,* 14, 419–430.

Kuiper, G. P. 1951. *Astrophysics.* McGraw-Hill.

Kuiper, T., and M. Morris. 1977. "Searching for extraterrestrial civilizations." *Science,* 196, 616. (Reading 34 here.)

Lalonde, L. M. 1974. "The upgraded Arecibo Observatory." *Science,* 186, 213–218.

Lappe, F. M. 1971. *Diet for a Small Planet.* Ballantine Books.

Latham, G., J. Dorman, F. Duennebier, M. Ewing, D. Lammlein, and Y. Nakamura. 1973a. "Moonquakes, meteorites, and the state of the lunar interior." *Abstracts of the Fourth Lunar Science Conference,* Lunar Science Institute, Houston, Texas.

———. 1973b. "Lunar seismology." *Abstracts of the Fourth Lunar Science Conference,* Lunar Science Institute, Houston, Texas.

Lederberg, J. 1960a. "Exobiology: Approaches to life beyond the Earth." *Science,* 132, 393. (Reading 11 here.)

———. 1960b. "A view of genetics." *Science,* 131, 269.

Lederberg, J., and D. B. Cowie. 1958. "Moondust." *Science,* 127, 1473.

Levin, B. 1964. *The Origin of the Earth and Planets.* 3d. ed. Moscow: Foreign Languages Publishing House.

Levin, G. V., A. H. Helm, J. R. Glendenning, and M. F. Thompson. 1962. "Gulliver: A quest for life on Mars." *Science,* 138, 114.

Lilly, J. C. 1961. *Man and Dolphin.* Doubleday.

Lowell, P. 1908. *Mars as the Abode of Life.* Macmillan. (Reading 14 here.)

Lunar Science Institute. 1973. *Abstracts of the Fourth Lunar Science Conference,* Houston.

Machol, R. 1976. "An ear to the universe." *IEEE Spectrum,* March. (Reading 31 here.)

MacLeish, K. 1973. "Australia's Wild." *National Geographic,* 143 (no. 2), 168.

Maeder, A. 1976. "Stellar evolution, V: Evolutionary models of population I stars with or without overshooting from convective cores." *Astron. and Astrophys.,* 47, 389.

Maiman, T. H. 1960a. *Brit. Commun. and Electronics,* 7, 674.

———. 1960b. *Nature,* 187, 493.

Mallove, E. F., and R. L. Forward. 1972. "Bibliography of interstellar travel and communication." Research Report 460, 16–21. Malibu, Calif.: Hughes Research Laboratories.

Markowitz, W. 1963. In *Air, Space, and Instruments,* ed. S. Lees. McGraw-Hill.

———. 1967. "The physics and metaphysics of unidentified flying objects." *Science,* 157, 1274. (Reading 49 here.)

Marx, G. 1960 *Astronautica Acta,* 6, 366.

———. 1963. *Astronautica Acta,* 9, 131.

———. 1966. *Nature,* 211, 22.

Matthew, W. D. 1921. "Life in other worlds," *Science,* 54, 239.

McCampbell, J. M. 1973. *Ufology*. San Francisco: Hollmann.

McCarthy, S. G. 1976. "Shuttle infrared telescope facility," in Society of Photo-optical Instrumentation Engineers (SPIE) 95, *Modern Utilization of Infrared Technology*, II, 2.

McCrosky, R. E. 1968. "Distributions of large meteoric bodies." *Smithsonian Astrophysical Observatory Special Report* No. 280.

Mendis, D. A., and N. C. Wickramasinghe. 1975. *Astrophys. Space Sci.*, 37, L13.

Menzel, D. H., and L. G. Boyd. 1963. *The World of Flying Saucers*. Doubleday.

Metz, W. D. 1972. *Science*, 178, 600.

Michael, W. H., Jr. 1963. "Considerations of the motion of a small body in the vicinity of the stable libration points of the Earth-Moon system." NASA TR-160.

Miller, S. L. 1955. "Production of some organic compounds under possible primitive Earth conditions." *J. Amer. Chem. Soc.*, 77, 2351.

Miller, S. L, and H. C. Urey. 1959. *Science*, 130, 245.

Molton, P. W. 1978. *J. Brit. Interplanetary Soc.*, 31, June, 203.

Morrison, P. 1962. "Interstellar communication." *Bull. Phil. Soc. Washington*, 16, 58. (Reading 26 here.)

———. 1974. "Entropy, life, and communication." In Ponnamperuma and Cameron (1974).

Morrison, P., J. Billingham, and J. Wolfe. 1977. "The search for extraterrestrial intelligence." NASA-Ames Report SP-419.

Nagy, B., W. G. Meinschein, and D. J. Hennessy. 1961. "Mass spectroscopic analyses of the Orgueil meteorite: Evidence for biogenic hydrocarbons." *Ann. N.Y. Acad. Sci.*, 93, 25.

Nandy, K., G. I. Thompson, C. Jamar, A. Monfils, and R. Wilson. 1975. *Astron. Astrophys.*, 44, 195.

NASA. 1969. "Meteoroid environment model, 1969: Near Earth to lunar surface." NASA SP-8013.

NASA. 1976. "Outlook for space." SP-386.

NASA. 1977. *The Search for Extraterrestrial Intelligence. Final Report of the Science Workshops on Interstellar Communication.* NASA Special Publ. 419.

National Astronomy and Ionosphere Center. 1975. "The Arecibo message of November 1974." *Icarus*, 26, 462. (Reading 58 here.)

Newcomb, S. 1895. *Fundamental Constants of Astronomy*. Washington, D.C.: U.S. Government Printing Office.

———. 1905. "Life in the Universe." *Harper's Mag.*, III, 404. (Reading 7 here.)

Node, H., ed. 1978. *Origin of Life*. Tokyo: Center for Academic Publications.

O'Connell, D. J. K., ed. 1958. *Stellar Population*. Amsterdam: North Holland Publishing Co.

Oke, J. B. 1963. *Nature*, 197, 1042.

Oliver, B. M. 1962. *International Science and Technol.*, no. 10, p. 55.

———. 1975. "Proximity of galactic civilizations." *Icarus*, 25, 360. (Reading 35 here.)

Oliver, B. M., and J. Billingham. 1972. "Project Cyclops: A design study of a system for detecting extraterrestrial intelligent life." NASA report CR114445. Moffet Field, Calif.: NASA-Ames Research Center.

O'Neill, G. 1974. "The colonization of space." *Physics Today*, 27 (no. 9), 32. (Reading 57 here.)

———. 1977. *The High Frontier (Human Colonies in Space)*. William Morrow.

Oort, J. H. 1958. In O'Connell (1958), p. 415.

———. 1959. "The origin of life on Earth." In Academy of Sciences of the U.S.S.R. (1959).

Oparin, A. I. 1938. *The Origin of Life*. 2d ed. Trans. by S. Morgulis. Macmillan. 1957. 3d ed. Oliver & Boyd.

Oparin, A. I., and F. Fesenkov. 1960. *The Universe*. 2d ed. Moscow: Foreign Languages Publishing House.

Oparin, A. I., A. G. Pasynskii, A. E. Braunshtein, and T. E. Pavloskaya, eds. 1959. *The Origin of Life on the Earth*. Pergamon Press.

Organic Electronic Spectral Data. 1946–1961. Vols. 1–5. Interscience.

Oro, J. 1962. *Trans. N.Y. Acad. Sci.*

Papagiannis, M. 1977. Int. Conf. on the Origin of Life, Japan; Astron. Contrib. Boston Univ. Series II, No. 61.

———. 1978a. "Are we all alone, or could they be in the asteroid belt?" *Quart. J. Roy. Astron. Soc.*, 19, 277. (Reading 46 here.)

———. 1978b. "Could we be the only advanced technological civilization in our galaxy?" In Node, H. (1978).

Peschka, W. 1956. *Astronautica Acta,* 2, 191.

Pierce, J. R. 1959. *Proc. Inst. Radio Engrs.,* 47, 1053.

Pipher, J. L., J. R. Houck, B. W. Jones, and M. Harwit. 1971. "Submillimeter observations of the night-sky emission above 120 kilometers." *Nature,* 231, 375.

Ponnamperuma, C., and A. G. W. Cameron, eds. 1974. *Interstellar Communication: Scientific Perspectives.* Boston: Houghton Mifflin.

Population Reference Bureau, 1970. *1970 World Population Data Sheet.* Population Reference Bureau, 1955 Massachusetts Av. N.W., Washington, D.C. 20036.

Powell, C. 1972. *Spaceflight,* 14, 442.

Powers, W. T. 1967. *Science,* 156, 11.

Purcell, E. "Radioastronomy and communication through space." USAEC Report BNL-658. (Reading 36 here.)

———. 1959. "Talk at Professional Group on Microwave Theory and Techniques." Inst. Rad. Eng., June.

Putnam, P. C. 1948. *Energy in the Future.* New York.

Rasool, L. I., and C. De Bergh. 1970. "The runaway greenhouse and the accumulation of CO_2 in the Venus atmosphere." *Nature,* 226, 1037.

Sagan, C. 1960. "Biological contamination of the Moon." *Proc. Nat. Acad. Sci.,* 46, 396.

———. 1961a. "Origin and planetary distribution of life." *Radiation Res.,* 15, 174.

———. 1961b. "The planet Venus." *Science,* 133, 849.

———. 1963. "Direct contact among galactic civilizations by relativistic interstellar spaceflight." *Planetary and Space Science,* 11, 485. (Reading 38 here.)

———. 1972. "UFOs: The extraterrestrial and other hypotheses." In Sagan and Page (1972), pp. 267–275.

———, ed. 1973a. *Communication with Extraterrestrial Intelligence.* Cambridge, Mass.: MIT Press.

———. 1973b. *The Cosmic Connection.* Dell.

———. 1973c. "On the detectivity of advanced galactic civilizations." *Icarus,* 19, 350. (Reading 29 here.)

———. 1977. *The Dragons of Eden.* Random House.

Sagan, C., and F. Drake. 1975. *Scientific American,* 233 (May) 83.

Sagan, C., F. Drake, A. Druyan, T. Ferris, J. Lomberg, and L. S. Sagan. 1978. *Murmurs of Earth.* Random House.

Sagan, C., and T. Page, eds. 1972. *UFOs: A Scientific Debate.* Norton.

Sagan, C., C. Ponnamperuma, and R. Mariner. 1963. *Nature 199,* 222.

Sagan, C., L. S. Sagan, and F. Drake. 1972. "A message from Earth." *Science,* 175, 881. (Reading 54 here.)

Sagan, C., and L. Whitehall. 1973. (unpublished).

Sakata, A., N. Nakagawa, T. Iguchi, S. Isobe, M. Morimoto, F. Hoyle, and N. C. Wickramasinghe. 1977. *Nature,* 266, 241.

Sampson, A. 1977. *The Arms Bazaar.* Viking.

Sanger, E. 1957. *Astronautica Acta,* 3, 89.

———. 1962. *J. Brit. Interplanetary Soc.,* 18, 273.

Schawlow, A. L., and C. H. Townes. 1958. *Phys. Rev.,* 112, 1940.

Schmidt, M. 1959. "The rate of star formation." *Astrophys. J.,* 129, 243.

———. 1963a. *Nature,* 197, 1040.

———. 1963b. "The rate of star formation, II." *Astrophys. J.,* 137, 759.

Schnabel, P., 1923. *Berossos und die Babylonisch-Hellenistische Literatur.* Leipzig: Teubner.

Schneider, S. H., and T. Gal-Chen. 1973. "Numerical experiments in climate stability." *J. Geophys. Res.,* 78, 6182.

Schwartz, R. M., and C. H. Townes. 1961. "Interstellar and interplanetary communication by optical masers." *Nature,* 190, 205. (Reading 23 here.)

Schwartzman, D. 1977. "The absence of extraterrestrials on Earth and the prospects for CETI." *Icarus,* 32, 473. (Reading 51 here.)

Sellers, W. D. 1969. "A global climate model based on the energy balance of the Earth-atmosphere system." *J. Appl. Meterol.,* 8, 392.

Serkowski, K. 1976. "Feasibility of a search for planets around solar-type stars with a polarimetric radial-velocity meter." *Icarus,* 27, 13.

———. 1978. "Should we search for planets around stars?" *Astronomy Quart.*, 1, 5.

Shain, C. A. 1956. *Aust. J. Phys.*, 9, 61.

Shannon, C. E., and W. Weaver. 1964. *The Mathematical Theory of Communication*. Urbana: Univ. of Illinois Press.

Shapley, H. 1958. *Of Stars and Men*. Boston: Beacon.

Shklovskii, I. S. 1977. "Man and Space: Conference in the U.S.S.R." *Astronomy*, 5 (no. 1), 56.

———. 1978. *The New York Times*, May 18.

Shklovskii, I. S., and C. Sagan. 1966. *Intelligent Life in the Universe*. San Francisco: Holden-Day.

Simpson, G. G. 1949. *The Meaning of Evolution*. New Haven, Conn.: Yale Univ. Press.

———. 1960. "The history of life." In *The Evolution of Life*, S. Tax, ed. (University of Chicago Press).

———. 1962. "Some cosmic aspects of organic evolution." In *Evolution und Hominisation*. G. Kurth, ed. (Stuttgart: Fischer).

———. 1964. "The nonprevalence of humanoids." *Science*, 143, 769. (Reading 39 here.)

Singer, S. F. 1970. *Scientific American*, Sept., p. 174.

Sinton, W. M. 1959. "Further evidence of vegetation on Mars." *Science*, 130, 1234.

Sinton, W. M., and J. Strong. 1960. *Astrophys. J.*, 131, 459.

Slysh, V. I. 1963. *Nature*, 199, 682.

Sneath, P. H. A. 1962. "Longevity of microorganisms." *Nature* (London), 195, 643.

Soifer, B. T., J. R. Houck, and M. Harwit. 1971. *Astrophys. J.*, 168, L73.

Spencer, D. W., and L. D. Jaffe. 1962. *Jet Propulsion Laboratory Technical Report*, 32–233.

———. 1963a. *Advanced Propulsion Concepts: Proceedings of the Third Symposium*. Gordon & Breach.

———. 1963b. *Astronautica Acta*, 9, 48.

———. 1976. *J. Brit. Interplanetary Soc.*, 29, 12.

Steinman, D. B. 1929. *A Practical Treatise on Suspension Bridges*. John Wiley.

Stephenson, D. G. 1977. *J. Brit. Interplanetary Soc.*, 30, March, 105.

———. 1979a. "Extraterrestrial cultures within the solar system?" *Quart. J. of the Roy. Aston. Soc.*, 20, 422. (Reading 47 here.)

———. 1979b. "Extraterrestrial intelligence." *Quart. J. of the Roy. Astron. Soc.*, 20, 481. (Reading 50 here.)

Størmer, C. 1928. *Nature*, 122, 681.

Struve, O. 1950. *Stellar Evolution*. Princeton University Press.

———. 1959. K. T. Compton Lecture, Mass. Institute of Technology, November.

———. 1960. *Sky and Telescope*, 19, 154.

Sturrock, P. 1977. "Survey of the membership of the American Astronomical Society concerning the UFO problem." Stanford University Institute for Plasma Research, Rept. no. 681.

Sullivan, W. 1966. *We Are Not Alone*. McGraw-Hill.

Temple, R. K. G. 1976. *The Sirius Mystery*. St. Martin's Press.

Thomas, L. 1972. "Notes of a biology watcher: Ceti." *New England J. Medicine*, 286, 306. (Reading 55 here.)

Tinsley, B. 1976. "Effect of main-sequence brightening on the luminosity evolution of elliptical galaxies." *Astrophys. J.*, 203, 63.

Townes, C. H. 1971. Jansky Lecture at the National Radio Astronomy Observatory, Charlottesville, Virginia, October 4.

Troitskii, V. S., A. M. Starodubtsev, L. I. Gershtein, V. L Rakhlin. 1971. "Search for monochromatic 927-MHz radio emission from nearby stars." *Soviet Astron.-AJ*, 15, 508.

Turtle, A. J., J. F. Pugh, S. Kenderine, and J. J. K. Pauliny-Toth. 1962. *Monthly Notices Roy. Astron. Soc.*, 124, 297.

Ulrich, R. K. 1975. "Solar neutrinos and variations in the solar luminosity." *Science*, 190, 619.

Ultraviolet Reference Spectra. Philadelphia: Sadtler Research Laboratories.

Urey, H. C. 1951. *The Planets*. New Haven, Conn.: Yale University Press.

———. 1960. "Lines of evidence regarding composition of the moon." *Proc. First Intern. Space Science Symposium, Nice*.

U.S. Government. 1959. *Space Handbook: Astronautics*

and Its Applications. Washington, D.C.: U.S. Government Printing Office.

———— . 1966. General Information Concerning Patents. Washington, D.C.: U.S. Government Printing Office.

U.S. Patent Office. 1947. Mimeographed statement POL-49.

Vallee, J. 1965. *Anatomy of a Phenomenon*. Ace Books.

———— . 1966. *Challenge to Science*. Regnery.

———— . 1975. *The Invisible College*. Dutton.

Van de Kamp, P. 1963. "Astrometric study of Barnard's Star from plates taken with the 24-inch Sproul refractor." *Astron. J.,* 68, 515.

———— . 1969a. "Parallax, proper motion, acceleration, and orbital motion of Barnard's Star." *Astron. J.,* 74, 238.

———— . 1969b. "Alternate dynamical analysis of Barnard's Star." *Astron. J.,* 74, 757.

———— . 1975. "Astrometric study of Barnard's Star." *Astron. J.,* 80, 667.

Van der Pol, B. 1928. *Nature,* 122, 878.

Verschuur, G. L. 1973a. "High-velocity clouds and normal galactic structure." *Astron. Astrophys.,* 22, 139.

———— . 1973b. "A search for narrow-band 21-cm wavelength signals from ten nearby stars." *Icarus,* 19, 329. (Reading 30 here.)

Viewing, D., 1975. *J. Brit. Interplanetary Soc.,* 28, 735.

Van Däniken, E. 1969. *Chariots of the Gods?* London: Souvenir Press.

Von Hoerner, S. 1961. "The search for signals from other civilizations." *Science,* 134, 1839.

———— . 1962. "The general limits of space travel." *Science,* 137, 18. (Reading 37 here.)

———— . 1972. In *Einheit und Veilheit,* E. Scheibe and G. Sussman, eds. (Göttingen: Vandenhoek and Ruprecht).

———— . 1975a. *J. Brit. Interplanetary Soc.* 28, 691.

———— . 1975b. In Wilson and Downes (1975).

———— . 1978. "Where is everybody?" *Naturwissenschaften,* 65, 553. (Reading 48 here.)

Wickramasinghe, N. C. 1974. *Nature,* 252, 462.

———— . 1975. *Monthly Notices Roy. Astron. Soc.,* 170, P11.

Wickramasinghe, N. C., F. Hoyle, and K. Nandy. 1977. *Astrophys. Space Sci.* (in press)

Wilson, R. W. 1963. *Observations Owens Valley Radio Observ., Cal Tech,* no. 3.

Wilson, T. L., and D. Downes, eds, 1975. *H II Regions and Related Topics.* Springer.

Witteborn, F. C., and L. S. Young. 1976. "The Spacelab infrared telescope facility (SIRTF)." *J. Spacecraft and Rockets,* 13, 667.

Worley, C. E. 1962. *Astron. J.,* 67, 590.

Index